Endothelial Cell Plasticity in the Normal and Injured Central Nervous System

Endothelial Cell Plasticity in the Normal and Injured Central Nervous System

Editors

Esperanza Meléndez Herrera
Profesor e Investigador Titular A
Laboratorio de Eco-fisiología Animal
Departamento de Zoología
Instituto de Investigaciones Sobre los Recursos Naturales
Universidad Michoacana de San Nicolas de Hidalgo (UMSNH)
Morelia, Michoacán
México

Bryan V. Phillips-Farfán
Investigador en Ciencias Médicas C
Laboratorio de Nutrición Experimental
Instituto Nacional de Pediatría
Coyoacán, D.F.
México

Gabriel Gutiérrez Ospina
Investigador Titular B
Instituto de Investigaciones Biomédicas
Universidad Nacional Autónoma de México (UNAM)
Coyoacán, D.F.
México

CRC Press
Taylor & Francis Group
Boca Raton London New York

CRC Press is an imprint of the
Taylor & Francis Group, an **informa** business
A SCIENCE PUBLISHERS BOOK

Cover Illustrations reproduced by kind courtesy of Dannia Colin Castelan.

CRC Press
Taylor & Francis Group
6000 Broken Sound Parkway NW, Suite 300
Boca Raton, FL 33487-2742

Version Date: 20150105

International Standard Book Number-13: 978-1-4665-9923-9 (eBook - PDF)

Visit the Taylor & Francis Web site at
http://www.taylorandfrancis.com

and the CRC Press Web site at
http://www.crcpress.com

Preface

It was once thought that blood vessels were passive elements that only provided metabolic and structural support for the central nervous system (CNS). However, the vasculature and the cells that comprise it are dynamic and may participate in many, if not all, aspects of CNS function. Endothelial cells, in particular, are highly heterogeneous and very plastic entities well suited and ideally placed to mediate the interactions between the vascular tree and the rest of the nervous system. There was a need to produce this volume to make the reader conscious of the fact that blood vessels have been long neglected with regard to their importance for assuring CNS function, development and maintenance. We think the chapters in this book are proficient in transmitting this point of view.

The title of this book emphasizes the very important fact that endothelial cells are capable of plastic changes. There are many examples of this throughout the book, some of which will be mentioned here. The basilar artery derives from migration of vein endothelial cells, showing that their phenotype is modifiable (Chapter 1). The microenvironment where endothelial cells reside in the forebrain determines the expression of homeobox transcription factors, which in turn determines their response to environmental signals (Chapter 1). The cells of the endothelium show extensive phenotypic heterogeneity and also functional diversity (Chapter 2). This variety favors the establishment and maintenance of compartmentalized microenvironments such as: neurogenic niches (SGZ and SVZ, Chapter 2), migratory routes (RMS, Chapter 2), sites that extensively contact blood-borne molecules (circumventricular organs, Chapter 2), regions where neurons differentiate (olfactory bulb) and locations that allow neuro-endocrine communication (median eminence, Chapter 4). Vascular cells, in particular the endothelium, are responsible for the dynamic and complex changes in blood flow and blood-brain barrier transport efficiency as a function of the local synaptic activity (neurovascular and neurobarrier coupling, Chapter 3). Plasticity of endothelial cells in the median eminence is ultimately in control of female reproduction (Chapter 4). Whether a cause or consequence of pathologies (drug-resistant epilepsy), the expression of transporter or carrier proteins is altered in endothelial cells (Chapter 5). Tumor neovascularization is extremely plastic; the vessels can grow by intussusceptive or proliferating angiogenesis, differentiation of cancer stem cells and vasculogenic mimicry (Chapter 6). Plasticity or heterogeneity of endothelial cells causes the differences between the blood-brain barrier and blood-spinal cord barrier which render the spinal cord more susceptible to certain lesion types (Chapter 7). Changes in the phenotype of endothelial cells participate importantly in the damage and recovery from stroke (Chapter 8). Evidently, the plasticity of endothelial cells has a limit, revealed by the fact that they are as vulnerable as neurons to the diverse types of damage observed in Alzheimer´s disease (Chapter 9). Endothelial cells also crucially participate in neuro-immune communication in both health and disease. For example, leukocyte entry into the CNS during inflammation requires their interaction with endothelial cells via adhesion ligand molecules and chemokine receptors. Under non-inflammatory conditions the sites of leukocyte migration into the CNS are the choroid plexus and the meninges. The chemokines and adhesion molecules also underlie leukocyte-endothelial cell communication at these sites.

Chapter 1 shows that vascular development in the CNS is stereotyped but nonetheless requires active interaction between all cells comprising the neurovascular unit. Thus, the formation and preservation of the CNS and its vessels are mutually dependent to ensure their proper operation. Chapter 2 discusses the idea that endothelial cells contribute to the generation of new neurons from undifferentiated neural progenitors, which originate from multi-potent neural stem cells. Endothelial cells may participate in neural stem cell proliferation, neuroblast migration and differentiation of neurons during development and in the adult. Moreover, the phenotypic heterogeneity and functional diversity of endothelial cells, which arise with age, seem to be important for the spatial and temporal restriction of neurogenesis to specialized microenvironments known as neurogenic niches. Chapter 3 reveals the complexity of the diverse blood-neural barriers and their importance for the homeostasis of the nervous system. Even blood flow to the brain and blood-brain barrier permeability are dynamic and complex processes, since both change as a function of local neural activity. Chapter 4 explains that the vasculature of the median eminence is similar to a neurogenic niche. Additionally, endothelial cells importantly participate in female reproduction by stimulating the withdrawal of the tanycyte outgrowths that cover the nerve terminals of neurons that secrete gonadotropin-releasing hormone and by promoting axonal sprouting from these neurons in the direction of the pituitary portal blood vessels. Chapter 5 reveals that over-expression of drug transporter proteins in the endothelial cells which comprise the blood-brain barrier may be responsible for drug-resistant epilepsy. Chapter 6 discusses the diverse molecules that participate in tumor growth, primary and secondary tumor cell invasion as well as tumor blood vessel angiogenesis. Chapter 7 expands upon the vessels of the spinal cord as well as vascular dysfunction, blood-spinal cord disruption, endothelial cell death and angiogenesis that follow spinal cord injury. Chapter 8 also touches upon angiogenesis and remodeling of the vasculature after stroke, detailing their molecular mediators. Finally, Chapter 9 highlights the importance of energy metabolism and blood vessel function in Alzheimer's disease.

Esperanza Meléndez Herrera
Bryan V. Phillips-Farfán
Gabriel Gutiérrez Ospina

Contents

CHAPTER 1

Early Development of the Vascular System Supplying the Brain

Bryan V. Phillips Farfán,[1,*] *Alma Lilia Fuentes Farías,*[2,a] *Gabriel Gutiérrez Ospina*[3] and *Esperanza Meléndez Herrera*[2,b,*]

1.1 Introduction

Vascular development in the embryonic and fetal central nervous system (CNS) is a highly stereotypical process that begins at the cervical levels of the neural tube (NT) and progresses towards the caudal and cephalic poles of the embryo following the process of neurulation. Until a few years ago, the ontogeny of blood vessels within the nervous system was believed to be a "passive" process subordinated to the metabolic demands of the neural tissue (Park et al. 2003). An increasing body of evidence, however, shows that vascularization in the developing brain proceeds through a relatively autonomous process. The vascular network guarantees proper neuronal

[1] Laboratorio de Nutrición Experimental, Instituto Nacional de Pediatría, Secretaria de Salud, México D.F., 04530, México.
Email: bvphillips@yahoo.com
[2] Laboratorio de Eco-fisiología animal, Departamento de Zoología, Instituto de Investigaciones Sobre los Recursos Naturales, Universidad Michoacana de San Nicolás de Hidalgo. Av. San Juanito Itzícuaro s/n. Col. Nueva Esperanza C.P. 58337 Morelia, Michoacán, México.
[a] Email: almafuentes70@hotmail.com
[b] Email: oazul74@yahoo.com.mx
[3] Laboratorio de Biología de Sistemas, Departamento de Biología Celular y Fisiología, Instituto de Investigaciones Biomédicas; Coordinación de Psicobiología, Facultad de Psicología, Universidad Nacional Autónoma de México, México, D.F. 04510, México.
Email: gabo@biomedicas.unam.mx
* Corresponding authors

development by providing nutrients, migratory guidance and trophic modulatory factors that allow neural cells to proliferate and differentiate (Sun et al. 2010). Blood vessels also modulate synaptic plasticity (Hopper and Garthwaite 2006) and the interactions among neural, immune and endocrine cells during development (Matsumura and Kobayashi 2004). In addition, it is now widely recognized that the patterning of the brain's vascular and neural networks is coordinated by the same set of molecular signals (Zacchigna et al. 2008). Early neurovascular communication thus establishes a life-long relationship between these systems that guarantees the proper function of the CNS. On the other hand, deficient neurovascular communication leads to several pathological conditions (Girouard and Iadecola 2006).

The formation of blood vessels throughout the nervous system, as in the rest of the body, may occur *de novo* (i.e., vasculogenesis) or from pre-existing vessels (i.e., angiogenesis). In turn, angiogenesis encompasses two mechanisms (Risau 1997). The first one, known as non-sprouting angiogenesis, is characterized by blood vessel elongation and intussusception in the absence of cell proliferation (Burri et al. 2004). The second one, termed sprouting angiogenesis, leads to the growth of blood vessels following the proliferation of endothelial cell precursors (Heinke et al. 2012). A combination of these processes produces the intricate vascular patterns observed in the adult brain.

Vascular development in the vertebrate nervous system is a complex process. Throughout this chapter, information obtained from a variety of animal models is used to elaborate a general picture in an effort to provide the reader with a comprehensive view on how the vertebrate nervous system is vascularized.

1.2 Vascular Patterning in the Neural Tube

During embryonic development, the brain and spinal cord are formed by morphogenetic movements combined with the processes of differential growth and remodeling that shape the NT. Once formed, the NT enlarges at its rostral-most segments giving rise to the primary brain vesicles: the caudal-most rhombencephalon, the mesencephalon and cephalic-most prosencephalon. Each of these vesicles is transiently divided along its anterio-posterior axis by metameric units called neuromeres. These sub-divisions represent morphogenetic fields and hence define cell migration and connectivity patterns, as well as cell fate within them (Puelles and Rubenstein 2003, Kiecker and Lumsden 2005). As development proceeds, differential proliferation in the rhombencephalon and prosencephalon leads to the formation of the mielencephalon/metencephalon and diencephalon/secondary prosencephalon, respectively, thus generating the so-called secondary brain vesicles (Puelles et al. 2013).

Early vascular patterning of the NT resembles that of the neural tissue during neurulation (Copp et al. 2003); however, angiogenesis precedes neurogenesis (Gotz 2013). The formation of blood vessels is triggered by NT-derived signals that induce the elaboration of the perineural vascular plexus (PNVP) around it. In the brain *primordium*, the PNVP is formed through vasculogenesis and angiogenesis. Angioblasts originating from head paraxial mesoderm (i.e., vasculogenesis) and from the cephalic vascular plexus (i.e., angiogenesis) are incorporated within the PNVP vascular network (Walls et al. 2008). At the spinal cord, the PNVP is formed by angioblasts coming from the somites and the lateral mesoderm (i.e., vasculogenesis) (Klessinger and Christ 1996, Pardanaud et al. 1996, Pardanaud and Dieterlen-Lievre 1999, Ambler et al. 2001) and from the vertebral artery and dorsal longitudinal anastomotical vessel (i.e., angiogenesis) (Walls et al. 2008). Later, the PNVP originates the intraneural vascular plexus (INVP), which supplies the CNS with nutrients and oxygen (James and Mukouyama 2011) through a process of angiogenesis (Lee et al. 2009). The development of the PNVP and INVP follows a stereotypical pattern proceeding rostrally from the rhombencephalon (Walls et al. 2008), caudally from the cervical segments of the spinal cord and from ventral to dorsal along the NT (Kurz et al. 2004) (Fig. 1A).

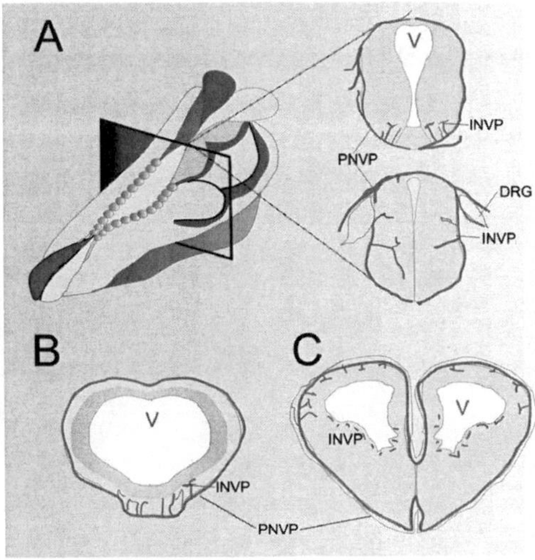

Figure 1. Early brain vascularization follows a stereotypical pattern. Neural tube closure instructs perineural vascular plexus (PNVP) formation, which starts at the rhombencephalic-cervical boundary (A, *left*). Intraneural vascular plexus (INVP) development is modulated by neuromere-specific local cues as demonstrated by its entry points into the central nervous system. PNVP-derived vessels sprout into the rhombencephalon (A, *top right*) and mesencephalon (B) ventrally while they enter the spinal cord laterally (A, *bottom right*). In the prosencephalon, pial and ventricular vessels develop independently (C). DRG= dorsal root ganglion, V= ventricle.

Color image of this figure appears in the color plate section at the end of the book.

1.2.1 Ontogeny of the Perineural Vascular Plexus

In mice, the cardiovascular system begins its formation by embryonic day (E) 7.3. At this age, the *endocardium primordia* appears in the thoracic segment of the embryo and a few hours later (E7.8), the aortic *primordia* can be clearly identified. A day later, the *endocardium* of the heart as well as the ventral and dorsal aortic segments are in place. A complete circulatory system is observed between E8–8.5 (Drake and Fleming 2000). Roughly by the same time, the cephalic vasculature begins the formation of the PNVP (Walls et al. 2008). At first, disconnected clusters of endothelial cells are identified throughout the cephalic mesenchyme and lateral mesoderm. Shortly afterwards, these cell clusters aggregate to form a rudimentary plexus at the level of the occipital somites. Then, capillaries grow towards the rostral and caudal aspects of the embryo. The latter reach the cervical somites a bit later in development. The NT is completely encased by the PNVP by E9.5 (Walls et al. 2008), an age that corresponds to incubation day (ID) four in chicken embryos (Kurz 2009).

The ontogeny of the PNVP is regulated by NT-derived signals, most important among them vascular endothelial growth factor (VEGF)-A. This factor induces differentiation and migration of presomitic mesoderm-derived endothelial progenitors after binding to the VEGF-2 receptor (VEGFR-2; Hogan et al. 2004). In addition, the PNVP in each side of the embryo develop independently from each other (Kurz and Christ 2002, Kurz 2009) due to the presence of NT-released chordin and noggin that work as endothelial cell repellent signals along the midline (Reese et al. 2004) (Fig. 2).

Figure 2. Neural tube-derived cues regulate perineural vascular plexus (PNVP) formation and vessel entry into the central nervous system (i.e., intraneural vascular plexus development). Some of the molecules expressed in angioblasts or endothelial cells are shown on the left, while neural tissue-derived signals are shown on the right (color code indicates the specific cell type that produces the protein). FP= floor plate, V= ventricle. For abbreviations of the molecules see text.

Color image of this figure appears in the color plate section at the end of the book.

1.2.2 Ontogeny of the Intraneural Vascular Plexus

In mouse embryos between E9.5 and E10, the PNVP forms branches that penetrate the developing NT from the *pia mater* deep into the ventricles (Kurz et al. 1996, Fantin et al. 2013). In vertebrates, this process starts at the rhombencephalic floor plate (Kurz et al. 1996) and then moves towards the cephalic and caudal poles of the embryo (Rovainen and Kakarala 1989, Kurz 2009, Marin-Padilla 2012) (Fig. 1A, *top right*). Blood vessels entering the spinal cord branch from primitive arteries (Kurz et al. 1996). These blood vessels reach the transition zone located between the ependymal and mantle layers, where they give rise to arch-like ventral sprouts (Kurz and Christ 2002) that later join lateral venous sprouts originating from the PNVP (Kurz et al. 2004). INVP angiogenesis at the forelimb bud level in mouse embryos follows a slightly different pattern to that seen in the chick embryo at cervical levels. It begins in the lateral border of the NT at E9.8 and then grows following a medial-ventral gradient until reaching the arch-like vessels in the floor plate at E10.2 (Nagase et al. 2005) (Fig. 1A, *bottom right*).

Since the INVP is formed by angiogenesis, it involves endothelial cell proliferation, migration and maturation. All these processes are known to be modulated by NT-derived growth factors (Risau 1997). Accordingly, in the chick spinal cord, the initial angiogenic sprouts are induced by the VEGF-VEGFR-2 pathway (James et al. 2009). In mice, this same process seems to be regulated by angiopoietin (Ang)-1, sonic hedgehog (Nagase et al. 2005) and epidermal growth factor (Sato et al. 1995). In addition, the orphan G protein-coupled receptor-124 (Anderson et al. 2011), transforming growth factor (TGF)-β (Mu et al. 2008) and wingless-related integration site (Wnt)-7a or 7b (Stenman et al. 2008) are required for the occurrence of normal angiogenic sprouting from the PNVP along the neural axis (Fig. 2).

Lastly, the final architectural features of the CNS vasculature require tightly controlled processes that involve a myriad of factors. For example, blood vessel fine tuning within the hindbrain requires the presence of neuropilin (Npn)-1 (Gerhardt et al. 2004), delta-like (Dll)-4 ligand (Suchting et al. 2007) and netrin-1 (Lu et al. 2004) to either facilitate or inhibit angiogenic sprout formation and branching (Fig. 2).

1.3 Vascularization in the Developing Brain

As previously mentioned, the process of vascularization throughout the vertebrate embryonic-fetal CNS generally follows specific gradients that are determined by multiple cues, such as NT closure (Copp et al. 2003, Kurz et al. 2004) and segmentation (Vasudevan et al. 2008), VEGF expression levels (Breier et al. 1992, Kawasaki et al. 1999) and local instructive signals released by discrete cellular domains (Hashimoto et al. 2001, Sure et al.

2001). Hence, this section of the chapter will review literature that clearly illustrate these general principles.

1.3.1 Development of the Hindbrain Vascular Network

During early vertebrate development, the full extent of the hindbrain is divided into eight units called rhombomeres (Kiecker and Lumsden 2005). In the zebrafish, the first vascular buds emerge from the primordial hindbrain veins. These buds ramify through the rostral aspect of the hindbrain most frequently at the levels of rhombomeres two, three or five, less often at rhombomeres four and six and never at rhombomere one, seven or eight (Fujita et al. 2011). Blood vessels first invade the center of each rhombomere to then grow dorsally until reaching the halfway point of the dorso-ventral axis. At that point, they turn laterally and ventrally forming a loop that allows blood vessels of both sides to form anastomoses with each other, with posterior communicating vessels at cephalic segments and with the basilar artery at the caudal segments (Ulrich et al. 2011). Thus, it is clear that vascularization along the zebrafish hindbrain does not occur simultaneously. Even though hindbrain vascularization exhibits a metameric pattern, rhombencephalic vessel sprouts do not occur at inter-rhombomere boundaries (Ulrich et al. 2011). Finally, it is worth emphasizing that blood vessel patterning shows spatial congruence with axonal tracts and neuronal nuclei that develop before them. For instance, the basilar artery runs along the medial longitudinal fascicle. Interestingly, some of its branches establish close associations with the branchiomotor and reticulospinal neurons (Ulrich et al. 2011), but avoid oligodendrocytes and neurons of the abducens motor nucleus in rhombomeres five and six (Fujita et al. 2011).

The pattern of hindbrain vascularization is controlled by several factors. In the zebrafish, NT-derived VEGF is necessary for formation of the primordial veins, the basilar artery (Covassin et al. 2006) and the INVP (Fujita et al. 2011, Ulrich et al. 2011). In addition, the chemokine C-X-C motif ligand-12b is required to generate hindbrain vascular buds and to orient basilar artery sprouts to the hindbrain midline, following the activation of phosphatidylinositide 3-kinase in endothelial cells after binding its receptor cxcr-4a (Fujita et al. 2011). The venous markers VEGFR-4 and disabeled (dab)-2 are present in the PNVP during early development. In contrast, dab-2 expression is reduced during the formation of the basilar artery. The basilar artery seems to originate primarily, if not entirely, from migration of primordial vein-derived endothelial cells (Fujita et al. 2011, Ulrich et al. 2011); suggesting that the endothelial cell phenotype is rather plastic during development, with some venous markers being expressed in cells that will later form part of the arterial walls (Fujita et al. 2011).

In mice, the pattern of parenchymal vascularization of the hindbrain follows the same basic sequence described before. At E9.5, pial vessels sprout radially into the ventricular zone and then turn laterally just below the ependymal layer to form the subventricular vascular plexus by E10.25–E12.5 (Gerhardt et al. 2004) (Fig. 1A, *top right*). The initial growth of these blood vessels is guided by radial glia and promoted by VEGF released from cells in the subventricular zone (Fantin et al. 2013). Subventricular lateral branching involves Npn-1 (Fantin et al. 2013) and the late formation of anastomosis is regulated by yolk sac-derived macrophages that enter the brain parenchyma independently from blood vessels (Fantin et al. 2010). Additionally, hindbrain vessel branching is negatively regulated by Dll-4 located in endothelial cells (Suchting et al. 2007) (Fig. 2).

1.3.2 Development of the Midbrain Vascular Network

During embryonic development, the mesencephalon does not subdivide as the other regions of the brain. Its alar plate gives rise to the *tectum*, whereas its basal plate differentiates into the *tegmentum*. In chick embryos, midbrain vascularization proceeds sequentially following a ventral to dorsal gradient starting at the *tegmentum*. At the initial stages, only a handful of vessels derived from the PNVP enter the midbrain orthogonally through the mantle layer (Roncali and Ambrosi 1982) (Fig. 1B). A day later, new vessels invade the ventrolateral *tectum* from the pial surface. At ID6, scant radial vessels sprout into the dorsomedial *tectum* (Roncali et al. 1985a). During the following days (ID 7–14), vessel density augments since radial vessels increase in number and length, branch and anastomose with one another to form the midbrain subventricular plexus. Later in development (ID 14–21), all vessels become surrounded by astrocytic endfeet, promoting endothelial cell maturation and the formation of the blood-brain barrier (BBB) (Bertossi et al. 1993). It has been suggested that vascular and neuronal development are somehow linked since the first vascular sprouts enter the ventrolateral *tectum* at a time when peak proliferative activity in the neuroepithelium is taking place. Also, branching activity of blood vessels entering the posterior *tectum* coincides with intense neuronal differentiation (Roncali and Ambrosi 1982, Roncali et al. 1985a).

In *Xenopus laevis* embryos, the process of hindbrain vascularization (Rovainen and Kakarala 1989) is fully comparable to that described for the chick. However, neither the removal of tectal visual afferents in this model (Rovainen and Kakarala 1989) nor chronic hypoxia in the chick embryo (Roncali et al. 1985b) significantly disturbs the progression of hindbrain angiogenesis, challenging the notion that angiogenesis and neurogenesis are interdependent.

In the zebrafish midbrain, oxygenated blood arrives through the basal communicating artery and deoxygenated blood departs through the choroidal vascular plexus (Isogai et al. 2001). Non-sprouting and sprouting angiogenesis contribute to create a highly intricate and anastomosed midbrain vascular network (Chen et al. 2012). Vessel regression and remodeling is commonly observed in the vascular bed that derives from the choroidal vascular plexus (Chen et al. 2012). Vascular pruning, promoted by diminished mechanical (i.e., hemodynamic) forces, improves the efficiency of the blood flow throughout the vascular network. Reduced blood flow activates the rat sarcoma (Ras)-related C3 botulinum toxin substrate (Rac)-1, resulting in endothelial cell migration from pre-existing vessels into adjacent vessels (Chen et al. 2012).

In the rat midbrain, as in other parts of the CNS, VEGF promotes angiogenesis by increasing the proliferation of endothelial cells. Interestingly, VEGF also stimulates astrocyte proliferation and improves neuronal survival, particularly of dopaminergic neurons, in mesencephalic explants (Silverman et al. 1999).

1.3.3 Development of the Forebrain Vascular Network

Early in vertebrate development, the forebrain is divided into two vesicles: the diencephalon and the secondary prosencephalon. According to recent cell fate mapping and gene expression studies, the diencephalon gives rise to the pretectum and thalamic nuclei, whereas the secondary prosencephalon forms the hypothalamus, retina, telencephalon and olfactory bulbs (Puelles et al. 2013). In this context it is worth noting that, with the exception of the telencephalic structures and the retina, we lack information on the vascular development of many structures derived from the prosencephalon. Hence, this part of the chapter will comment mostly on aspects of the development of the telencephalic and retinal vascular beds.

In mouse embryos, the initiation of telencephalic vascularization occurs by E10 to E11. During this time frame, the PNVP generates sprouts that invade the dorsolateral aspect of the telencephalic vesicle and ventral ventricular vessels appear (Vasudevan et al. 2008). In contrast to what happens in caudal vesicles, telencephalic structures are vascularized following a precise ventro-dorsal ventricular gradient that is independent from pial-derived vessels (Fig. 1C). Furthermore, as is the case of neuronal populations, endothelial cell fate and migration in telencephalic structures is regulated by region-exclusive homeobox proteins. Endothelial cells in the ventral telencephalon express the thyroid transcription factor-1 and distal-less like-1 and -2 proteins. In contrast, endothelial cells migrating through the dorsal telencephalon produce the paired box (Pax)-6 protein. These region-exclusive homeobox transcription factors differentially modify the expression of molecules critical for endothelial development, such as VEGF and brain derived neurotrophic factor, and guide endothelial cell migration through cell autonomous region-specific signals (Vasudevan et al. 2008).

Glial cells contribute a great deal to forebrain vascularization. Radial glia prevent regression of the vasculature irrigating the cerebral cortex by inhibiting Wnt signaling and the expression of matrix metalloproteinases (Ma et al. 2013). On the other hand, astrocytes and their precursors expand their processes onto vessel walls, covering and making widespread contacts with the vasculature. Thus, glia lend structural support in addition to guiding growing vessels and migrating neuroblasts. Additionally, glial cells participate in BBB development, since only vessels associated with glia express BBB markers (Virgintino et al. 1998).

Interestingly, neurogenesis follows a progression pattern similar to that seen for the vasculature during telencephalic development. However, vascularization takes place first (Vasudevan et al. 2008), suggesting that endothelial cells importantly participate in corticogenesis (Li et al. 2013). In support, VEGF gene deletion in endothelial cells alters neuronal proliferation and migration, as well as axonal tract formation in the developing cortex (Li et al. 2013).

Finally, a dynamic analysis of hypophyseal vascularization was recently reported in zebrafish embryos (Gutnick 2011). Hypophyseal vascular formation begins 24 hours after the formation of the hypothalamic-neurohypophyseal tract. Arteries irrigating the neurohypophysis form within the hypophysis itself, lengthen forward and attach to the palatocerebral arteries. Afterwards, the main lateral vein sinuses lengthen anteriorly, develop medial offshoots and come round the posterior neurohypophysis. The hypophyseal veins connect bilaterally to each other and to the hypophyseal arteries to form a loop. Interestingly, neural-derived oxytocin and its receptor, located in endothelial cells, control the development of the neurohypophyseal vessels (Gutnick 2011).

1.3.4 Development of the Retinal Vascular Network

As mentioned earlier, the secondary prosencephalon provides the anatomical *substratum* necessary for the retina to be formed. In a three-week old human embryo, bilateral symmetrical evaginations called optic cups emerge from the anterolateral surface of the prosencephalon; the retina proper arises from the ventral region of these structures. As the retina *primordium* evaginates by the fourth week of human gestation, the central hyaloid artery arises from the choroidal veins that surround the optic cup (Anand-Apte and Hollyfield 2011). However, this blood vessel is a transient structure. In all species so far studied, it disappears as ontogeny proceeds

(Fruttiger 2002, Saint-Geniez and D'Amore 2004). The process of hyaloid artery regression involves Wnt-7b (released by macrophages) binding to the frizzled (Fzd)-4 receptor. This process also requires the participation of the low-density lipoprotein receptor related protein (Lrp)-5 (Kato et al. 2002, Lobov et al. 2005), Norrin (a Fzd-4 receptor ligand) (Richter et al. 1998), Ang-2 acting on tyrosine kinases with immunoglobulin-like and epithelial growth factor-like domains (Tie)-2 (Hackett et al. 2002) and fibroblast growth factor receptors (Rousseau et al. 2003).

As soon as the hyaloid artery degenerates, a vascular plexus forms at the distal-most segment of the optic nerve. In the mouse, this vascular network is called the primary vessel plexus (PVP) and it spreads over the nerve fiber layer during the first ten days after birth (Dorrell et al. 2002). Newly formed vessels grow towards the outer side of the retina and, between days seven and nine, constitute a secondary vascular plexus at the outer edge of the inner nuclear layer (INL). During the third postnatal week, an intermediate vascular plexus is formed between the superficial and deep plexuses (Dorrell et al. 2002). The formation of the PVP is likely promoted by retinal astrocytes that experience hypoxia and release VEGF (Pierce et al. 1996, West et al. 2005) by the day of birth. Retinal astrocytes originate from Pax-6 positive optic nerve astroblasts (Mi et al. 1999, Chu et al. 2001), are present in the optic nerve by E15 (Dorrell et al. 2002) and enter the retina days later (Ling and Stone 1988, Dorrell et al. 2002, Fruttiger 2002). In addition to releasing VEGF, these cells also express platelet derived growth factor receptor (PDGFR)-α (Mudhar et al. 1993) and vimentin (Chu et al. 2001). Once the PVP is formed, the central retinal artery and vein develop first followed by their branches. This vascular developmental pattern seems to respond to the graded availability of VEGF across the retina (Ruhrberg et al. 2002, Stalmans et al. 2002, Gerhardt et al. 2003). When retinal oxygen concentrations are adequate, blood vessel branching, astrocyte proliferation and VEGF release stop (West et al. 2005).

Diverse genes and their protein products participate in modeling the retinal angio-architecture. For instance, the tailless homologue promotes astrocyte proliferation (Miyawaki et al. 2004). Apelin, Dll-4, platelet derived growth factor (PDGF)-B, uncoordinated (Unc)-5b and VEGFR-2 support angiogenesis during the formation of the primary inner vascular plexus (Saint-Geniez et al. 2002, Gerhardt et al. 2003, Claxton and Fruttiger 2004, Lu et al. 2004, Suchting et al. 2007). Other ligand-receptor families that also participate in retinal angiogenesis are Ephrin-Eph, Semaphorin-Plexin, Slit-Roundabout and Netrin-Unc (Eichmann 2005). Müller glial cells appear by E18-P12 in the rat (Rapaport et al. 2004). These cells produce Norrin which, together with tetraspanin-12, interacts with Fzd-4 and Lrp-5 to induce angiogenesis by activating the β-catenin and sex-determining region Y-related high mobility group box-17 signaling pathway in retinal endothelial cells (Junge et al. 2009, Ye et al. 2009). Additionally, mutual signaling among endothelial cells, smooth muscle cells and pericytes promotes retinal vascular maturation. Indeed, smooth muscle cells and pericytes secrete Ang-1 that acts on endothelial Tie-2 receptors; while endothelial cells release PDGF that acts on PDGFR-β in pericytes or smooth muscle cells (Suri et al. 1996, Hellstrom et al. 1999). In addition, VEGF released by INL glial cells promotes angiogenesis in the outer vascular plexus (Stone et al. 1995, although see Fruttiger 2007). Similar to the primary plexus, the center matures first and then the periphery. Veins of the primary plexus grow angiogenic sprouts that enter the retina following Müller cell extensions, perpendicularly to the primary plexus. They make a 90° turn upon reaching the inner and outer boundaries of the INL; therefore, two vessel systems are laid down parallel to the primary plexus (Fruttiger 2007). Ang-2 produced by horizontal INL cells might stimulate angiogenesis by inhibiting Ang-1-Tie-2 signaling (Hackett et al. 2000, Hackett et al. 2002). Norrin and Fzd-4 also participate in outer plexus vascular development (Xu et al. 2004b, Luhmann et al. 2005).

1.4 Brain Vascular Malformations

The term brain vascular malformation (BVM) groups a series of disorders that affect the anatomical and cytological arrangement of the blood vessels that nourish the brain. Arterio-venous (AVMs; tangles of abnormal and poorly formed blood vessels) and cavernous (CaV; clusters of abnormal dilated vessels also called cavernoma) malformations are the two most prevalent types of BVMs. Although BVMs were first thought to arise during embryonic development, it is now known that they can appear at any age throughout the organism's life (Minakawa et al. 1989, Clatterbuck et al. 2000). BVMs arise at sites where capillaries would normally be located. In an AVM, arteries and veins seem to coalesce forming a meshwork of direct, high-flow shunts. In a CaV, the vessels are thin-walled, dilated, low-flow and may contain thrombi. Capillary dysplasia might be the first event that leads to BVMs (Braverman et al. 1990, Rigamonti et al. 1991). A single lesion probably results in the production of multiple malformations (Sato et al. 2004, Shenkar et al. 2008).

Mutations in a single copy of the endoglin or activin-like kinase receptor (ALK)-1 genes result in loss of protein function and subsequent hereditary hemorrhagic telangiectasia (HHT) type 1 or 2, respectively (Goumans et al. 2009). Similarly, mutations in the SMAD (small, mothers against decapentaplegic)-4 gene are also associated with HHT (Goumans et al. 2009). On the other hand, mutations in Kirsten-Ras-revertant interaction trapped (KRIT)-1, osmosensing scaffold for mitogen activated protein kinase kinase kinase-3 (OSM) and programmed cell death (PDCD)-10 genes are the most likely factors that cause CaVs (Uhlik et al. 2003, Glading et al. 2007, Ma et al. 2007, Crose et al. 2009).

The prevailing hypothesis regarding development of BVMs is that they arise as a result of two hits. The first hit corresponds to a hereditary mutation in a copy of the above genes, while the second hit can be a somatic mutation in the second copy or a localized environmental alteration. This hypothesis might explain why BVMs are confined to a small area of the vasculature. Evidence supporting the idea comes from CaV patient brain tissue showing germline and somatic biallelic mutations in a subpopulation of endothelial cells within the lesion site (Gault et al. 2005, Akers et al. 2009, Gault et al. 2009, Pagenstecher et al. 2009). Further evidence comes from AVMs in which hemodynamic stress enhances local inflammation and/or the expression of angiogenic factors (Hashimoto et al. 2006, Goettsch et al. 2008). In support of this possibility, previous studies have shown that venous hypertension increases the expression of hypoxia-inducible factor-1 and VEGF (Zhu et al. 2006). Also, patients suffering BVMs have augmented serum VEGF levels and up-regulated expression of VEGF and Ang-2 in the lesion (Jung et al. 2003, Hashimoto et al. 2005). Local increases in the expression of these angiogenic factors induce endothelial cell proliferation (Hashimoto et al. 2001, Sure et al. 2001); thus increasing the occurrence of vascular dysplasia, which is aggravated by increments of local blood-flow in endoglin and ALK-1 mutant mice (Xu et al. 2004a, Hao et al. 2008).

Endoglin, ALK-1 and SMAD-4 constitute part of the TGF-β signaling cascade (ten Dijke et al. 2008). All these proteins are expressed in endothelial cells during development and their expression is reduced in adulthood, unless needed for vascular repair (Ma et al. 2000, Jonker and Arthur 2002, Seki et al. 2003, van Laake et al. 2006, ten Dijke et al. 2008). Both ALK-1 and SMAD-4 are essential for vascular recruitment of smooth muscle cells (Lan et al. 2007, Park et al. 2008). Thus, these data suggest that the principal problem in AVMs and HHTs is a defective function of the endothelial TGF-β pathway which results in secondary deregulation of smooth muscle development. This is supported by the finding that bone morphogenetic protein-9 or 10, also members of the TGF-β family, are the physiological ALK-1 ligands (David et al. 2007, Scharpfenecker et al. 2007). Interestingly, mice with reduced expression of ALK-1 or endoglin have dilated, thin-walled arteries with reduced smooth muscle cells and lower expression of ephrin B2, an arterial marker (Urness et al. 2000, Sorensen et al. 2003). Thus, defects in differentiation between arteries and veins (or their specification) may also participate in the pathogenesis of AVMs. In addition, monocytes might be also involved in BVMs since they express endoglin. Furthermore, even though endoglin or ALK-1 knockout mice show no major defects of vasculogenesis, they lack secondary capillary beds and pericyte vascular coverage. These mice die during gestation likely from severe impairments in vascular development (Li et al. 1999, Urness et al. 2000). Alterations in a single copy of endoglin or ALK-1 in mice result in spontaneous vascular dysplasia, telangiectasia and random hemorrhages. Small vessel malformations are observed in these mice together with disappearance of vascular smooth muscle cells (Bourdeau et al. 1999, Satomi et al. 2003, Srinivasan et al. 2003, Torsney et al. 2003, Park et al. 2008). Also, endothelial nitric oxide synthase is dysfunctional in mice with an altered copy of endoglin, resulting in production of superoxide and impaired vascular function (Toporsian et al. 2005).

KRIT-1, OSM and PDCD-10 are expressed by endothelial cells, glia and neurons (Petit et al. 2006, Plummer et al. 2006). Endothelial, but not glial or neuronal, expression of KRIT-1 and OSM is needed for normal cerebral vascular formation and function (Hogan et al. 2008, Boulday et al. 2009, Whitehead et al. 2009). All these proteins constitute a single complex that interacts with cytoskeletal, signaling, cell junction and extracellular matrix proteins necessary to maintain vascular structural stability, such as: Ras-related protein-1 (Bos 2005, Carmona et al. 2009), heart of glass-1 (Whitehead et al. 2004, Beraud-Dufour et al. 2007, Glading et al. 2007, Kleaveland et al. 2009), β1-integrin and integrin cytoplasmic domain-associated protein-1α (Zhang et al. 2008, Faurobert and Albiges-Rizo 2010, Hilder et al. 2007). KRIT-1 also promotes β1-integrin action, such as endothelial adhesion to the extracellular matrix (Rupp and Little 2001), lumen formation (Iruela-Arispe and Davis 2009) and recruitment of pericytes and smooth muscle cells to vessels (Astrof and Hynes 2009).

Finally, endothelial cells with altered OSM function have cytoskeletal defects, diminished junctions between them and impaired barrier function (Hogan et al. 2008, Whitehead et al. 2009). Accordingly, OSM knockout mice show lesions that resemble those observed in cerebral CaV patients (Plummer et al. 2006). OSM is named due to its ability to restore cell shape and volume after hyperosmotic shock. It acts as a scaffold protein for Rac-1 and upstream mitogen activated protein kinases (Uhlik et al. 2003). Thus, OSM regulates cytoskeletal dynamics and cell shape via mitogen activated protein kinase signaling. Cytoskeletal dynamics are also important for maintaining cell polarity; KRIT-1, OSM and PDCD-10 maintain cell polarity by modulating Ras homolog family member A (RhoA) activity on the cytoskeleton. Indeed, OSM increases RhoA degradation by binding the SMAD specific E3 ubiquitin protein ligase-1 (Crose et al. 2009). KRIT-1 and PDCD-10 also inhibit RhoA activity. Although the mechanism of KRIT-1 action is unknown, PDCD-10 binds to protein kinases of the germinal center kinase sub-family to stimulate Golgi apparatus assembly and activate moesin (Preisinger et al. 2004, Fidalgo et al. 2010, Zheng et al. 2010).

1.5 Conclusions

1. The early establishment of blood vessels in the CNS follows a highly stereotyped pattern that is conserved throughout vertebrate evolution, suggesting a genetic underpinning. Nonetheless, CNS vascularization is an active process requiring communication between all cells of the neurovascular unit. This is particularly true for intraneural sprouting angiogenesis which follows a very specific ventro-dorsal pattern highly regulated by local signals throughout the entire antero-posterior axis.
2. Multi-directional communication among resident neuronal and glial cell populations with the newly formed endothelial cells is very important for brain function. Endothelial cells are perfectly located between the blood and the neural tissue, giving them the capacity to regulate molecular trafficking and consequently neural function. Similarly, neurons and glia can help in determining the phenotype of endothelial cells endowing them with a high specificity and heterogeneity. Although the metabolic demands of the neural tissue do not participate during early vascularization, the establishment and maintenance of the CNS and its vasculature are interdependent to guarantee their normal function.
3. The close interdependence between neural and endothelial cells is demonstrated by their response to the same signals. For example, VEGF and its receptors participate in nearly all aspects of CNS vascular development. This trophic factor also plays a prominent role in astrocyte proliferation, neuronal survival and neural cell differentiation.
4. Endothelial cells are likely to exert functions besides those commonly attributed to them (e.g., nutrient delivery) during CNS development. Exploring this possibility awaits the development of experimental designs capable of providing an adequate milieu for neural cells in the absence of blood vessels. Once these conditions are established, the re-introduction of controlled numbers of endothelial cells will undoubtedly provide scenarios in which to test their role on various aspects of neuronal and glial biology during brain development.

Acknowledgements

Authors are grateful to Jesús Ramírez, Ivonne Mora and Raymundo Reyes for their administrative and technical support. This work was funded by grants from CONACyT (grant No. 180762) and PROMEP (UMSNH-PTC-336) to EMH and by the INP (grant 055-2012) to BVPF.

References

Akers, A.L., E. Johnson, G.K. Steinberg, J.M. Zabramski and D.A. Marchuk. 2009. Biallelic somatic and germline mutations in cerebral cavernous malformations (CCMs): evidence for a two-hit mechanism of CCM pathogenesis. Hum. Mol. Genet. 18: 919–930.

Ambler, C.A., J.L. Nowicki, A.C. Burke and V.L. Bautch. 2001. Assembly of trunk and limb blood vessels involves extensive migration and vasculogenesis of somite-derived angioblasts. Dev. Biol. 234: 352–364.

Anand-Apte, B. and J.G. Hollyfield. 2010. Developmental anatomy of the retinal and choroidal vasculature. pp. 9–15. *In*: D.A. Dartt, J. Besharse and R. Dana [eds.]. Encyclopedia of the Eye. Elsevier, Oxford, GB.

Anderson, K.D., L. Pan, X.M. Yang, V.C. Hughes, J.R. Walls, M.G. Dominguez, M.V. Simmons, P. Burfeind, Y. Xue, Y. Wei, L.E. Macdonald, G. Thurston, C. Daly, H.C. Lin, A.N. Economides, D.M. Valenzuela, A.J. Murphy, G.D. Yancopoulos and N.W. Gale. 2011. Angiogenic sprouting into neural tissue requires Gpr124, an orphan G protein-coupled receptor. Proc. Natl. Acad. Sci. USA 108: 2807–2812.

Astrof, S. and R.O. Hynes. 2009. Fibronectins in vascular morphogenesis. Angiogenesis 12: 165–175.

Beraud-Dufour, S., R. Gautier, C. Albiges-Rizo, P. Chardin and E. Faurobert. 2007. Krit1 interactions with microtubules and membranes are regulated by Rap1 and integrin cytoplasmic domain associated protein-1. FEBS J. 274: 5518–5532.

Bertossi, M., L. Roncali, B. Nico, D. Ribatti, L. Mancini, D. Virgintino, G. Fabiani and A. Guidazzoli. 1993. Perivascular astrocytes and endothelium in the development of the blood-brain barrier in the optic tectum of the chick embryo. Anat. Embryol. (Berl.) 188: 21–29.

Bos, J.L. 2005. Linking Rap to cell adhesion. Curr. Opin. Cell Biol. 17: 123–128.

Boulday, G., A. Blecon, N. Petit, F. Chareyre, L.A. Garcia, M. Niwa-Kawakita, M. Giovannini and E. Tournier-Lasserve. 2009. Tissue-specific conditional CCM2 knockout mice establish the essential role of endothelial CCM2 in angiogenesis: implications for human cerebral cavernous malformations. Dis. Model. Mech. 2: 168–177.

Bourdeau, A., D.J. Dumont and M. Letarte. 1999. A murine model of hereditary hemorrhagic telangiectasia. J. Clin. Invest. 104: 1343–1351.

Braverman, I.M., A. Keh and B.S. Jacobson. 1990. Ultrastructure and three-dimensional organization of the telangiectases of hereditary hemorrhagic telangiectasia. J. Invest. Dermatol. 95: 422–427.

Breier, G., U. Albrecht, S. Sterrer and W. Risau. 1992. Expression of vascular endothelial growth factor during embryonic angiogenesis and endothelial cell differentiation. Development 114: 521–532.

Burri, P.H., R. Hlushchuk and V. Djonov. 2004. Intussusceptive angiogenesis: its emergence, its characteristics, and its significance. Dev. Dyn. 231: 474–488.

Carmona, G., S. Gottig, A. Orlandi, J. Scheele, T. Bauerle, M. Jugold, F. Kiessling, R. Henschler, A.M. Zeiher, S. Dimmeler and E. Chavakis. 2009. Role of the small GTPase Rap1 for integrin activity regulation in endothelial cells and angiogenesis. Blood 113: 488–497.

Chen, Q., L. Jiang, C. Li, D. Hu, J.W. Bu, D. Cai and J.L. Du. 2012. Haemodynamics-driven developmental pruning of brain vasculature in zebrafish. PLoS Biol. 10: e1001374.

Chu, Y., S. Hughes and T. Chan-Ling. 2001. Differentiation and migration of astrocyte precursor cells and astrocytes in human fetal retina: relevance to optic nerve coloboma. FASEB J. 15: 2013–2015.

Clatterbuck, R.E., J.L. Moriarity, I. Elmaci, R.R. Lee, S.N. Breiter and D. Rigamonti. 2000. Dynamic nature of cavernous malformations: a prospective magnetic resonance imaging study with volumetric analysis. J. Neurosurg. 93: 981–986.

Claxton, S. and M. Fruttiger. 2004. Periodic Delta-like 4 expression in developing retinal arteries. Gene Expr. Patterns 5: 123–127.

Copp, A.J., N.D. Greene and J.N. Murdoch. 2003. The genetic basis of mammalian neurulation. Nat. Rev. Genet. 4: 784–793.
Couly, G.F., P.M. Coltey and N.M. Le Douarin. 1993. The triple origin of skull in higher vertebrates: a study in quail-chick chimeras. Development 117: 409–429.
Covassin, L.D., J.A. Villefranc, M.C. Kacergis, B.M. Weinstein and N.D. Lawson. 2006. Distinct genetic interactions between multiple Vegf receptors are required for development of different blood vessel types in zebrafish. Proc. Natl. Acad. Sci. USA 103: 6554–6559.
Crose, L.E., T.L. Hilder, N. Sciaky and G.L. Johnson. 2009. Cerebral cavernous malformation 2 protein promotes smad ubiquitin regulatory factor 1-mediated RhoA degradation in endothelial cells. J. Biol. Chem. 284: 13301–13305.
David, L., C. Mallet, S. Mazerbourg, J.J. Feige and S. Bailly. 2007. Identification of BMP9 and BMP10 as functional activators of the orphan activin receptor-like kinase 1 (ALK1) in endothelial cells. Blood 109: 1953–1961.
Dorrell, M.I., E. Aguilar and M. Friedlander. 2002. Retinal vascular development is mediated by endothelial filopodia, a preexisting astrocytic template and specific R-cadherin adhesion. Invest. Ophthalmol. Vis. Sci. 43: 3500–3510.
Drake, C.J. and P.A. Fleming. 2000. Vasculogenesis in the day 6.5 to 9.5 mouse embryo. Blood 95: 1671–1679.
Eichmann, A., T. Makinen and K. Alitalo. 2005. Neural guidance molecules regulate vascular remodeling and vessel navigation. Genes Dev. 19: 1013–1021.
Fantin, A., J.M. Vieira, A. Plein, C.H. Maden and C. Ruhrberg. 2013. The embryonic mouse hindbrain as a qualitative and quantitative model for studying the molecular and cellular mechanisms of angiogenesis. Nat. Protoc. 8: 418–429.
Fantin, A., J.M. Vieira, G. Gestri, L. Denti, Q. Schwarz, S. Prykhozhij, F. Peri, S.W. Wilson and C. Ruhrberg. 2010. Tissue macrophages act as cellular chaperones for vascular anastomosis downstream of VEGF-mediated endothelial tip cell induction. Blood 116: 829–840.
Faurobert, E. and C. Albiges-Rizo. 2010. Recent insights into cerebral cavernous malformations: a complex jigsaw puzzle under construction. FEBS J. 277: 1084–1096.
Fidalgo, M., M. Fraile, A. Pires, T. Force, C. Pombo and J. Zalvide. 2010. CCM3/PDCD10 stabilizes GCKIII proteins to promote Golgi assembly and cell orientation. J. Cell Sci. 123: 1274–1284.
Fruttiger, M. 2002. Development of the mouse retinal vasculature: angiogenesis versus vasculogenesis. Invest. Ophthalmol. Vis. Sci. 43: 522–527.
Fruttiger, M. 2007. Development of the retinal vasculature. Angiogenesis 10: 77–88.
Fujita, M., Y.R. Cha, V.N. Pham, A. Sakurai, B.L. Roman, J.S. Gutkind and B.M. Weinstein. 2011. Assembly and patterning of the vascular network of the vertebrate hindbrain. Development 138: 1705–1715.
Gault, J., I.A. Awad, P. Recksiek, R. Shenkar, R. Breeze, M. Handler and B.K. Kleinschmidt-DeMasters. 2009. Cerebral cavernous malformations: somatic mutations in vascular endothelial cells. Neurosurgery 65: 138–145.
Gault, J., R. Shenkar, P. Recksiek and I.A. Awad. 2005. Biallelic somatic and germ line CCM1 truncating mutations in a cerebral cavernous malformation lesion. Stroke 36: 872–874.
Gerhardt, H., C. Ruhrberg, A. Abramsson, H. Fujisawa, D. Shima and C. Betsholtz. 2004. Neuropilin-1 is required for endothelial tip cell guidance in the developing central nervous system. Dev. Dyn. 231: 503–509.
Gerhardt, H., M. Golding, M. Fruttiger, C. Ruhrberg, A. Lundkvist, A. Abramsson, M. Jeltsch, C. Mitchell, K. Alitalo, D. Shima and C. Betsholtz. 2003. VEGF guides angiogenic sprouting utilizing endothelial tip cell filopodia. J. Cell Biol. 161: 1163–1177.
Girouard, H. and C. Iadecola. 2006. Neurovascular coupling in the normal brain and in hypertension, stroke, and Alzheimer disease. J. Appl. Physiol. 100: 328–335.
Glading, A., J. Han, R.A. Stockton and M.H. Ginsberg. 2007. KRIT-1/CCM1 is a Rap1 effector that regulates endothelial cell cell junctions. J. Cell Biol. 179: 247–254.

Goettsch, W., C. Gryczka, T. Korff, E. Ernst, C. Goettsch, J. Seebach, H.J. Schnittler, H.G. Augustin and H. Morawietz. 2008. Flow-dependent regulation of angiopoietin-2. J. Cell. Physiol. 214: 491–503.

Gotz, M. and W.B. Huttner. 2005. The cell biology of neurogenesis. Nat. Rev. Mol. Cell Biol. 6: 777–788.

Goumans, M.J., Z. Liu and P. ten Dijke. 2009. TGF-beta signaling in vascular biology and dysfunction. Cell Res. 19: 116–127.

Gutnick, A., J. Blechman, J. Kaslin, L. Herwig, H.G. Belting, M. Affolter, J.L. Bonkowsky and G. Levkowitz. 2011. The hypothalamic neuropeptide oxytocin is required for formation of the neurovascular interface of the pituitary. Dev. Cell 21: 642–654.

Hackett, S.F., H. Ozaki, R.W. Strauss, K. Wahlin, C. Suri, P. Maisonpierre, G. Yancopoulos and P.A. Campochiaro. 2000. Angiopoietin 2 expression in the retina: upregulation during physiologic and pathologic neovascularization. J. Cell. Physiol. 184: 275–284.

Hackett, S.F., S. Wiegand, G. Yancopoulos and P.A. Campochiaro. 2002. Angiopoietin-2 plays an important role in retinal angiogenesis. J. Cell. Physiol. 192: 182–187.

Hao, Q., H. Su, D.A. Marchuk, R. Rola, Y. Wang, W. Liu, W.L. Young and G.Y. Yang. 2008. Increased tissue perfusion promotes capillary dysplasia in the ALK1-deficient mouse brain following VEGF stimulation. Am. J. Physiol. Heart Circ. Physiol. 295: H2250–H2256.

Hashimoto, T., H. Meng and W.L. Young. 2006. Intracranial aneurysms: links among inflammation, hemodynamics and vascular remodeling. Neurol. Res. 28: 372–380.

Hashimoto, T., R. Mesa-Tejada, C.M. Quick, A.W. Bollen, S. Joshi, J. Pile-Spellman, M.T. Lawton and W.L. Young. 2001. Evidence of increased endothelial cell turnover in brain arteriovenous malformations. Neurosurgery 49: 124–132.

Hashimoto, T., Y. Wu, M.T. Lawton, G.Y. Yang, N.M. Barbaro and W.L. Young. 2005. Coexpression of angiogenic factors in brain arteriovenous malformations. Neurosurgery 56: 1058–1065.

Heinke, J., C. Patterson and M. Moser. 2012. Life is a pattern: vascular assembly within the embryo. Front. Biosci. (Elite Ed.) 4: 2269–2288.

Hellstrom, M., M. Kalen, P. Lindahl, A. Abramsson and C. Betsholtz. 1999. Role of PDGF-B and PDGFR-beta in recruitment of vascular smooth muscle cells and pericytes during embryonic blood vessel formation in the mouse. Development 126: 3047–3055.

Hilder, T.L., M.H. Malone, S. Bencharit, J. Colicelli, T.A. Haystead, G.L. Johnson and C.C. Wu. 2007. Proteomic identification of the cerebral cavernous malformation signaling complex. J. Proteome Res. 6: 4343–4355.

Hogan, B.M., J. Bussmann, H. Wolburg and S. Schulte-Merker. 2008. ccm1 cell autonomously regulates endothelial cellular morphogenesis and vascular tubulogenesis in zebrafish. Hum. Mol. Genet. 17: 2424–2432.

Hogan, K.A., C.A. Ambler, D.L. Chapman and V.L. Bautch. 2004. The neural tube patterns vessels developmentally using the VEGF signaling pathway. Development 131: 1503–1513.

Hopper, R.A. and J. Garthwaite. 2006. Tonic and phasic nitric oxide signals in hippocampal long-term potentiation. J. Neurosci. 26: 11513–11521.

Iruela-Arispe, M.L. and G.E. Davis. 2009. Cellular and molecular mechanisms of vascular lumen formation. Dev. Cell. 16: 222–231.

Isogai, S., M. Horiguchi and B.M. Weinstein. 2001. The vascular anatomy of the developing zebrafish: an atlas of embryonic and early larval development. Dev. Biol. 230: 278–301.

James, J.M. and Y.S. Mukouyama. 2011. Neuronal action on the developing blood vessel pattern. Semin. Cell Dev. Biol. 22: 1019–1027.

James, J.M., C. Gewolb and V.L. Bautch. 2009. Neurovascular development uses VEGF-A signaling to regulate blood vessel ingression into the neural tube. Development 136: 833–841.

Jonker, L. and H.M. Arthur. 2002. Endoglin expression in early development is associated with vasculogenesis and angiogenesis. Mech. Dev. 110: 193–196.

Jung, K.H., K. Chu, S.W. Jeong, H.K. Park, H.J. Bae and B.W. Yoon. 2003. Cerebral cavernous malformations with dynamic and progressive course: correlation study with vascular endothelial growth factor. Arch. Neurol. 60: 1613–1618.

Junge, H.J., S. Yang, J.B. Burton, K. Paes, X. Shu, D.M. French, M. Costa, D.S. Rice and W. Ye. 2009. TSPAN12 regulates retinal vascular development by promoting Norrin- but not Wnt-induced FZD4/beta-catenin signaling. Cell 139: 299–311.

Kato, M., M.S. Patel, R. Levasseur, I. Lobov, B.H. Chang, D.A. Glass 2nd, C. Hartmann, L. Li, T.H. Hwang, C.F. Brayton, R.A. Lang, G. Karsenty and L. Chan. 2002. Cbfa1-independent decrease in osteoblast proliferation, osteopenia, and persistent embryonic eye vascularization in mice deficient in Lrp5, a Wnt coreceptor. J. Cell Biol. 157: 303–314.

Kawasaki, T., T. Kitsukawa, Y. Bekku, Y. Matsuda, M. Sanbo, T. Yagi and H. Fujisawa. 1999. A requirement for neuropilin-1 in embryonic vessel formation. Development 126: 4895–4902.

Kiecker, C. and A. Lumsden. 2005. Compartments and their boundaries in vertebrate brain development. Nat. Rev. Neurosci. 6: 553–564.

Kleaveland, B., X. Zheng, J.J. Liu, Y. Blum, J.J. Tung, Z. Zou, S.M. Sweeney, M. Chen, L. Guo, M.M. Lu, D. Zhou, J. Kitajewski, M. Affolter, M.H. Ginsberg and M.L. Kahn. 2009. Regulation of cardiovascular development and integrity by the heart of glass-cerebral cavernous malformation protein pathway. Nat. Med. 15: 169–176.

Klessinger, S. and B. Christ. 1996. Axial structures control laterality in the distribution pattern of endothelial cells. Anat. Embryol. (Berl.) 193: 319–330.

Kurz, H. 2009. Cell lineages and early patterns of embryonic CNS vascularization. Cell Adh. Migr. 3: 205–210.

Kurz, H. and B. Christ. 2002. Vascular development of the brain and spinal cord. pp. 157–191. *In*: R.J. Tomanek [ed.]. Assembly of the vaculature and its regulation. Birkhauser, Boston, USA.

Kurz, H., T. Gartner, P.S. Eggli and B. Christ. 1996. First blood vessels in the avian neural tube are formed by a combination of dorsal angioblast immigration and ventral sprouting of endothelial cells. Dev. Biol. 173: 133–147.

Kurz, H., J. Korn and B. Christ. 2004. Morphogenesis of embryonic CNS vessels. Cancer Treat. Res. 117: 33–50.

Lan, Y., B. Liu, H. Yao, F. Li, T. Weng, G. Yang, W. Li, X. Cheng, N. Mao and X. Yang. 2007. Essential role of endothelial Smad4 in vascular remodeling and integrity. Mol. Cell. Biol. 27: 7683–7692.

Lee, H.S., J. Han, H.J. Bai and K.W. Kim. 2009. Brain angiogenesis in developmental and pathological processes: regulation, molecular and cellular communication at the neurovascular interface. FEBS J. 276: 4622–4635.

Li, D.Y., L.K. Sorensen, B.S. Brooke, L.D. Urness, E.C. Davis, D.G. Taylor, B.B. Boak and D.P. Wendel. 1999. Defective angiogenesis in mice lacking endoglin. Science 284: 1534–1537.

Li, S., K. Haigh, J.J. Haigh and A. Vasudevan. 2013. Endothelial VEGF sculpts cortical cytoarchitecture. J. Neurosci. 33: 14809–14815.

Ling, T.L. and J. Stone. 1988. The development of astrocytes in the cat retina: evidence of migration from the optic nerve. Brain Res. Dev. Brain Res. 44: 73–85.

Lobov, I.B., S. Rao, T.J. Carroll, J.E. Vallance, M. Ito, J.K. Ondr, S. Kurup, D.A. Glass, M.S. Patel, W. Shu, E.E. Morrisey, A.P. McMahon, G. Karsenty and R.A. Lang. 2005. WNT7b mediates macrophage-induced programmed cell death in patterning of the vasculature. Nature 437: 417–421.

Lu, X., F. Le Noble, L. Yuan, Q. Jiang, B. De Lafarge, D. Sugiyama, C. Breant, F. Claes, F. De Smet, J.L. Thomas, M. Autiero, P. Carmeliet, M. Tessier-Lavigne and A. Eichmann. 2004. The netrin receptor UNC5B mediates guidance events controlling morphogenesis of the vascular system. Nature 432: 179–186.

Luhmann, U.F., J. Lin, N. Acar, S. Lammel, S. Feil, C. Grimm, M.W. Seeliger, H.P. Hammes and W. Berger. 2005. Role of the Norrie disease pseudoglioma gene in sprouting angiogenesis during development of the retinal vasculature. Invest. Ophthalmol. Vis. Sci. 46: 3372–3382.

Ma, S., H.J. Kwon, H. Johng, K. Zang and Z. Huang. 2013. Radial glial neural progenitors regulate nascent brain vascular network stabilization via inhibition of Wnt signaling. PLoS Biol. 11: e1001469.

Ma, X., H. Zhao, J. Shan, F. Long, Y. Chen, Y. Chen, Y. Zhang, X. Han and D. Ma. 2007. PDCD10 interacts with Ste20-related kinase MST4 to promote cell growth and transformation via modulation of the ERK pathway. Mol. Biol. Cell 18: 1965–1978.

Ma, X., M. Labinaz, J. Goldstein, H. Miller, W.J. Keon, M. Letarte and E. O'Brien. 2000. Endoglin is overexpressed after arterial injury and is required for transforming growth factor-beta-induced inhibition of smooth muscle cell migration. Arterioscler. Thromb. Vasc. Biol. 20: 2546–2552.

Marin-Padilla, M. 2012. The human brain intracerebral microvascular system: development and structure. Front. Neuroanat. 6: 38.

Matsumura, K. and S. Kobayashi. 2004. Signaling the brain in inflammation: the role of endothelial cells. Front. Biosci. 9: 2819–2826.

Mi, H. and B.A. Barres. 1999. Purification and characterization of astrocyte precursor cells in the developing rat optic nerve. J. Neurosci. 19: 1049–1061.

Minakawa, T., R. Tanaka, T. Koike, S. Takeuchi and O. Sasaki. 1989. Angiographic follow-up study of cerebral arteriovenous malformations with reference to their enlargement and regression. Neurosurgery 24: 68–74.

Miyawaki, T., A. Uemura, M. Dezawa, R.T. Yu, C. Ide, S. Nishikawa, Y. Honda, Y. Tanabe and T. Tanabe. 2004. Tlx, an orphan nuclear receptor, regulates cell numbers and astrocyte development in the developing retina. J. Neurosci. 24: 8124–8134.

Mu, Z., Z. Yang, D. Yu, Z. Zhao and J.S. Munger. 2008. TGFbeta1 and TGFbeta3 are partially redundant effectors in brain vascular morphogenesis. Mech. Dev. 125: 508–516.

Mudhar, H.S., R.A. Pollock, C. Wang, C.D. Stiles and W.D. Richardson. 1993. PDGF and its receptors in the developing rodent retina and optic nerve. Development 118: 539–552.

Nagase, T., M. Nagase, K. Yoshimura, T. Fujita and I. Koshima. 2005. Angiogenesis within the developing mouse neural tube is dependent on sonic hedgehog signaling: possible roles of motor neurons. Genes Cells 10: 595–604.

Pagenstecher, A., S. Stahl, U. Sure and U. Felbor. 2009. A two-hit mechanism causes cerebral cavernous malformations: complete inactivation of CCM1, CCM2 or CCM3 in affected endothelial cells. Hum. Mol. Genet. 18: 911–918.

Pardanaud, L. and F. Dieterlen-Lievre. 1999. Manipulation of the angiopoietic/hemangiopoietic commitment in the avian embryo. Development 126: 617–627.

Pardanaud, L., D. Luton, M. Prigent, L.M. Bourcheix, M. Catala and F. Dieterlen-Lievre. 1996. Two distinct endothelial lineages in ontogeny, one of them related to hemopoiesis. Development 122: 1363–1371.

Park, J.A., K.S. Choi, S.Y. Kim and K.W. Kim. 2003. Coordinated interaction of the vascular and nervous systems: from molecule- to cell-based approaches. Biochem. Biophys. Res. Commun. 311: 247–253.

Park, S.O., Y.J. Lee, T. Seki, K.H. Hong, N. Fliess, Z. Jiang, A. Park, X. Wu, V. Kaartinen, B.L. Roman and S.P. Oh. 2008. ALK5- and TGFBR2-independent role of ALK1 in the pathogenesis of hereditary hemorrhagic telangiectasia type 2. Blood 111: 633–642.

Pierce, E.A., E.D. Foley and L.E. Smith. 1996. Regulation of vascular endothelial growth factor by oxygen in a model of retinopathy of prematurity. Arch. Ophthalmol. 114: 1219–1228.

Petit, N., A. Blecon, C. Denier and E. Tournier-Lasserve. 2006. Patterns of expression of the three cerebral cavernous malformation (CCM) genes during embryonic and postnatal brain development. Gene Expr. Patterns 6: 495–503.

Plummer, N.W., T.L. Squire, S. Srinivasan, E. Huang, J.S. Zawistowski, H. Matsunami, L.P. Hale and D.A. Marchuk. 2006. Neuronal expression of the Ccm2 gene in a new mouse model of cerebral cavernous malformations. Mamm. Genome 17: 119–128.

Preisinger, C., B. Short, V. De Corte, E. Bruyneel, A. Haas, R. Kopajtich, J. Gettemans and F.A. Barr. 2004. YSK1 is activated by the Golgi matrix protein GM130 and plays a role in cell migration through its substrate 14-3-3zeta. J. Cell Biol. 164: 1009–1020.

Puelles, L. and J.L. Rubenstein. 2003. Forebrain gene expression domains and the evolving prosomeric model. Trends Neurosci. 26: 469–476.

Puelles, L., M. Harrison, G. Paxinos and C. Watson. 2013. A developmental ontology for the mammalian brain based on the prosomeric model. Trends Neurosci. 36: 570–578.

Rapaport, D.H., L.L. Wong, E.D. Wood, D. Yasumura and M.M. LaVail. 2004. Timing and topography of cell genesis in the rat retina. J. Comp. Neurol. 474: 304–324.

Reese, D.E., C.E. Hall and T. Mikawa. 2004. Negative regulation of midline vascular development by the notochord. Dev. Cell 6: 699–708.

Richter, M., J. Gottanka, C.A. May, U. Welge-Lussen, W. Berger and E. Lutjen-Drecoll. 1998. Retinal vasculature changes in Norrie disease mice. Invest. Ophthalmol. Vis. Sci. 39: 2450–2457.

Rigamonti, D., P.C. Johnson, R.F. Spetzler, M.N. Hadley and B.P. Drayer. 1991. Cavernous malformations and capillary telangiectasia: a spectrum within a single pathological entity. Neurosurgery 28: 60–64.

Risau, W. 1997. Mechanisms of angiogenesis. Nature 386: 671–674.

Roncali, L. and G. Ambrosi. 1982. Development of the vascular supply and morphogenesis of the nervous substratum in the chick embryo mesencephalic tegmentum. Boll. Soc. Ital. Biol. Sper. 58: 1102–1107.

Roncali, L., D. Ribatti and G. Ambrosi. 1985a. Ultrastructural basis of the vessel wall differentiation in the chick embryo optic tectum. J. Submicrosc. Cytol. 17: 83–88.

Roncali, L., D. Ribatti, B. Nico and M. Bertossi. 1985b. Blood vessel ultrastructure in the chick optic tectum developing under chronic hypoxia conditions. Acta Neuropathol. 67: 242–246.

Rousseau, B., F. Larrieu-Lahargue, A. Bikfalvi and S. Javerzat. 2003. Involvement of fibroblast growth factors in choroidal angiogenesis and retinal vascularization. Exp. Eye Res. 77: 147–156.

Rovainen, C.M. and M.H. Kakarala. 1989. Angiogenesis on the optic tectum of albino Xenopus laevis tadpoles. Brain Res. Dev. Brain Res. 48: 197–213.

Ruhrberg, C., H. Gerhardt, M. Golding, R. Watson, S. Ioannidou, H. Fujisawa, C. Betsholtz and D.T. Shima. 2002. Spatially restricted patterning cues provided by heparin-binding VEGF-A control blood vessel branching morphogenesis. Genes Dev. 16: 2684–2698.

Rupp, P.A. and C.D. Little. 2001. Integrins in vascular development. Circ. Res. 89: 566–572.

Saint-Geniez, M. and P.A. D'Amore. 2004. Development and pathology of the hyaloid, choroidal and retinal vasculature. Int. J. Dev. Biol. 48: 1045–1058.

Saint-Geniez, M., B. Masri, F. Malecaze, B. Knibiehler and Y. Audigier. 2002. Expression of the murine msr/apj receptor and its ligand apelin is upregulated during formation of the retinal vessels. Mech. Dev. 110: 183–186.

Sato, S., N. Kodama, T. Sasaki, M. Matsumoto and T. Ishikawa. 2004. Perinidal dilated capillary networks in cerebral arteriovenous malformations. Neurosurgery 54: 163–170.

Sato, T.N., Y. Tozawa, U. Deutsch, K. Wolburg-Buchholz, Y. Fujiwara, M. Gendron-Maguire, T. Gridley, H. Wolburg, W. Risau and Y. Qin. 1995. Distinct roles of the receptor tyrosine kinases Tie-1 and Tie-2 in blood vessel formation. Nature 376: 70–74.

Satomi, J., R.J. Mount, M. Toporsian, A.D. Paterson, M.C. Wallace, R.V. Harrison and M. Letarte. 2003. Cerebral vascular abnormalities in a murine model of hereditary hemorrhagic telangiectasia. Stroke 34: 783–789.

Scharpfenecker, M., M. van Dinther, Z. Liu, R.L. van Bezooijen, Q. Zhao, L. Pukac, C.W. Lowik and P. ten Dijke. 2007. BMP-9 signals via ALK1 and inhibits bFGF-induced endothelial cell proliferation and VEGF-stimulated angiogenesis. J. Cell Sci. 120: 964–972.

Seki, T., J. Yun and S.P. Oh. 2003. Arterial endothelium-specific activin receptor-like kinase 1 expression suggests its role in arterialization and vascular remodeling. Circ. Res. 93: 682–689.

Shenkar, R., P.N. Venkatasubramanian, A.M. Wyrwicz, J.C. Zhao, C. Shi, A. Akers, D.A. Marchuk and I.A. Awad. 2008. Advanced magnetic resonance imaging of cerebral cavernous malformations: part II. Imaging of lesions in murine models. Neurosurgery 63: 790–798.

Silverman, W.F., J.M. Krum, N. Mani and J.M. Rosenstein. 1999. Vascular, glial and neuronal effects of vascular endothelial growth factor in mesencephalic explant cultures. Neuroscience 90: 1529–1541.

Sorensen, L.K., B.S. Brooke, D.Y. Li and L.D. Urness. 2003. Loss of distinct arterial and venous boundaries in mice lacking endoglin, a vascular-specific TGFbeta coreceptor. Dev. Biol. 261: 235–250.

Srinivasan, S., M.A. Hanes, T. Dickens, M.E. Porteous, S.P. Oh, L.P. Hale and D.A. Marchuk. 2003. A mouse model for hereditary hemorrhagic telangiectasia (HHT) type 2. Hum. Mol. Genet. 12: 473–482.

Stalmans, I., Y.S. Ng, R. Rohan, M. Fruttiger, A. Bouche, A. Yuce, H. Fujisawa, B. Hermans, M. Shani, S. Jansen, D. Hicklin, D.J. Anderson, T. Gardiner, H.P. Hammes, L. Moons, M. Dewerchin, D. Collen, P. Carmeliet and P.A. D'Amore. 2002. Arteriolar and venular patterning in retinas of mice selectively expressing VEGF isoforms. J. Clin. Invest. 109: 327–336.

Stenman, J.M., J. Rajagopal, T.J. Carroll, M. Ishibashi, J. McMahon and A.P. McMahon. 2008. Canonical Wnt signaling regulates organ-specific assembly and differentiation of CNS vasculature. Science 322: 1247–1250.

Stone, J., A. Itin, T. Alon, J. Peer, H. Gnessin, T. Chan-Ling and E. Keshet. 1995. Development of retinal vasculature is mediated by hypoxia-induced vascular endothelial growth factor (VEGF) expression by neuroglia. J. Neurosci. 15: 4738–4747.

Suchting, S., C. Freitas, F. le Noble, R. Benedito, C. Breant, A. Duarte and A. Eichmann. 2007. The Notch ligand Delta-like 4 negatively regulates endothelial tip cell formation and vessel branching. Proc. Natl. Acad. Sci. USA 104: 3225–3230.

Sun, J., W. Zhou, D. Ma and Y. Yang. 2010. Endothelial cells promote neural stem cell proliferation and differentiation associated with VEGF activated Notch and Pten signaling. Dev. Dyn. 239: 2345–2353.

Sure, U., N. Butz, J. Schlegel, A.M. Siegel, J.P. Wakat, H.D. Mennel, S. Bien and H. Bertalanffy. 2001. Endothelial proliferation, neoangiogenesis, and potential *de novo* generation of cerebrovascular malformations. J. Neurosurg. 94: 972–977.

Suri, C., P.F. Jones, S. Patan, S. Bartunkova, P.C. Maisonpierre, S. Davis, T.N. Sato and G.D. Yancopoulos. 1996. Requisite role of angiopoietin-1, a ligand for the TIE2 receptor, during embryonic angiogenesis. Cell 87: 1171–1180.

ten Dijke, P., M.J. Goumans and E. Pardali. 2008. Endoglin in angiogenesis and vascular diseases. Angiogenesis 11: 79–89.

Toporsian, M., R. Gros, M.G. Kabir, S. Vera, K. Govindaraju, D.H. Eidelman, M. Husain and M. Letarte. 2005. A role for endoglin in coupling eNOS activity and regulating vascular tone revealed in hereditary hemorrhagic telangiectasia. Circ. Res. 96: 684–692.

Torsney, E., R. Charlton, A.G. Diamond, J. Burn, J.V. Soames and H.M. Arthur. 2003. Mouse model for hereditary hemorrhagic telangiectasia has a generalized vascular abnormality. Circulation 107: 1653–1657.

Uhlik, M.T., A.N. Abell, N.L. Johnson, W. Sun, B.D. Cuevas, K.E. Lobel-Rice, E.A. Horne, M.L. Dell'Acqua and G.L. Johnson. 2003. Rac-MEKK3-MKK3 scaffolding for p38 MAPK activation during hyperosmotic shock. Nat. Cell Biol. 5: 1104–1110.

Ulrich, F., L.H. Ma, R.G. Baker and J. Torres-Vazquez. 2011. Neurovascular development in the embryonic zebrafish hindbrain. Dev. Biol. 357: 134–151.

Urness, L.D., L.K. Sorensen and D.Y. Li. 2000.Arteriovenous malformations in mice lacking activin receptor-like kinase-1. Nat. Genet. 26: 328–331.

van Laake, L.W., S. van den Driesche, S. Post, A. Feijen, M.A. Jansen, M.H. Driessens, J.J. Mager, R.J. Snijder, C.J. Westermann, P.A. Doevendans, C.J. van Echteld, P. ten Dijke,

H.M. Arthur, M.J. Goumans, F. Lebrin and C.L. Mummery. 2006. Endoglin has a crucial role in blood cell-mediated vascular repair. Circulation 114: 2288–2297.

Vasudevan, A., J.E. Long, J.E. Crandall, J.L. Rubenstein and P.G. Bhide. 2008. Compartment-specific transcription factors orchestrate angiogenesis gradients in the embryonic brain. Nat. Neurosci. 11: 429–439.

Virgintino, D., E. Maiorano, M. Errede, A. Vimercati, P. Greco, L. Selvaggi, L. Roncali and M. Bertossi. 1998. Astroglia-microvessel relationship in the developing human telencephalon. Int. J. Dev. Biol. 42: 1165–1168.

Walls, J.R., L. Coultas, J. Rossant and R.M. Henkelman. 2008. Three-dimensional analysis of vascular development in the mouse embryo. PLoS One 3: e2853.

West, H., W.D. Richardson and M. Fruttiger. 2005. Stabilization of the retinal vascular network by reciprocal feedback between blood vessels and astrocytes. Development 132: 1855–1862.

Whitehead, K.J., A.C. Chan, S. Navankasattusas, W. Koh, N.R. London, J. Ling, A.H. Mayo, S.G. Drakos, C.A. Jones, W. Zhu, D.A. Marchuk, G.E. Davis and D.Y. Li. 2009. The cerebral cavernous malformation signaling pathway promotes vascular integrity via Rho GTPases. Nat. Med. 15: 177–184.

Whitehead, K.J., N.W. Plummer, J.A. Adams, D.A. Marchuk and D.Y. Li. 2004. Ccm1 is required for arterial morphogenesis: implications for the etiology of human cavernous malformations. Development 131: 1437–1448.

Xu, B., Y.Q. Wu, M. Huey, H.M. Arthur, D.A. Marchuk, T. Hashimoto, W.L. Young and G.Y. Yang. 2004a. Vascular endothelial growth factor induces abnormal microvasculature in the endoglin heterozygous mouse brain. J. Cereb. Blood Flow Metab. 24: 237–244.

Xu, Q., Y. Wang, A. Dabdoub, P.M. Smallwood, J. Williams, C. Woods, M.W. Kelley, L. Jiang, W. Tasman, K. Zhang and J. Nathans. 2004b. Vascular development in the retina and inner ear: control by Norrin and Frizzled-4, a high-affinity ligand-receptor pair. Cell 116: 883–895.

Ye, X., Y. Wang, H. Cahill, M. Yu, T.C. Badea, P.M. Smallwood, N.S. Peachey and J. Nathans. 2009. Norrin, frizzled-4, and Lrp5 signaling in endothelial cells controls a genetic program for retinal vascularization. Cell 139: 285–298.

Zacchigna, S., C. Ruiz de Almodovar and P. Carmeliet. 2008. Similarities between angiogenesis and neural development: what small animal models can tell us. Curr. Top. Dev. Biol. 80: 1–55.

Zhang, J., S. Basu, D. Rigamonti, H.C. Dietz and R.E. Clatterbuck. 2008. Krit1 modulates beta 1-integrin-mediated endothelial cell proliferation. Neurosurgery 63: 571–578.

Zheng, X., C. Xu, A. Di Lorenzo, B. Kleaveland, Z. Zou, C. Seiler, M. Chen, L. Cheng, J. Xiao, J. He, M.A. Pack, W.C. Sessa and M.L. Kahn. 2010. CCM3 signaling through sterile 20-like kinases plays an essential role during zebrafish cardiovascular development and cerebral cavernous malformations. J. Clin. Invest. 120: 2795–2804.

Zhu, Y., M.T. Lawton, R. Du, Y. Shwe, Y. Chen, F. Shen, W.L. Young and G.Y. Yang. 2006. Expression of hypoxia-inducible factor-1 and vascular endothelial growth factor in response to venous hypertension. Neurosurgery 59: 687–696.

CHAPTER 2

Endothelial Cell Heterogeneity and Maintenance of Neurogenic Niches During Development and Adult Life

Dannia Colín Castelán

2.1 Introduction

Neurogenesis was considered to be limited to embryonic life and to early postnatal stages. However, in recent years, it has been demonstrated that this phenomenon occurs throughout mammalian life (Alvarez-Buylla and Garcia-Verdugo 2002). It has also been shown that neurogenesis is maintained in spatiotemporally restricted microenvironments called neurogenic niches (Luskin 1993, Stenman et al. 2003, Tucker et al. 2006, Riquelme et al. 2008).

In recent years, research has begun showing the importance of the vascular bed in the generation and maintenance of neurogenic niches. These specialized microenvironments are, at least partially, the result of the phenotypic heterogeneity of endothelial cells as well as their functional diversity (e.g., permeability, secretion and transport). This chapter addresses the issue of how endothelial cells support neurogenic niches throughout embryonic and adult life.

Laboratorio de Biología de Sistemas, Departamento de Biología Celular y Fisiología, Instituto de Investigaciones Biomédicas, Universidad Nacional Autónoma de México, 04510. México, D.F.
Email: danniacc13@gmail.com

2.2 What are Neurogenic Niches?

Neurogenesis, the process of generating new neurons from undifferentiated progenitors, is essential for brain formation. It also seems to be an important aspect of how the brain maintains its structural plasticity in adult life (Burghardt et al. 2012).

There are at least three different phases of neurogenesis. The first phase starts with the proliferation of multi-potent neural stem cells, giving rise to other neural stem cells (self-renewal) and transit-amplifying neural progenitors as daughter cells (Alvarez-Buylla and Lois 1995). In the second neurogenic phase, the transit-amplifying neural progenitors become neuroblasts and migrate towards their target location (Doetsch et al. 1997). Finally, in the third phase, neuroblasts specify their fate and differentiate into mature neurons (Bovetti et al. 2007).

Neurogenesis occurs since early embryonic development and continues throughout life. However, it takes place in different locations during different developmental stages. During early embryonic development neural stem cells are located around the ventricles where they proliferate. Afterwards, neuroblasts migrate along the radiating fibers, formed by radial glia, towards the surface of the central nervous system (Franco and Muller 2011). As embryonic development proceeds, the neurogenic areas become restricted to the hippocampus and areas around the ventricles, such as the lateral ganglionic eminence (Taupin 2006, Tucker et al. 2006). The latter structure gives rise to the striatum and the subventricular zone neurogenic compartment. During postnatal life, only two major neurogenic areas remain: (1) the forebrain neurogenic niche running tangentially from the subventricular zone of the lateral ventricle, through the rostral migratory stream to the olfactory bulb; and (2) the dentate gyrus subgranular zone of the hippocampus (Eriksson et al. 1998, Seaberg and van der Kooy 2002, Taupin 2006). In either case, it is noteworthy that neurogenesis is maintained in the adult brain only in restricted zones. It has been proposed that this spatiotemporal restriction of neurogenesis may be due to the specialized microenvironments that are maintained within the neurogenic regions, which allow the proliferation, migration and differentiation of the neuronal progenitors. These specialized microenvironments have been called neurogenic niches (Taupin 2006).

The exact composition of neurogenic niches and how they regulate neural progenitors is not known. However, some elements forming the neurogenic microenvironment are currently being elucidated; they include growth factors, guidance cues, extracellular matrix, cell-cell interactions and certain gases and other molecules, such as glucose or amino acids (Yang et al. 2011). All these components are provided by different cells residing within the niche; i.e., the neural stem cells, ependymal cells, astroglia, neuroblasts and endothelial cells (Doetsch et al. 1997). These cell types form the neurogenic microenvironment and some recent reports suggest that each of them contributes to a greater or lesser extent to the regulation of the neurogenic process. Further, the role of the vascular bed in the regulation of neurogenesis has recently generated special interest due to several reports that indicate a close association between neuronal progenitors and endothelial cells.

2.3 Role of Endothelial Cells in Neurogenic Niches During Development

2.3.1 Endothelial Cells and Neural Stem Cell Proliferation During Development

As described in the previous section, neurogenic niches are specialized microenvironments that allow or promote neurogenesis. Endothelial cells can contribute to these microenvironments in different ways, one of them being the production and secretion of trophic factors. Indeed, in the neonatal brain, endothelial cells induce an increase in neural stem cell proliferation and differentiation depending on the endothelial cell-derived vascular endothelial growth factor (Sun et al. 2010). Moreover, endothelial cells produce soluble factors that promote embryonic neural stem cell renewal and inhibit their differentiation *in vitro* (Shen et al. 2004). Interestingly, in this co-culture model, endothelial cell-derived factors induced neural stem cell expansion where there was no contact between the endothelial cells and the neural stem cells. However, in a co-culture model where there was contact between the endothelial cells and the neural stem cells, endothelial cells limited neural stem cell proliferation by inhibiting the G1-S phase transition through the secretion of bone morphogenetic proteins. The exit from the cell cycle induced by endothelial cells in this *in vitro* model was reversible (Mathieu et al. 2008). The contrast between these two studies is revealing because it suggests that endothelial cells not only regulate neural stem cell proliferation through soluble secreted factors, but that there is another level of regulation that resides in cell-cell interactions.

2.3.2 Endothelial Cells and Neuroblast Migration During Development

Endothelial cells also participate in neuroblast migration during embryonic development. During the first stages of brain formation, when neuroblasts migrate radially, blood vessels are oriented radially as well. In addition, at times when neuroblast migration becomes tangential in the rostral migratory stream, blood vessels show the same orientation (Colin-Castelan et al. 2011). Furthermore, a recent study shows that during postnatal development neuroblast migration in the rostral migratory stream is less efficient than in the adult, partially because of the not yet fully developed vasculature (Bozoyan et al. 2012). What is more, endothelial cells are essential for brain cortex assembly because they secrete endothelins, which in turn promote inter-kinetic nuclear migration, a process by which neural progenitor cell nuclei dynamically translocate from the ventricle towards the pia mater (Nishikawa et al. 2011).

Besides promoting neuroblast migration, blood vessels may be involved in maintaining the limits of the forebrain neurogenic niche since early embryonic development. In mice, at embryonic day 14.5, blood vessels already border the rostral migratory stream elbow at its dorso-central aspect. By postnatal day seven, the vascular bed that lines the borders of the entire rostral migratory stream is already formed. In contrast, at this age the core of the rostral migratory stream lacks blood vessels (Colin-Castelan et al. 2011). Interestingly, during postnatal development, endothelial cells secrete Semaphorin-3A along the borders of the rostral migratory stream. This molecule usually functions as a neuronal navigation guidance cue. However, in this case, endothelial cells are also the major targets of Semaphorin-3A (Melendez-Herrera et al. 2008). Taking this into account, Semaphorin-3A might modulate blood vessel formation, growth and remodeling along the borders of the rostral migratory stream. This may be one of the multiple mechanisms involved in delimiting the forebrain neurogenic niche.

We are just beginning to elucidate how endothelial cells participate in neurogenic niche function in the developing brain. So far the information indicates that the endothelium plays a crucial role in brain formation.

2.4 The Role of Endothelial Cells in Adult Neurogenesis

2.4.1 Endothelial Cells and Adult Neural Stem Cell Proliferation

Recently, the role of the vasculature as a support system for stem cell proliferation has become clear, even if the mechanisms are not well understood. One of the first clues which suggest that the vascular bed supports neural stem cell and neural progenitor cell proliferation is the close anatomical relationship between blood vessels and proliferating cells within adult stem cell niches.

There are several niches where stem cells are known to be located near the blood vessels, such as the muscular satellite cell niche, the bone marrow mesenchymal cell niche and the neurogenic niches (Palmer et al. 2000, Kopp et al. 2005, Christov et al. 2007, Shen et al. 2008, Tavazoie et al. 2008). Some studies have revealed that neural stem cells and neural progenitor cells form clusters of proliferating cells around the blood vessels within the two proliferative regions of the adult mammalian brain (Palmer et al. 2000, Shen et al. 2008). Moreover, neural stem cells and progenitor cells are in direct contact with endothelial cells in locations which lack astrocytic endfeet throughout the adult subventricular zone (Tavazoie et al. 2008). For some time it was thought that this association responded only to the metabolic requirements of proliferating stem cells and progenitors. Given the energy cost of cell proliferation, neural stem cells and neural progenitors would be expected to be located in a space where they would readily have access to oxygen, glucose and other nutrients. However, some recent studies indicate that the vascular bed not only delivers nutrients through the blood as previously described. In a recent report published by Shen et al. in 2004, they demonstrated that endothelial cell-derived soluble factors facilitate embryonic and adult neural stem cell proliferation and inhibit their differentiation, in a two-chamber co-culture system.

Interestingly, it seems that the vascular beds located in the proliferative regions within the adult brain have certain special characteristics that may be important in promoting neural stem cell and progenitor cell proliferation. For example, in the mammalian hippocampus, proliferating neuronal precursors are associated specifically with proliferating blood vessels (Palmer et al. 2000). On the other hand, testosterone-induced neurogenesis of the higher vocal center of adult songbirds is concomitant with testosterone-induced angiogenesis (Louissaint et al. 2002). Moreover, in an *in vitro* ischemia model, angiogenesis and neurogenesis are actually coupled processes (Teng et al. 2008). Using a co-culture system, the authors found that endothelial cells activated by ischemia enhance neural progenitor cell proliferation and that progenitor cells derived from the ischemic subventricular zone promote angiogenesis (Teng et al. 2008).

2.4.2 Endothelial Cells and Adult Neuroblast Migration

Regarding neuroblast migration, the concomitant orientation between blood vessels and migrating neuroblasts seen during development is maintained until adulthood (Colin-Castelan et al. 2011). Moreover, the vascular bed and neuroblasts are not only oriented in the same direction, but blood vessels actually serve as a scaffold for migrating neuroblasts along the adult rostral migratory stream, as a recent study shows (Whitman et al. 2009). Also, when neuroblasts exit the rostral migratory stream, they migrate radially within the olfactory bulb until reaching their final destination and this migration is also associated with the corresponding blood vessel orientation (Bovetti et al. 2007). The association between blood vessels and migrating neuroblasts has not only been seen in physiological conditions, it has been observed that ischemic insults increase neural stem cell proliferation and cause neuroblast recruitment to the site of the lesion. After middle cerebral artery occlusion, neuroblasts migrate to the ischemic striatum in close association with blood vessels (Yamashita et al. 2006, Thored et al. 2007). In addition, blood vessels generated through angiogenesis after cortical ischemia also serve as a scaffold for neuroblast migration and indeed, when angiogenesis is inhibited, neuroblasts are not able to reach the site of the lesion (Ohab et al. 2006). This implies that the vascular support is required for neuroblast migration at least in ischemia-generated conditions.

Apart from the structural support, blood vessels are able to provide molecular cues that guide or facilitate neuroblast migration. In the higher vocal center of the adult songbird, endothelial cells generated by testosterone-induced angiogenesis produce brain derived neurotrophic factor in a testosterone dependent manner; which, in turn, promotes neural migration and recruitment (Teng et al. 2008).

In addition to guidance molecules, endothelial cells can also provide the extracellular matrix required for neuroblast migration. Indeed, endothelial cells secrete laminin (Kramer et al. 1985). Interestingly, it has been reported that the laminin receptor $\alpha 6\beta 1$ integrin directs neuroblast migration in the adult mouse forebrain neurogenic niche (Emsley 2003). Moreover, endothelial cells produce extracellular matrix remodeling enzymes such as matrix metalloproteinase two and nine, as well as "a disintegrin and metalloproteinase" two and 21, all of which promote neuroblast migration by degrading the extracellular matrix and thus allow neuroblast motility (Yang et al. 2005, Wang et al. 2006, Murase et al. 2008).

2.4.3 Endothelial Cells and Adult Neuronal Differentiation

There is increasing evidence indicating that the vascular bed actively controls cell differentiation. *In vitro* inhibition of laminin degradation by matrix metalloproteinases impedes neural progenitor cell differentiation toward the neuronal lineage (Szymczak et al. 2010). This is especially interesting given that both laminin and matrix metalloproteinases are produced by the vasculature. It has also been shown that endothelial cells in culture are capable of maintaining neural stem cells in an undifferentiated state. Nonetheless, when these cells are allowed to differentiate, they give rise to neurons in greater proportions compared to neural stem cells without endothelial exposure (Shen et al. 2004). In a different study using ischemic conditions, neural progenitor cells isolated from the subventricular zone of normal adult rats co-cultured with cerebral endothelial cells isolated from stroke boundary vessels, substantially increased neuronal differentiation and reduced astrocytic differentiation (Teng et al. 2008). The same effect was seen when co-culturing normal neural stem cells or neural progenitor cells with endothelial cells deprived of oxygen and glucose (Plane et al. 2010). Furthermore, it has been shown that co-transplantation of neural stem cells with endothelium increased the survival of neural progenitors and also accelerated neuronal differentiation compared with transplantation of neural precursors alone in an *in vivo* model of cerebral ischemia (Nakagomi et al. 2009). Moreover, Plane et al. demonstrated *in vitro*, that media conditioned with control endothelial cells or ischemic endothelial cells promote different neurogenic phases. Untreated endothelium promotes neurosphere expansion, whereas ischemic endothelial cells promote neural differentiation and migration (Plane et al. 2010). The data suggest that diverse kinds of vasculature (in this case, endothelium from different physiological conditions) might provide neural precursors with different signals, thus generating heterogeneous responses.

Even if most of these results were obtained *in vitro*, they provide unequivocal evidence that blood vessels could be regulating neuronal differentiation *in vivo*. As mentioned before, neuroblasts migrate radially within the olfactory bulb in association with the vascular bed (Bovetti et al. 2007). Interestingly, by then neuroblasts have already begun their differentiation and show some typical characteristics of mature neurons (Luskin 1998), so it is likely that the blood vessels could be providing signals *in vivo* that promote neuronal differentiation.

Neurogenesis is a complex phenomenon that we are just beginning to understand. However, there is accumulating evidence indicating that endothelial cells are important elements within neurogenic niches. The vascular bed provides structural, molecular and trophic support for different processes underlying neurogenesis. Endothelial cells have proven to be more than just cells conforming blood vessels. Endothelial cells also provide soluble trophic factors, extracellular matrix and navigational guidance cues.

Taking into account that neurogenesis is widespread during embryonic development but becomes restricted with increasing age and the fact that endothelial cells are capable of supporting all the neurogenic phases, there may be different kinds of specialized endothelial cells helping in the formation and maintenance of the specialized microenvironments known as neurogenic niches.

2.5 Endothelial Cell Heterogeneity

In neuroscience, neurons have typically been the focus of most studies. Relatively recently, we have started to realize the importance of other cell types for the correct functioning of the brain. Since the discovery of the blood brain barrier by Edwin Goldmann in the beginning of the 20th century, scientists have studied the structure and function of the vasculature supplying the brain. Nonetheless, blood vessels have usually been regarded almost as static elements similar to pipelines through which nutrients and oxygen arrive while waste and carbon dioxide depart. However, it has been proven that the role of the vasculature goes beyond simply being a way to transport blood.

In fact, blood vessels have been shown to perform different tasks in different organs. For example, the capillaries that form the glomeruli of the kidney participate in primary filtration of the blood. The endothelial cells that form these capillaries have fenestrae that allow the filtration of plasma proteins and fluids, but not blood cells (Satchell and Braet 2009). In contrast, liver capillaries (sinusoids) are vascular channels that receive blood from the hepatic artery and portal vein and deliver it into the central veins. These sinusoids are discontinuous, allowing hepatocytes to be bathed in nutrient-rich plasma derived from the blood flow of the small intestine (Reviewed in Braet and Wisse 2002). In the brain, blood vessels form the blood brain barrier that restricts the flux of blood-derived substances to the brain parenchyma (Pardridge 1991, Rubin and Staddon 1999). These different functions are the consequence of inter-organ endothelial heterogeneity.

Although inter-organ vascular heterogeneity has been known for quite some time, recent findings have led us to think that there is also intra-organ vascular heterogeneity. Although for some organs this idea may not come as a surprise, the notion that there may be endothelial cell heterogeneity within the brain still remains largely unexplored.

Based on morphological data, brain endothelial cells have typically been considered homogeneous. Nevertheless, some studies have shown that brain endothelial cells differ in their molecular identity. In fact, transgenic mice carrying a chimeric human Von Willebrand factor construct fused with the coding sequence of lactose Z (LacZ), showed 5-bromo-4-chloro-3-indolyl-β-D-galactopyranoside reaction in a subpopulation of brain endothelial cells (Aird et al. 1995). Additionally, transgenic mice that have the LacZ reporter under the transcriptional control of the tyrosine kinase with immunoglobulin -like and epithelial growth factor-like domains-2 gene present LacZ activity only in some brain endothelial cells (Deutsch et al. 2008). In the human brain, capillary endothelial cells from blood type A individuals show heterogeneous expression of the A and H antigens (Wang et al. 2013). Another example of endothelial heterogeneity is the distribution of endothelial barrier antigen and P-glycoprotein. The endothelial barrier antigen is absent in cerebral arterioles, whereas its expression varies greatly among endothelial cells within venules and capillaries. In contrast, P-glycoprotein is highly and uniformly expressed in venules and capillaries, but it is absent in cerebral arterioles (Saubamea et al. 2012). In addition, the venous marker EphB4 is heterogeneously expressed by endothelial cells in different parts of the adult mouse brain (Colin-Castelan et al. 2011). Interestingly, developmental studies have shown that endothelial cells express molecular markers homogeneously in the embryonic brain; thus it appears that endothelial cell heterogeneity arises with age (Aird et al. 1995, Deutsch et al. 2008, Colin-Castelan et al. 2011). It is possible that the formation of specific subsets of endothelial cells is required for proper homeostasis of the mature tissue.

Besides molecular heterogeneity, some studies have also found functional differences within the brain vascular bed. For instance, cerebral blood flood presents diverse spatial responses even when all the tissue is subjected to the same stimulus (Khalili-Mahani et al. 2012); probably due to the uneven distribution of receptors among endothelial cells. Moreover, even the barrier function of blood vessels within the central nervous system is not homogeneous as normally assumed. For example, the circumventricular organs do not possess a complete blood brain barrier, thus allowing some substances in circulation to modify brain function (Fig. 1) (Reviewed in Horsburgh and Massoud 2013). Likewise, the vascular bed of the neurogenic subventricular zone has a modified blood brain barrier that lacks astrocytic endfeet and pericyte coverage (Fig. 1). The consequence of these changes is that the blood brain barrier of the subventricular zone is permeable to small molecules (Tavazoie et al. 2008). Interestingly, the analysis of tight junction proteins, such as claudin-5, in the spinal cord demonstrates that the density of tight junctions is greatest in capillaries and smaller venules, while it is less in larger venules. In addition, the response to inflammatory stimuli, such as experimental autoimmune encephalomyelitis, is also heterogeneous. During experimental autoimmune encephalomyelitis, the spinal venules show reduction in the density of tight junctions allowing focal leukocyte extravasation, whereas capillaries maintain their integrity (Paul et al. 2013).

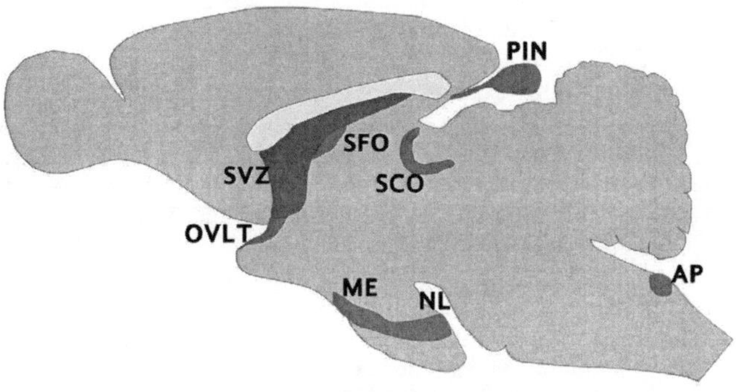

Figure 1. Schematic representation of a sagittal section through the rat brain. The areas highlighted with blue or purple lack tight junctions between endothelial cells (Blue: circumventricular organs, Purple: subventricular zone). AP = area postrema; ME = median eminence; NL = neural lobe of the pituitary; OVLT = vascular organ of the lamina terminalis; PIN = pineal organ; SCO = subcomissural organ; SFO = subfornical organ; SVZ = subventricular zone.

Color image of this figure appears in the color plate section at the end of the book.

We are just beginning to unravel the complexity of the vascular bed within the central nervous system. There is little but conclusive information demonstrating endothelial cell heterogeneity within the brain and spinal cord. Most of the studies have focused on the distribution of the barrier properties of the endothelial cells. However, it is possible that the heterogeneity also extends to non-barrier properties of the endothelial cells (i.e., growth factor secretion, extracellular matrix formation, leukocyte adhesion, surface antigens expression). We need to explore brain endothelial heterogeneity more in depth, because this may represent a new level of regulation of brain function and development that had never been explored before.

References

Aird, W.C., N. Jahroudi, H. Weiler-Guettler, H.B. Rayburn and R.D. Rosenberg. 1995. Human von Willebrand factor gene sequences target expression to a subpopulation of endothelial cells in transgenic mice. Proc. Natl. Acad. Sci. USA 92: 4567–4571.

Alvarez-Buylla, A. and J. Garcia-Verdugo. 2002. Neurogenesis in adult subventricular zone. J. Neurosci. 22: 629–634.

Alvarez-Buylla, A. and C. Lois. 1995. Neuronal stem cells in the brain of adult vertebrates. Stem Cells 13: 263–272.

Bovetti, S., Y.C. Hsieh, P. Bovolin, I. Perroteau, T. Kazunori and A.C. Puche. 2007. Blood vessels form a scaffold for neuroblast migration in the adult olfactory bulb. J. Neurosci. 27: 5976–5980.

Bovetti, S., P. Peretto, A. Fasolo and S. De Marchis. 2007. Spatio-temporal specification of olfactory bulb interneurons. J. Mol. Histol. 38: 563–569.

Bozoyan, L., J. Khlghatyan and A. Saghatelyan. 2012. Astrocytes control the development of the migration-promoting vasculature scaffold in the postnatal brain via VEGF signaling. J. Neurosci. 32: 1687–1704.

Braet, F. and E. Wisse. 2002. Structural and functional aspects of liver sinusoidal endothelial cell fenestrae: a review. Comp. Hepatol. 1: 1.

Burghardt, N.S., E.H. Park, R. Hen and A.A. Fenton. 2012. Adult-born hippocampal neurons promote cognitive flexibility in mice. Hippocampus 22: 1795–1808.

Christov, C., F. Chretien, R. Abou-Khalil, G. Bassez, G. Vallet, F.J. Authier, Y. Bassaglia, V. Shinin, S. Tajbakhsh, B. Chazaud and R.K. Gherardi. 2007. Muscle satellite cells and endothelial cells: close neighbors and privileged partners. Mol. Biol. Cell. 18: 1397–1409.

Colin-Castelan, D., B.V. Phillips-Farfan, G. Gutierrez-Ospina, A.L. Fuentes-Farias, A. Baez-Saldaña, P. Padilla-Cortes and E. Melendez-Herrera. 2011. EphB4 is developmentally and differentially regulated in blood vessels throughout the forebrain neurogenic niche in the mouse brain: Implications for vascular remodeling. Brain Res. 1383: 90–98.

Deutsch, U., T.M. Schlaeger, B. Dehouck, A. Doring, S. Tauber, W. Risau and B. Engelhardt. 2008. Inducible endothelial cell-specific gene expression in transgenic mouse embryos and adult mice. Exp. Cell Res. 314: 1202–1216.

Doetsch, F., J.M. Garcia-Verdugo and A. Alvarez-Buylla. 1997. Cellular composition and three-dimensional organization of the subventricular germinal zone in the adult mammalian brain. J. Neurosci. 17: 5046–5061.

Emsley, J. 2003. alpha6beta1 integrin directs migration of neuronal precursors in adult mouse forebrain. Exp. Neurol. 183: 273–285.

Eriksson, P.S., E. Perfilieva, T. Bjork-Eriksson, A.M. Alborn, C. Nordborg, D.A. Peterson and F.H. Gage. 1998. Neurogenesis in the adult human hippocampus. Nat. Med. 4: 1313–1317.

Franco, S.J. and U. Muller. 2011. Extracellular matrix functions during neuronal migration and lamination in the mammalian central nervous system. Dev. Neurobiol. 71: 889–900.

Horsburgh, A. and T.F. Massoud. 2013. The circumventricular organs of the brain: conspicuity on clinical 3T MRI and a review of functional anatomy. Surg. Radiol. Anat. 35: 343–349.

Khalili-Mahani, N., M.J. Osch, M.D. Rooij, C.F. Beckmann, M.A. Buchem, A. Dahan, J.M. Gerven and S.A. Rombouts. 2012. Spatial heterogeneity of the relation between resting-state connectivity and blood flow: An important consideration for pharmacological studies. Hum. Brain Mapp. doi: 10.1002/hbm.22224.

Kopp, H.G., S.T. Avecilla, A.T. Hooper and S. Rafii. 2005. The bone marrow vascular niche: home of HSC differentiation and mobilization. Physiology (Bethesda) 20: 349–356.

Kramer, R.H., G.M. Fuh, K.G. Bensch and M.A. Karasek. 1985. Synthesis of extracellular matrix glycoproteins by cultured microvascular endothelial cells isolated from the dermis of neonatal and adult skin. J. Cell Physiol. 123: 1–9.

Louissaint, A. Jr., S. Rao, C. Leventhal and S.A. Goldman. 2002. Coordinated interaction of neurogenesis and angiogenesis in the adult songbird brain. Neuron 34: 945–960.

Luskin, M.B. 1993. Restricted proliferation and migration of postnatally generated neurons derived from the forebrain subventricular zone. Neuron 11: 173–189.

Luskin, M.B. 1998. Neuroblasts of the postnatal mammalian forebrain: their phenotype and fate. J. Neurobiol. 36: 221–233.

Mathieu, C., K. Sii-Felice, P. Fouchet, O. Etienne, C. Haton, A. Mabondzo, F.D. Boussin and M.A. Mouthon. 2008. Endothelial cell-derived bone morphogenetic proteins control proliferation of neural stem/progenitor cells. Mol. Cell Neurosci. 38: 569–577.

Melendez-Herrera, E., D. Colin-Castelan, A. Varela-Echavarria and G. Gutierrez-Ospina. 2008. Semaphorin-3A and its receptor neuropilin-1 are predominantly expressed in endothelial cells along the rostral migratory stream of young and adult mice. Cell Tissue Res. 333: 175–184.

Murase, S., C. Cho, J.M. White and A.F. Horwitz. 2008. ADAM2 promotes migration of neuroblasts in the rostral migratory stream to the olfactory bulb. Eur. J. Neurosci. 27: 1585–1595.

Nakagomi, N., T. Nakagomi, S. Kubo, A. Nakano-Doi, O. Saino, M. Takata, H. Yoshikawa, D.M. Stern, T. Matsuyama and A. Taguchi. 2009. Endothelial cells support survival, proliferation, and neuronal differentiation of transplanted adult ischemia-induced neural stem/progenitor cells after cerebral infarction. Stem Cells 27: 2185–2195.

Nishikawa, K., K. Ayukawa, Y. Hara, K. Wada and S. Aoki. 2011. Endothelin/endothelin-B receptor signals regulate ventricle-directed interkinetic nuclear migration of cerebral cortical neural progenitors. Neurochem. Int. 58: 261–272.

Ohab, J.J., S. Fleming, A. Blesch and S.T. Carmichael. 2006. A neurovascular niche for neurogenesis after stroke. J. Neurosci. 26: 13007–13016.

Palmer, T., A. Willhoite and F.H. Gage. 2000. Vascular niche for adult hippocampal neurogenesis. J. Comp. Neurol. 425: 479–494.

Pardridge, W.M. 1991. Advances in cell biology of blood-brain barrier transport. Semin. Cell Biol. 2: 419–426.

Paul, D., A.E. Cowan, S. Ge and J.S. Pachter. 2013. Novel 3D analysis of Claudin-5 reveals significant endothelial heterogeneity among CNS microvessels. Microvasc. Res. 86: 1–10.

Plane, J.M., A.V. Andjelkovic, R.F. Keep and J.M. Parent. 2010. Intact and injured endothelial cells differentially modulate postnatal murine forebrain neural stem cells. Neurobiol. Dis. 37: 218–227.

Riquelme, P.A., E. Drapeau and F. Doetsch. 2008. Brain micro-ecologies: neural stem cell niches in the adult mammalian brain. Philos. Trans. R. Soc. Lond. B Biol. Sci. 363: 123–137.

Rubin, L.L. and J.M. Staddon. 1999. The cell biology of the blood-brain barrier. Annu. Rev. Neurosci. 22: 11–28.

Satchell, S.C. and F. Braet. 2009. Glomerular endothelial cell fenestrations: an integral component of the glomerular filtration barrier. Am. J. Physiol. Renal Physiol. 296: F947–F956.

Saubamea, B., V. Cochois-Guegan, S. Cisternino and J.M. Scherrmann. 2012. Heterogeneity in the rat brain vasculature revealed by quantitative confocal analysis of endothelial barrier antigen and P-glycoprotein expression. J. Cereb. Blood Flow Metab. 32: 81–92.

Seaberg, R.M. and D. van der Kooy. 2002. Adult rodent neurogenic regions: the ventricular subependyma contains neural stem cells, but the dentate gyrus contains restricted progenitors. J. Neurosci. 22: 1784–1793.

Shen, Q., S.K. Goderie, L. Jin, N. Karanth, Y. Sun, N. Abramova, P. Vincent, K. Pumiglia and S. Temple. 2004. Endothelial cells stimulate self-renewal and expand neurogenesis of neural stem cells. Science 304: 1338–1340.

Shen, Q., Y. Wang, E. Kokovay, G. Lin, S.M. Chuang, S.K. Goderie, B. Roysam and S. Temple. 2008. Adult SVZ stem cells lie in a vascular niche: a quantitative analysis of niche cell-cell interactions. Cell Stem Cell 3: 289–300.

Stenman, J., H. Toresson and K. Campbell. 2003. Identification of two distinct progenitor populations in the lateral ganglionic eminence: implications for striatal and olfactory bulb neurogenesis. J. Neurosci. 23: 167–174.

Sun, J., B. Sha, W. Zhou and Y. Yang. 2010. VEGF-mediated angiogenesis stimulates neural stem cell proliferation and differentiation in the premature brain. Biochem. Biophys. Res. Commun. 394: 146–152.

Sun, J., W. Zhou, D. Ma and Y. Yang. 2010. Endothelial cells promote neural stem cell proliferation and differentiation associated with VEGF activated Notch and Pten signaling. Dev. Dyn. 239: 2345–2353.

Szymczak, P., L. Wojcik-Stanaszek, J. Sypecka, A. Sokolowska and T. Zalewska. 2010. Effect of matrix metalloproteinases inhibition on the proliferation and differentiation of HUCB-NSCs cultured in the presence of adhesive substrates. Acta Neurobiol. Exp. (Wars.) 70: 325–336.

Taupin, P. 2006. Adult neural stem cells, neurogenic niches, and cellular therapy. Stem Cell Rev. 2: 213–219.

Tavazoie, M., L. Van der Veken, V. Silva-Vargas, M. Louissaint, L. Colonna, B. Zaidi, J.M. Garcia-Verdugo and F. Doetsch. 2008. A specialized vascular niche for adult neural stem cells. Cell Stem Cell 3: 279–288.

Teng, H., Z.G. Zhang, L. Wang, R.L. Zhang, L. Zhang, D. Morris, S.R. Gregg, Z. Wu, A. Jiang, M. Lu, B.V. Zlokovic and M. Chopp. 2008. Coupling of angiogenesis and neurogenesis in cultured endothelial cells and neural progenitor cells after stroke. J. Cereb. Blood Flow Metab. 28: 764–771.

Thored, P., J. Wood, A. Arvidsson, J. Cammenga, Z. Kokaia and O. Lindvall. 2007. Long-term neuroblast migration along blood vessels in an area with transient angiogenesis and increased vascularization after stroke. Stroke 38: 3032–3039.

Tucker, E.S., F. Polleux and A.S. LaMantia. 2006. Position and time specify the migration of a pioneering population of olfactory bulb interneurons. Dev. Biol. 297: 387–401.

Wang, L., Z.G. Zhang, R.L. Zhang, S.R. Gregg, A. Hozeska-Solgot, Y. LeTourneau, Y. Wang and M. Chopp. 2006. Matrix metalloproteinase 2 (MMP2) and MMP9 secreted by erythropoietin-activated endothelial cells promote neural progenitor cell migration. J. Neurosci. 26: 5996–6003.

Wang, M.M., S.J. Lee, J. Kim, J.J. Majersik, M. Blaivas and J. Borjigin. 2013. ABO blood antigens define human cerebral endothelial diversity. Neuroreport 24: 79–83.

Whitman, M.C., W. Fan, L. Rela, D.J. Rodriguez-Gil and C.A. Greer. 2009. Blood vessels form a migratory scaffold in the rostral migratory stream. J. Comp. Neurol. 516: 94–104.

Yamashita, T., M. Ninomiya, P. Hernandez Acosta, J.M. Garcia-Verdugo, T. Sunabori, M. Sakaguchi, K. Adachi, T. Kojima, Y. Hirota, T. Kawase, N. Araki, K. Abe, H. Okano and K. Sawamoto. 2006. Subventricular zone-derived neuroblasts migrate and differentiate into mature neurons in the post-stroke adult striatum. J. Neurosci. 26: 6627–6636.

Yang, P., K.A. Baker and T. Hagg. 2005. A disintegrin and metalloprotease 21 (ADAM21) is associated with neurogenesis and axonal growth in developing and adult rodent CNS. J. Comp. Neurol. 490: 163–179.

Yang, X.T., Y.Y. Bi and D.F. Feng. 2011. From the vascular microenvironment to neurogenesis. Brain Res. Bull. 84: 1–7.

CHAPTER 3

Neuronal-Glial-Endothelial Interactions Regulate Central Nervous System Homeostasis

Beatriz Gómez-Gónzalez,[a,*] *Adrian M. Cabañas-Morales,*[b] *Gabriela Hurtado-Alvarado*[c] and *Javier Velázquez-Moctezuma*[d]

3.1 Energy and Nutrient Requirements of the Central Nervous System

Neural cells are open systems that constantly exchange energy with the microenvironment; this energy is used to perform the required work for vital cell functions such as endocytocis, synthesis and transport of macromolecules, exocytosis, cell structure maintenance, locomotion, membrane excitability, adhesion and, in some cases, proliferation. All these cellular functions require a stable microenvironment and conservation of an adequate metabolic rate (Riera et al. 2008). Homeostasis is the maintenance of constant conditions in the internal milieu in spite of changed external conditions (Cannon 1932). Homeostasis also requires large amounts of energy in the form of adenosine triphosphate (ATP) (Riera et al. 2008).

Area of Neurosciences, Dept. of Biology of Reproduction, CBS, Universidad Autónoma Metropolitana, Unidad Iztapalapa, Av. San Rafael Atlixco No. 186, Col. Vicentina, Deleg. Iztapalapa, México City, D.F., Mexico, 09340.
[a] Email: bgomezglez@gmail.com & bgomez@xanum.uam.mx
[b] Email: adrian.mi.c.m@gmail.com
[c] Email: gadasa313@yahoo.com.mx
[d] Email: jvm@xanum.uam.mx
* Corresponding author

Although the human brain accounts for only two per cent of the total body weight, it uses 20 per cent of the energy available; therefore, a specialized transport system is required to supply the required amount of energy to support neural activity (Riera et al. 2008). Energy supply to active synaptic regions involves the delivery of both glucose and lactate to neural and glial cells; in fact, glucose concentration in the brain interstitial fluid is reduced temporarily 10–52 per cent when synaptic activity increases (Chen et al. 1993, Lowry et al. 1998, Barros and Deitmer 2010). In addition to glucose, there are huge differences in the plasma concentration of amino acids, proteins, ions and other macromolecules compared both to cerebrospinal fluid and brain interstitial fluid (as shown in Table 1)

Table 1. Plasma and cerebrospinal fluid (CSF) concentration of some solutes.

Solute	**Units**	**Plasma**	**CSF**
Proteins			
Albumin	mg/ml	42	192
Inmunoglobulin G	mg/ml	9.87	0.012
Transferrin	mg/ml	2.6	0.014
Plasminogen	mg/ml	0.7	0.000025
Fibrinogen	mg/ml	325	0.00275
α2-macroglobulin	mg/ml	3	0.0046
Cistatin-C	mg/ml	0.001	0.004
Ions			
Sodium	mM	140	141
Potassium	mM	4.6	2.9
Calcium	mM	5	2.5
Magnesium	mM	1.7	2.4
Chlorine	mM	101	124
Amino acids			
Leucine	μM	109	10.1–14.9
Arginine	μM	80	14.2–21.6
Glicine	μM	249	4.7–8.5
Alanine	μM	330	23.2–32.7
Serine	μM	149	23.5–37.8
Glutamic acid	μM	83	1.79–14.7
Taurine	μM	78	5.3–6.8
Other factors			
Osmolarity	mOsmol	305.2	298.5
pH		7.4	7.3
Glucose	mM	5	3

Modified from Abbott et al. 2010.

(Abbott et al. 2010). These concentration differences are essential to ensure proper neural and glial functions. In addition, selective transport of specific highly demanded molecules (e.g., glucose, lactate, some amino acids and neurotrophic factors) is essential to maintain homeostasis and preserve proper brain function. Homeostasis of the nervous system is maintained by the coordinated action of blood-neural barriers (Fishman 1990, Spector 2010), including the blood-brain barrier, blood-spinal cord barrier, blood-cerebrospinal fluid barrier, blood-retinal barrier and blood-nerve barrier (Choi and Kim 2008).

3.2 Blood-Neural Barriers

As the keepers of homeostasis in the nervous system, blood-neural barriers provide protection to neural cells against potentially neurotoxic circulating molecules, while selectively transporting molecules required for proper neural function. Cells forming part of the blood-neural barriers express a highly specialized phenotype, which includes the presence of solute carriers and efflux transporters and the maintenance of restricted access through the paracellular route. Because of their localization in all neural structures (brain, spinal cord, retina and peripheral nerves), these blood-neural barriers could be considered as the main interface between the periphery and the nervous system.

3.2.1 Blood-Brain Barrier

The blood-brain barrier constitutes the main barrier in the central nervous system; 99 per cent of brain capillaries possess it (Nag and Begley 2005, Pardridge 2005). Brain capillaries are preferentially distributed in the gray matter instead of the white matter (Peters et al. 1991). In mammals, capillary density also varies as a function of regional synaptic activity; there are more capillaries per area in the cortical layers with more synaptic activity (e.g., layer four of primary sensory cortices, where thalamic afferents end) compared to the rest of the cortical layers (Patel 1983, Zheng et al. 1991, Woolsey et al. 1996). The olfactory bulb glomeruli also present a higher capillary density than the interglomerular spaces (Borowsky and Colins 1989) and capillaries are more numerous in the barrels of the somatosensory cortex than in the inter-barrel regions (Patel 1983, Woolsey et al. 1996). It is estimated that almost every neuron has its own capillary in the vicinity; in the mammalian gray matter, capillaries are separated from each other by approximately 50 μm and each capillary may perfuse to a surrounding area of 15–50 μm (Pardridge 2005, Hamilton et al. 2010).

Large blood vessels, like the arteries and veins, are mainly located in the external surface of the brain, while arterioles and venules are located in the brain parenchyma. Although penetrating arteries and arterioles do not constitute a proper blood-brain barrier, they are highly secluded from the neural environment; they have three coats, called *tunica adventitia, tunica media* and *tunica intima* (Peters et al. 1991). The outermost layer, the *tunica adventitia,* is composed of cells derived from the leptomeninges; the *tunica media* is formed by several interposing layers of smooth muscle cells; and the *tunica intima* is constituted of one layer of endothelial cells. Collagen fibers separate the *tunica adventitia* from the *tunica media* and a thick basal lamina divides the *tunica media* from the *tunica intima.* On the other hand, only endothelial cells and their surrounding basal lamina form the walls of venules and veins. In addition to their variable coatings, all medium and large blood vessels are isolated from the surrounding neuropil by the presence of the perivascular space and by a thick impermeable layer of astrocyte endfeet, called glia *limitans* (Frederickson and Low 1969, Peters et al. 1991).

The capillary wall, therefore, constitutes the main site for molecular exchange; it represents a 20 m^2 surface area (150–200 cm^2/g), emerging as the major interface between the periphery and the brain (Nag and Begley 2005, Pardridge 2005). The blood brain barrier is constituted by brain endothelial cells, astroglia, pericytes and microglia; in addition it includes two specialized extracellular matrix components, the basal lamina and the glycocalyx (Fig. 1A). All these components are essential to maintain the biochemical and physical properties of the blood-brain barrier, however, the main component is the endothelial cell due to its structure and function (Risau 1991, Abbott et al. 2006).

3.2.2 Blood-Spinal Cord Barrier

The blood-spinal cord barrier is localized in the capillaries irrigating the spinal cord (Choi and Kim 2008, Bartanusz et al. 2011). These capillaries derive from arteries of the pial plexus and progressive ramifications of the anterior and posterior spinal arteries; arterioles penetrate radially into the spinal cord and then ramify into microvessels. Capillaries in the mammalian spinal cord are preferentially distributed in the gray matter, as compared to the white matter (Dommise 1980, Qiu and Zhu 2004); moreover, capillary density is higher in the anterior horn than in the posterior horn throughout the spinal cord (Qiu and Zhu 2004).

Like the blood-brain barrier, the main cellular components of the blood-spinal cord barrier are the endothelial cells; other cellular components of the blood-spinal cord barrier include the astroglia, pericytes and microglia (see Fig. 1A) (Bartanusz et al. 2011). Endothelial cells conforming the

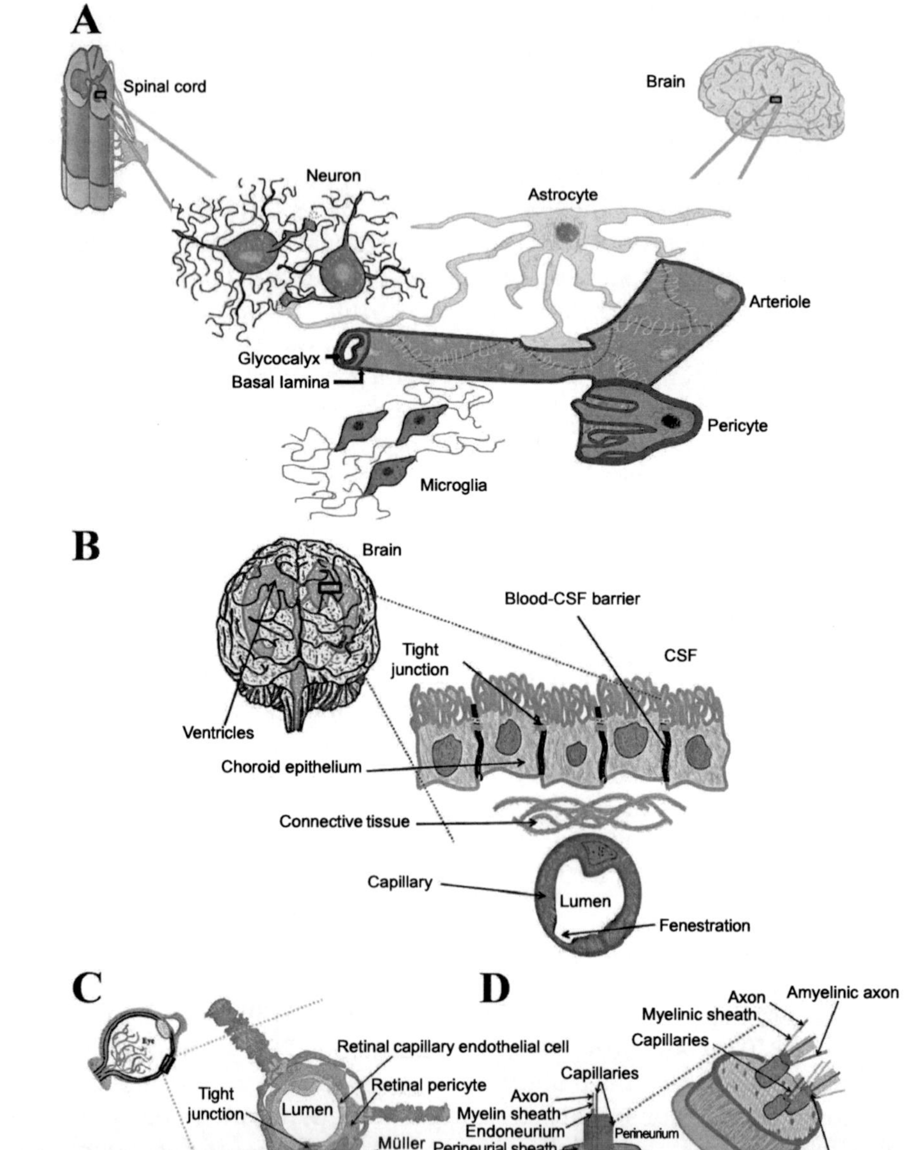

Figure 1. Blood-neural barriers. A) Blood-brain and blood-spinal cord barriers, B) Blood-cerebrospinal fluid barrier, C) Blood-retinal barrier and D) Blood-nerve barrier. CSF = cerebrospinal fluid.

Color image of this figure appears in the color plate section at the end of the book.

blood-spinal cord barrier differ from brain endothelial cells by several characteristics; they present glycogen deposits, higher permeability to some circulating molecules (e.g., the cytokines interferon α, interferon γ, tumor necrosis factor (TNF)-α and the tracer ^{14}C-carboxyl-inulin) and have reduced expression of some proteins conforming the specialized interendothelial junctions (Prockop et al. 1995, Pan et al. 1997, Sharma 2005, Ge and Patcher 2006).

3.2.3 Blood-Cerebrospinal Fluid Barrier

The blood-cerebrospinal fluid barrier is localized in the subarachnoid space and choroid plexus of the lateral, third and fourth ventricles (Johanson et al. 2011); however, the primary site for blood to cerebrospinal fluid exchange occurs in the 75 cm^2 surface area of the choroid plexus of the lateral ventricles (Keep and Jones 1990). In contrast to the blood-brain barrier and blood-spinal cord barrier, the blood-cerebrospinal fluid barrier is constituted exclusively by epithelial cells; the capillaries of the choroid plexus are fenestrated, thus allowing the free exchange of molecules from the blood to the interstitial fluid (Fig. 1B) (Hurley et al. 1981). Epithelial cells are responsible for the physical and chemical properties of the blood-cerebrospinal fluid barrier. The complete belt tight junctions between the epithelial cells restrict paracellular diffusion of molecules from the choroid plexus and interstitial space to the cerebrospinal fluid. A special system of carriers, present in apical and basolateral membranes of epithelial cells, guarantees that the concentration of water and solutes in the cerebrospinal fluid is maintained stable within the physiological parameters required for normal brain physiology (see Table 1) (Brightman and Reese 1969, Johanson et al. 2011).

These epithelial cells are specialized in the production and absorption of cerebrospinal fluid; the epithelial cells of the choroid plexus produce up to 0.4 ml/min/g of cerebrospinal fluid in the mammalian central nervous system (Cserr 1971). On the other hand, the epithelial cells of the subarachnoid space are pivotal for cerebrospinal fluid absorption and drainage to the venous circulation (Johanson et al. 2011). Cerebrospinal fluid circulates from the brain ventricles to the central canal of the spinal cord and to the subarachnoid space; once molecules traverse the blood-cerebrospinal fluid barrier they may enter into the brain interstitial fluid.

3.2.4 Other Blood-Neural Barriers

Other neural structures located outside the central nervous system also require a blood-barrier due to their high metabolic demands and the need for protection against potentially neurotoxic molecules circulating in the blood (Abbott 2005). Structures in which an endothelial or epithelial barrier is observed include the internal ear and peripheral nerves (Juhn et al. 2001, Choi and Kim 2008, Hosoya and Tachikawa 2009, Mizisin and Weerasuriya 2011).

3.2.4.1 Blood-Retinal Barrier

The blood-retinal barrier includes the outer and inner blood-retinal barrier (Choi and Kim 2008, Hosoya and Tachikawa 2009). The outer blood-retinal barrier is localized in the epithelial cells of the external pigmented layer of the retina; it protects the retinal cells against potentially neurotoxic molecules that might cross through the fenestrated endothelium of the choroid layer (Choi and Kim 2008). The inner blood-retinal barrier is an endothelial barrier localized in the capillaries irrigating the internal face of the retina (Choi and Kim 2008, Hosoya and Tachikawa 2009). The capillaries of the retina derive from the central retinal artery, which penetrates into the eye through the optic nerve as a ramification of the ophthalmic artery. The retina has two capillary beds: one located superficially, extending from the nerve fiber layer to the ganglion cell layer (called superficial capillary plexus) and the other located in the inner nuclear layer (named deep capillary plexus). Capillaries are also distributed heterogeneously along the retina, they are denser in the peripheral (lateral and nasal) retina and absent in the fovea (Anand-Apte and Hollyfield 2010).

In addition to endothelial cells, the inner retinal barrier is composed of Müller glia, pericytes, and extracellular matrix components (Fig. 1C) (Choi and Kim 2008, Hosoya and Tachikawa 2009, Anand-Apte and Hollyfield 2010). The endothelial cells that form the inner blood-retinal barrier are joined by tight junctions that preclude paracellular diffusion of circulating molecules. These cells also express numerous influx carriers and efflux transporters that allow selective passage of glucose and other nutrients into the retina and carry out xenobiotics and endobiotics (Hosoya and Tachikawa 2009).

3.2.4.2 Blood-Nerve Barrier

Peripheral nerves are insulated from variations in extracellular fluid composition owing to their epineurium, perineurium and endoneurium covering (Fig. 1D). The epineurium is formed by collagen fibers, scattered fibroblasts and some adipocytes; it represents an extension of the dura mater. The perineurium consists of numerous layers of flattened squamous cells, up to 15 in the vicinity of the root nerve and only one at the nerve terminal endings, which form intercellular tight junctions to restrict molecular diffusion from external to internal layers; it is considered an extension of the leptomeninges (Shanthaveerappa and Bourne 1962, Burkel 1967, Weerasuriya and Mizisin 2011). Finally, the endoneurium forms a space which contains the connective tissue attached to myelin sheaths and the endoneurial fluid. The blood-nerve barrier extends from the roots of the spinal and cranial nerves to their sensory and motor terminal endings; the blood-nerve barrier is located in the endothelial cells forming the capillaries of the most internal nerve coat, the endoneurium and the flattened squamous cells of the perineurium (Weerasuriya et al. 1979, 1980, Mizisin and Weerasuriya 2011, Weerasuriya and Mizisin 2011). In the endoneurium, continuous non-fenestrated capillaries establish tight junctions at the interendothelial contacts; pericytes also participate in the formation of the blood-nerve barrier in endoneurial capillaries (Bell and Wedell 1984).

3.3 Components of the Blood-Brain Barrier

3.3.1 Endothelial Cells

The main cellular components of the blood-brain barrier are the endothelial cells (Risau 1991, Abbott et al. 2006). In the adult mammalian central nervous system, brain endothelial cells are arranged in a single cell layer, about 500 nm thick, forming the wall of the brain capillaries (Abbott et al. 2010, Peters et al. 1991, Ueno 2007). Brain endothelial cells are characterized by the absence of fenestrations, presence of tight junctions at interendothelial cell contacts and the membrane expression of influx and efflux carriers in a polarized pattern (see below for details). As shown in the following sections, these cells present reduced transport of macromolecules by receptor-mediated and adsorptive endocytosis routes. Although they are the major cell types that determine the physical and chemical properties of the blood-brain barrier, endothelial cells need contacts with surrounding cells and factors released by astroglia and pericytes to maintain their barrier phenotype (Abbott et al. 2006, Risau 1991).

3.3.2 Astrocytes

Astrocytes are named after their morphology, which is characterized by multiple processes that look like a cartoon star. Astroglia in the mammalian brain can be divided into nine types; tanycytes, radial cells, Bergmann glia, fibrous astrocytes, protoplasmatic astrocytes, marginal glia, perivascular glia, ependymal glia and Müller cells (Emsley and Macklis 2006, Matyash and Kettenmann 2010). Astroglia numbers vary from one species to another, it is calculated that glia outnumber neurons in the human brain hemispheres;

specifically in the human neocortex there are 1.5–1.65 astrocytes per neuron, while in the rat neocortex the ratio is 0.3–0.5 astrocytes per neuron (Sherwood et al. 2006, Taber and Hurley 2008). Moreover, in the human neocortex, protoplasmic astroglia are 2.6 times larger in diameter and extend 10 times longer their primary processes than their counterparts in rodents (Matyash and Kettenmann 2010).

Gray matter astrocytes form syncytia, highly communicating networks which control synapse plasticity and remodeling (Houades et al. 2006, Taber and Hurley 2008). In the rodent neocortex and hippocampus astroglial networks may be conformed by more than 50 cells. They establish gap junctions, ensuring direct intracellular communication (Houades et al. 2006). Astrocyte syncytia are established according to the information processing characteristics of each region; for example in the barrel neocortex, astrocytes located in layers one, two and three form horizontal syncytia, while those located in layer four establish gap junctions within the same cortical column (Houades et al. 2006).

There is a symbiotic relationship between astroglia and neurons in the mammalian brain; the phenotype of astrocytes is adapted to the local neuronal environment to optimize information processing (Hewet 2009). Astrocytes extend thousands of fine membranous processes that envelop synapses and capillaries (Barrers 2008), dividing the neuropil into small independent functional unities (Fig. 1A) (Nedergaard et al. 2003). Gray matter astrocyes may contact 20,000 to 120,000 synapses in the rodent; while in the human gray matter, the numbers rise up to two million synapses per astrocyte, constituting the so-called tripartite synapse (Barrers 2008, Oberheim et al. 2009). On the other hand, perivascular astrocyte endfeet constitute a 1.2 μm thick layer that covers around 85–99 per cent of the brain capillary wall (McCaslin et al. 2011, Peters et al. 1991, Risau 1991).

Astrocyte functions include homeostasis of fluids, ions, pH and neurotransmitter extracellular content (Parpura et al. 2012, Sofroniev and Vinters 2010). They are rich in Na^+-dependent glutamate transporters, which control excitatory neurotransmission and prevent excitotoxicity (Matyash and Kettenmann 2010, Parpura et al. 2012). Astrocytes are also responsible for a wide variety of complex and essential functions in the central nervous system, including genesis and maintenance of synapses, participation in information processing (Sofroniev and Vinters 2010) and sensing of changes in synaptic activity with the consequent modification of cerebral blood flow (CBF, see the last section for a detailed description) (Hewet 2009). Astrocytes produce and release many mediators; such as nitric oxide (NO) (Li et al. 2003), taurine, aspartate, eicosanoids (Martin 1992), cyclooxygenase and epoxygenase derivatives (Amruthesh et al. 1992), ATP (Queiroz et al. 1999), angiopoietin-1 and transforming growth factor β (Abbott et al. 2006), that can modify capillary diameter and thus control local blood flow.

3.3.3 Pericytes

Pericytes are smooth-muscle analogs located in the walls of arterioles, venules and capillaries. They are distributed throughout the whole organism; however, they are preferentially located in brain and retinal capillaries (Kamouchi et al. 2011, Shepro and Morel 1993). In the brain, pericytes cover around 20–30 per cent of the capillary wall (Balanov and Dore-Duffy 1998, Kim et al. 2006). They extend primary process along the vessel axis and secondary processes that surround the capillary wall (Kamouchi et al. 2011). Because of their location and since they express the α-muscle actin, pericytes have been deemed as contractile cells (Balanov and Dore-Duffy 1998, Bandopadhyay et al. 2001, Shepro and Morel 1993). In fact, recent *in vivo* studies have shown that pericytes can modify capillary diameter in response to changes in local synaptic activity (Peppiatt et al. 2006, see Hamilton et al. 2010 for an extensive review).

Pericytes establish gap, adherens and tight junctions with brain endothelial cells; these junctions are required to maintain the phenotype of endothelial cells (Allt and Lawrenson 2001, Lai and Kuo 2005, Shimizu et al. 2008). The presence of gap junctions with brain endothelial cells and surface fragment crystallizable receptor expression indicate that pericytes exert phagocytic functions (Guillemin and Brew 2004, Sims 1986, Thomas 1999). On the other hand, tight junctions between pericytes and brain endothelial cells provide structural support for interendothelial tight junctions (Lai and Kuo 2005, Shimizu et al. 2008).

Under normal physiological conditions, pericytes are embedded in the basal lamina surrounding brain capillaries; however under pathological conditions, such as neovascularization secondary to hypoxia or tumorigenesis, pericytes detach from the vessel wall and migrate into the brain parenchyma. It has been postulated that pericytes guide neovascularization both during early development and in the pathological adult central nervous system (Kamouchi et al. 2011). Pericytes have surface molecules, such as platelet-derived growth factor β receptor and transforming growth factor β receptor II; their expression seems to be important for angiogenesis and the adequate establishment of brain endothelial cell phenotypes (Andrae et al. 2008, Dohgu et al. 2005, Hellstrom et al. 2001, Kamouchi et al. 2011). Pericytes also secrete numerous vascular factors that promote or maintain the barrier properties of brain endothelial cells; including angiopoietin-1 (Fiedler and Augustin 2006, Hori et al. 2007), fibroblast growth factor and vascular endothelial growth factor (VEGF) (Shepro and Morel 1993).

3.3.4 Microglia

Microglia constitute around 10 per cent of all central nervous system cells and are mainly involved in the immune response. Microglia have two phenotypes, resting and activated microglia; resting microglia have small bodies and long thin processes while activated microglia have phagocytic morphology with protruding short thick processes (Streit et al. 1999). Microglia continuously survey the neural microenvironment, they activate in response to foreign antigens or changes in brain homeostasis (Yang et al. 2010); therefore, they are deemed as the first defense line against pathogens (Skaper et al. 2012). Reactive microglia may migrate and surround damaged or dead cells, subsequently clearing cellular debris from the local area (Fetler and Amigorena 2005). Under normal physiological conditions, resting microglia are involved in immune surveillance, synapse maintenance and trophic support; due to their highly motile cytoplasmic processes that provide extensive coverage of the entire interstitial space (Wong et al. 2012).

To sense alterations within their environment, microglia express a large number of cell surface receptors (e.g., cytokine or chemokine receptors, toll-like receptors, scavenger receptors), ion channels (e.g., Na^+ channel, Cl^- intracellular channel 1) and adhesion molecules; all of which may activate complex and dynamic signaling pathways that finally result in microglial motility to the damaged region and differentiation into reactive microglia (Black and Waxman 2012, Schomberg and Olson 2012). A key molecule that modulates microglial activity is ATP, an endogenous ligand of the P2-purinoceptor family (Inoue and Tsuda 2012, Trang et al. 2012); the microglial chemoattractive response is triggered by ATP released at the injury site and the subsequent activation of the P2Y12 receptor on microglia (Ohsawa and Kohsaka 2011). Once activated, microglia release pro-inflammatory cytokines, such as interleukin (IL)-1, IL-6 and TNF-α, chemokines (e.g., fractalkine), extracellular proteases (e.g., lysosomal cysteine protease cathepsin S), NO, prostaglandins, growth factors and superoxide species (Clark and Malcangio 2011, Loane and Byrnes 2010, Schomberg and Olson 2012, Smith et al. 2012).

Under normal physiological conditions, microglia are located in the brain parenchyma near capillaries and their interactions with endothelial cells contribute to maintain the properties of the blood-brain barrier (Ransohoff and Perry 2009). Similar to the other blood-brain barrier components, microglia have recently been associated with other important processes in the brain; microglia located within the neurogenic niches may modulate neurogenesis in both the healthy and injured brain. Indeed, a decrease in the survival of new neurons follows the presence of reactive microglia (Ekdahl 2012). It has been proposed that microglial phagocytosis of damaged neurons can be beneficial by preventing the release of pro-inflammatory intracellular contents, but under certain conditions, such as inflammation, microglia can promote the death of viable neurons and phagocytose them (Neher et al. 2012).

3.3.5 Extracellular Matrix Components

The interactions between the cellular components of the blood-brain barrier are mediated by the extracellular components. Both the glycocalyx and basal lamina allow the communication from a specific cell type with other cells through diffusion of soluble signaling molecules (Reitsma et al. 2007, Ueno 2009).

3.3.5.1 Glycocalyx

The glycocalyx is a carbohydrate-rich layer lining the luminal endothelium of blood vessels (Fig. 1A), in which soluble plasma- or endothelium-derived molecules are incorporated. This network of membrane-bound proteoglycans and glycoproteins is considered the first site where water and nutrients are exchanged between circulating blood and the brain tissue. Its thickness depends on the vessel diameter; in capillaries, it is around 0.2–0.5 µm, while in arterioles and venules it is around 2–4.5 µm. The major components of the glycocalyx are the proteoglycans, the glycosaminoglycans (e.g., chondroitin or dermatan sulphate, keratan sulphate and hyaluronic acid), the glycoproteins (e.g., integrins) and soluble components (e.g., growth factors) (Palte and Raines 2012, Reitsma et al. 2007, Ueno 2009).

The composition of the glycocalyx differs between the blood-contact region and the endothelial cell-contact region; glycoproteins such as integrins are in contact mainly with the endothelial cell surface, while soluble molecules and or some-glycosaminoglycans are located in the blood-contact region (Job et al. 2012, Yen et al. 2012). The most important "backbone" molecules of the glycocalyx are the proteoglycans, which consist of a core protein and glycosaminoglycan chains; the major core proteins are syndecan, glypican, versican, decorin, biglycan and mimecan (Reitsma et al. 2007). Glycosaminoglycans are linear polysaccharides bound to proteoglycans; of these heparan sulfate is the most abundant (it represents 50–90 per cent of all glycosaminoglycans found in the blood-brain barrier). All these molecules confer a negative charge to the glycocalyx (Palte and Raines 2012); glycocalyx neutralization increases blood-brain barrier permeability (Gomez-Gonzalez et al. 2011, Huxley and Curry 1995, Noble et al. 1996).

The glycocalyx is the first barrier between the blood and brain because it restricts direct contact between blood elements and endothelial cells.

Size, steric hindrance and electrostatic charge of the glycocalyx and the permeating substances play a role in glycocalyx-dependent permeability (Gao and Lipowsky 2010). The glycocalyx can also modify blood-brain barrier permeability by mechanical forces and by interactions between its components with diverse signaling molecules (Reitsma et al. 2007). Interactions between blood-born molecules and glycocalyx elements may include: (1) binding of receptors, enzymes or their ligands to the endothelial glycocalyx causing a local increase in the concentration of these substances; (2) binding of plasma-derived molecules leading to a local gradient concentration, which is often seen in growth factor-regulated gene transcription and developmental processes; and (3) docking of enzymes and their agonists or inhibitors gives a vasculoprotective role to the glycocalyx (Lawrenson et al. 2000, Reitsma et al. 2007).

3.3.5.2 Basal Lamina

The basal lamina surrounds the brain endothelial cells and pericytes, producing a layer of variable thickness (0.5–1 μm). Its molecules are distributed in three apposed layers: structural, specialized and proteoglycan layers; collagen IV and elastin form the structural layer, whereas laminin and fibronectin form the specialized layer (Weiss et al. 2009, Lourenco et al. 2010). Basal lamina proteins are regulated by matrix metalloproteinases (MMPs) and plasmin; MMP-2 digests collagen IV, fibronectin and laminin, while MMP-9 only digests collagen IV. Endothelial cells, pericytes and astrocytes release the proteins that form the basal lamina; astrocytes express high levels of collagen IV and MMP-2 messenger ribonucleic acid (mRNA), while fibronectin and MMP-9 mRNA are high in pericytes. Other important molecules contained in the basal lamina are integrins, which anchor the extracellular matrix to the endothelial cytoskeleton (Hynes 1992, Lourenco et al. 2010, Persidsky et al. 2006).

Basal lamina functions include cell adhesion, maintenance of capillary vessel morphology, mechanical support for cell attachment, a stratum for the migration of pericytes and endothelial cells, separation of adjacent tissues, regulating passage of macromolecules and permitting communication between cellular elements of the blood-brain barrier. An uncoupling of the basal lamina with other cellular components may alter the endothelial cell cytoskeleton and lead to blood-brain barrier hyperpermeability (Carvey et al. 2009, Lourenco et al. 2010, Persidsky et al. 2006, Zlokovic 2008).

3.4 Specializations of Brain Endothelial Cells

The blood-brain barrier is the critical physical, metabolical and immunological barrier that maintains homeostasis in the central nervous system. The blood-brain barrier regulates entry of blood-borne molecules into the brain. This regulation requires close communication between its cellular and extracellular components; however, endothelial cells function as the main physical and chemical barriers (Ueno 2007). We have previously described the general characteristics of brain endothelial cells, now we will describe their specializations; which preclude paracellular transport of molecules and, on the other hand, provide selective transport of required molecules through specialized carrier systems.

3.4.1 Interendothelial Junctional Complexes

Brain endothelial cells are distinguished from peripheral endothelial cells by the presence of an interendothelial junctional complex constituted by tight and adherens junctions. Because of their functionality and localization, endothelial junctional complexes and especially tight junctions, constitute the gate and fence of the brain. The gate function is given by the high restriction to paracellular diffusion of water-soluble molecules from the blood to the brain at the tight junctions. The fence function is given by the establishment of cell polarity, dividing endothelial cells into apical and basolateral poles (Wolburg et al. 2009). On the other hand, adherens junctions are critical for tight junction formation during early development and for tight junction maintenance during adulthood (McCaffrey et al. 2007, Wolburg and Lippoldt 2002).

3.4.1.1 Tight Junctions

Tight junctions are responsible for some of the key properties of the blood-brain barrier, including its high transendothelial electric resistance (1500–2000 Ω/cm^2) and restricted paracellular permeability of endothelial cells (Butt et al. 1990). Tight junctions are large protein complexes localized in the apical region of brain endothelial cells; which constitute a continuous network of parallel, interconnected, transmembrane protein strands. They maintain brain endothelial cell polarity by promoting the asymmetric allocation of the luminal and abluminal solute transporters and efflux carriers (Deli 2009). Additionally, tight junctions are incorporated within cholesterol-enriched plasma membrane regions associated to caveolin 1 (Nusrat et al. 2000). Interactions between cholesterol and the hydrocarbon chains of sphingolipids generate a liquid-ordered membrane domain that resists detergent solubilization and promotes tight juntion protein oligomerization (Lambert et al. 2005).

The integral proteins that constitute tight junctions are occludin, claudins (Furuse et al. 1993, 1998) and junctional adhesion molecules (JAM)

(Fig. 2) (Chiba et al. 2008, Martín-Padura et al. 1998). The first protein identified was occludin, a ≈60 kDa protein with four transmembrane domains plus amino (N)- and carboxy (C)-terminal cytoplasmic domains (Fig. 2). The N-terminal domain is constituted mostly by tyrosine and

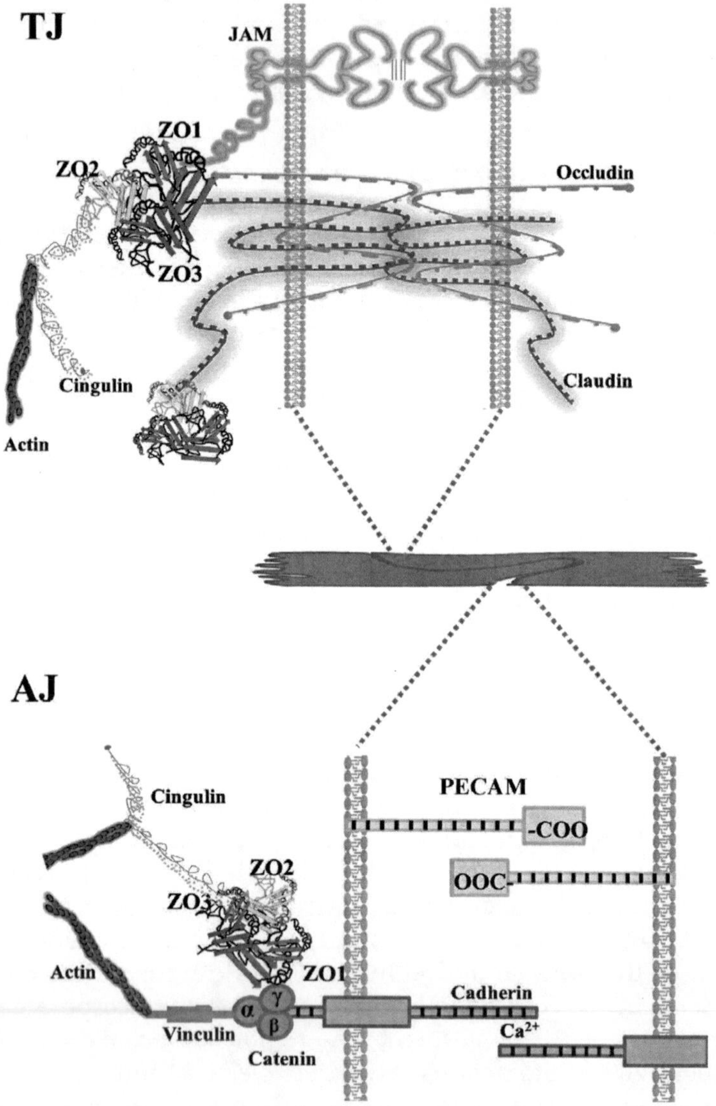

Figure 2. The interendothelial junctional complex. The image shows tight junctions (TJ) in the apical region and adherens junctions (AJ) in the basolateral region of brain endothelial cells. Abbreviations: ZO-1 to 3 = Zonula occludens 1, 2 and 3; JAM = junctional adhesion molecules; PECAM = platelet endothelial cell adhesion molecule.

Color image of this figure appears in the color plate section at the end of the book.

glycine residues (≈60 per cent), as well as many charged amino acids. The C-terminal domain is rich in threonine, serine and tyrosine residues, which are targets for phosphorylation by protein kinases (Schneeberger and Lynch 2004). The C-terminal region is directly bound to cytoplasmic anchoring proteins, such as zona occludens protein 1 (ZO-1) (Furuse et al. 1993), which anchors occludin to the actin cytoskeleton (Furuse et al. 1994). The first extracellular loop is involved in intercellular adhesion and the second loop generates the high transendothelial electric resistance characteristic of brain endothelial cells (Nusrat et al. 2005). Occludin also participates in apoptosis and cytoskeleton maintenance of brain endothelial cells (Chiba et al. 2008).

Claudins are a family of 20–24 kDa transmembrane proteins that include up to 24 members (Schneeberger and Lynch 2004). Claudin-1, 3, 4, 5, 8, 11, 14 and 19 contribute to the tightness of paracellular barriers; specifically, brain endothelial cells express claudin-1, 3, 5, 12 and 18 (Deli 2009, Krause et al. 2008). The C-terminal domain binds to the PDZ domains of ZO-1, 2 and 3 (Fig. 2) (Furuse et al. 1999). The extracellular loops generate interendothelial pores of variable sizes (≈10 Å) that allow bidirectional water flux along the tight junction (Stamatovic et al. 2008, Tang and Goodenough 2003). In fact, tracer experiments and magnetic resonance imaging (MRI) studies show that the blood-brain barrier loses its selective impermeability to 800 Da molecules in the absence of claudin-5 (Nitta et al. 2003). The first extracellular loop has several charged amino acids; in claudin-5, the substitution of negatively charged residues in positions 53 and 64 by positive charged residues converts the protein from cation- to anion-permissive (Colegio et al. 2002).

JAM are transmembrane proteins and members of the immunoglobulin (Ig) superfamily (Fig. 2). The JAM family includes five members, of which only JAM-A is present in brain endothelial cells (Nag 2011). JAM are 45 kDa glycosylated proteins that have two extracellular V-type Ig domains, a single transmembrane domain and an intracellular C-terminal domain. The C-terminal domain interacts with the PDZ3 domain of ZO-1. The first V-type Ig domains of opposing JAM molecules interact with each other (Martín-Padura et al. 1998). JAM regulate leukocyte migration in association with platelet endothelial cellular adhesion molecule 1 (Itoh et al. 2001). In addition, they are involved in the development of brain endothelial cell polarity (Rehder et al. 2006).

Transmembrane proteins forming tight junctions are linked to the cell cytoskeleton by cytoplasmic plaques, which are protein complexes adhered to the internal membrane surface in the apical pole of brain endothelial cells. The proteins of the cytoplasmic plaque anchor the transmembrane proteins of tight junctions to the actin cytoskeleton (Wolburg et al. 2009). Among the proteins forming the cytoplasmic plaque are ZO-1, 2 and 3, 7H6 antigen, cingulin, vinculin and symplekin (Fig. 2) (Furuse et al. 1993). ZOs are 130–220 kDa phosphoproteins of the membrane-associated guanylate kinase family that participate in tight junction assembly; indeed, dissociation or loss of ZO-1 from the cytoplasmic plaque increases paracellular permeability of the blood-brain barrier (Mark and Davis 2002). Consequently, tight junction assembly depends on the structural organization of the actin cytoskeleton of brain endothelial cells (Deli 2009); it was shown that actin-disrupting substances, such as cytochalasin D, cytokines and phalloidin, disrupt tight junction structure and function (Stevenson and Begg 1994).

Several cytoplasmic and extracellular molecules regulate the signaling cascades that control the assembly and disassembly of tight junctions. Extracellular Ca^{2+} is involved in many processes, such as heterotrimeric G protein and protein kinase C signaling pathways. It is necessary for the interactions between vascular endothelial (VE)-cadherins, which are related to the initial events of junctional complex assembly. Furthermore, when extracellular Ca^{2+} is removed, there is a decrease in the transendothelial electric resistance and an increase in paracellular permeability. On the contrary, a rise in intracellular Ca^{2+} increases transendothelial electric resistance, ZO-1 migration from intracellular sites to the plasma membrane and tight junction assembly in brain endothelial cells (Stuart et al. 1994). In the cytoskeleton, Ca^{2+} activates Ca^{2+} calmodulin-dependent myosin light chain kinase, which phosphorylates myosin light chain and promotes cross-bridge movement between actin and myosin. Ca^{2+} calmodulin also activate many signaling cascades mediated by proteins kinases that induce phosphorylation of proteins from the tight junctions (Kumar et al. 2009). Protein kinases regulate tight junction properties, adhesion strength among cells and interactions between the plasma membrane and cytoskeleton of brain endothelial cells (Kovbasnjuk et al. 1998).

3.4.1.2 Adherens Junctions

Adherens junctions are dynamic complexes spread among tight junctions that maintain cellular adhesion between adjacent cells; indeed, they are a pre-requisite for tight junction formation (Zlokovic 2008). At the electron microscopic level, they are characterized by apposed parallel membranes of adjacent endothelial cells forming an intercellular space of ≈200 Å that extends 0.2–0.5 μm (Farquhar and Palade 1963, Niessen and Gottardi 2008). Adherens junctions are composed of cadherins, catenins, vinculin and actinin (Fig. 2) (Lourenco et al. 2010). The cadherins expressed in brain endothelial cell membranes include epithelial, placental, neural and VE-cadherin. They are single-pass transmembrane glycoproteins that need extracellular Ca^{2+} to interact with other cadherins of the adjacent membrane. Their cytoplasmic domain mediates crucial structural and signaling activities required for adhesion, through its association with distinct proteins known as catenins (Navarro et al. 1998).

Catenins have two isoforms, α-catenin and β-catenin, which bind F-actin ensuring polymerization in or near adherens junctions (Vasioukhin et al. 2000). Monomeric α-catenin binds β-catenin in adherens junctions and, upon release, forms dimers that promote actin bundle formation. The transition from branched actin networks to bundled actin filaments correlates with the creation of mature adherens junctions. Both VE-cadherin and β-catenins are implicated in endothelial hyperpermeability, since they can be phosphorylated by receptor tyrosine kinases when VEGF is present; which activates p21-activated kinase and focal adhesion kinase, thereby triggering integrin signaling (Stockton et al. 2004). Due to their close association with the cytoskeleton of brain endothelial cells, any alteration in the integrity of tight or adherens junctions may modify the dynamics of transcellular transport. Indeed, both vesicle-mediated transport and differential allocation of specific carriers systems depend on the stability of the cytoskeleton of brain endothelial cells (Lourenco et al. 2010, Zlokovic 2008).

3.4.2 Specialized Carrier Systems

Macromolecule movement across brain endothelial cells by the paracellular route is restricted due to the presence of tight junctions and simple diffusion across the endothelial cell membrane is limited to small lipophilic molecules (e.g., O_2, CO_2 and ethanol) (Jones et al. 1992). Hence, hydrophilic and large hydrophobic molecules require specific carrier mechanisms. Thus, brain endothelial cells transport nutrients, vitamins and metabolic precursors from the blood-to-brain through specialized carrier systems. Transcellular bidirectional transport across the blood-brain barrier can be classified into five main categories: carrier-mediated passage, ion transfer, active efflux conveyance, receptor-mediated movement and caveolae-mediated transport (Fig. 3) (Zlokovic et al. 2008).

3.4.2.1 Carrier-Mediated Transport

Carrier systems present in brain endothelial cells transport nutrients such as hexoses (glucose and galactose), neutral, basic and acidic amino acids, monocarboxylic acids (lactate, pyruvate and ketone bodies), nucleosides, purines, amines and vitamins from the blood-to-brain and vice versa (Zlokovic et al. 2008). The influx and efflux rates of molecules vary as a function of their concentration and requirement. The major carrier systems present in the

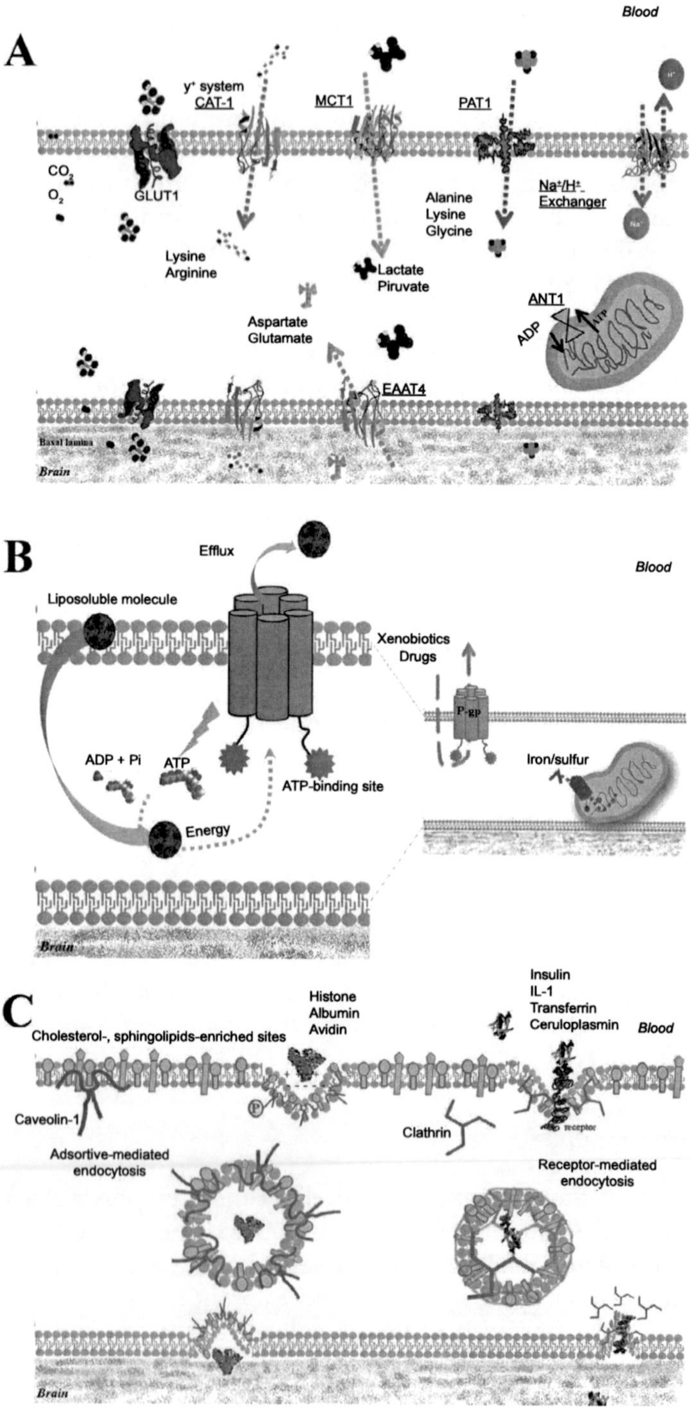

Figure 3. contd....

blood-brain barrier are glucose transporter (GLUT)-1, large neutral amino acid transporters (L1) and the monocarboxylic acid transporter 1 (MCT1) (Fig. 3) (Pardridge 2007).

GLUT-1 is the main glucose carrier present in the blood-brain barrier (Fig. 3A); an isoform of 55 kDa is selectively expressed in brain endothelial cells; while another isoform of 45 kDa is expressed in astrocytes (Dwyer et al. 2002). GLUT-1 also transports mannose, 2-deoxyglucose, galactose, ascorbic acid and other glycoconjugates (Lourenco et al. 2010). It is a bidirectional Na^+-independent transporter that carries glucose as a function of the gradient concentration. GLUT-1 is predominantly expressed in the abluminal membrane of brain endothelial cells, compared to the luminal membrane (Cornford and Hyman 2005, Dobrogowska and Vorbrodt 1999, Farrell and Pardridge 1991, Pardridge et al. 1990). This differential expression assures that the mean glucose concentration in the interstitial fluid is kept constant at one mM under resting conditions, despite plasma variations in concentration (Barros and Deitmer 2010, Leybaert 2005). Although glucose transport rate into the brain is kept at 0.37 mM/g/min under basal conditions, which is enough to support neural and glial activity (de Graaf et al. 2001); it was shown that glycogen polymers accumulate inside brain astrocytes and in spinal cord endothelial cells (Cataldo and Broadwell 1986, Sharma 2005). Glycogen is an energy reservoir of the central nervous system that can be used in emergencies, such as energetic stress caused by hypoglycemia or cerebral ischemia (Riera et al. 2008).

MCT also carries energy substrates into the brain, including ketone bodies (β-hydroxybutyrate and acetoacetate), lactate and pyruvate (Fig. 3A); in endothelial cells the transport of those molecules is facilitated by the concentration gradient (Simpson et al. 2007, Zlokovic et al. 2008). Brain endothelial cells express both MCT-1 and MCT-2 (Tsuji 2005); however, MCT-2 is also expressed in the membranes of astrocytes and neurons (Bergersen et al. 2001, Cornford and Hyman 2005, Simpson et al. 2007). Both carriers localize at luminal and abluminal membranes of brain endothelial cells

Figure 3. contd.

Figure 3. Specialized carrier systems in brain endothelial cells. A) Solute carrier systems for glucose, amino acids and ion exchange; B) ATP-binding cassette transporters for the efflux of xenobiotics and drugs; C) adsorptive and receptor-mediated endocytosis. Abbreviations: GLUT-1 = Glucose transporter 1; CAT-1 = cationic amino acid transporter (y+ system); MCT1 = monocarboxylic acid transporter; LAT-1 = large amino acid transporter 1; EAAT-4 = excitatory amino acid transporter; ANT-1 = mitochondrial adenine nucleotide translocator; Na^+/H^+ = sodium proton pump; P-gp = P glycoprotein; IL-1 = interleukin 1; ATP = adenosine triphosphate; ADP = adenosine biphosphate; Pi = inorganic phosphate.

Color image of this figure appears in the color plate section at the end of the book.

(Abbott et al. 2010). The mean rate of substrate transport is ≈ 91 nmol/g/min in brain endothelial cells; thus, guaranteeing enough supply of energy substrates and maintenance of proper brain function (Pardridge 1983).

Amino acids are also essential for brain metabolism, especially for neurotransmitter synthesis. The main amino acid influx carriers in brain endothelial cells are L1 carriers which transport large neutral amino acids and y^+ carriers that transport cationic amino acids (Fig. 3A) (Abbott et al. 2010, Hawkins et al. 2006). The L1 and y+ carrier systems are expressed on both luminal and abluminal membranes of brain endothelial cells; they exert bidirectional, Na^+-independent facilitative transport of essential amino acids. L1 mediates transport of the large neutral amino acids leucine, isoleucine, tryptophan, valine, tyrosine, phenylalanine, methionine and threonine. Under normal physiological conditions these amino acids compete; indeed, tryptophan has the lowest affinity for the L1 system. Meanwhile, the y+ system mediates transport of cationic aminoacids (e.g., lysine, arginine and ornithine) (Hawkins et al. 2006).

Additionally, brain endothelial cells also express efflux amino acid transporters; among them the most important is the excitatory amino acid transporter (EAAT) (Fig. 3A). It is a Na^+-dependent carrier that provides a mechanism for net removal of potentially neurotoxic amino acids (Zlokovic 2008). EAAT extrudes glutamate and aspartate from the brain interstitial fluid into endothelial cells and the x_G and n carrier systems extrude them from the endothelial cells. Brain endothelial cells also express a Na^+-dependent small neutral amino acid carrier system, which transports alanine, cysteine, serine, glycine, threonine, valine, leucine, isoleucine and methionine unidirectionally from the brain (Abbott et al. 2010, Pardridge 2005, Zlokovic 2008). The efflux carrier systems maintain the amino acid concentration in the brain interstitial fluid at 10 per cent of the plasma concentration (Zlokovic 2008). Other carrier systems which are expressed in the membrane of brain endothelial cells are shown in Table 2.

3.4.2.2 ABC Transporters

The chemical barrier function of brain endothelial cells includes extruding mechanisms for lipid soluble molecules. Under normal physiological conditions, molecules <400–500 Da in size, high lipid solubility and <8–10 hydrogen bonds freely cross the blood-brain barrier (Pardridge 2005). Large liposoluble molecules have low concentrations in the central nervous system because they are substrates for the ATP-binding cassette (ABC) transporter family. These energy-dependent carriers, located in the luminal membrane of brain endothelial cells, keep xenobiotics at low concentrations in the brain interstitial fluid. ABC transporters play a protective role, but also

Table 2. Solute carriers expressed in brain endothelial cells.

Solute carrier	Substrate function
SGLT-1	Glucose Na^+-dependent transporter
GLUT-3	Glucose transporter
Choline transporter	Creatinine and choline Na^+ (and Cl^-)-dependent transport
Oatp-1	Thyroid hormone and organic anions transporter
GAT-1	Gamma-amino butyric acid Na^+ (and Cl^-)-dependent transporter
FATP-1	Long chain fatty acids transporter
PiT-1	Phosphate symporter
ANT-1	ADP ATP exchange across mitochondrial inner membrane
PiT-2	Na^+-dependent phosphate transporter
AE-3	Anion (Cl^- HCO_3^-) exchanger
Xtrp-3	Na^+ and Cl^- coupled transport
NHERF	Na^+ H^+ exchange regulatory factor
IREG-1	Proton-coupled divalent metal ion transporter
SAT-2	Na^+-dependent neutral amino acid transporter
NHE-1	Na^+ H^+ exchanger
NIS	Na^+ iodide symporter
ArsA	Arsenite efflux
Znt-1	Zinc efflux

Modified from Enerson and Drewes (2006). Abbreviations: ADP = adenosine diphosphate; AE-3 = anion exchanger 3; ANT-1 = mitochondrial adenine nucleotide translocator; ArsA = arsenite transporter; ATP = adenosine triphosphate; FATP-1 = fatty acid transporter 1; GAT-1 = gamma-amino butyric acid transporter; GLUT-3 = glucose transporter 3; IREG-1 = iron-regulated transporter; NHERF = sodium hydrogen exchanger regulator; NHE-1 = sodium hydrogen exchanger 1; NIS = sodium iodide symporter; Oatp-1 = organic anion transporter; PiT-1 PiT-2 = phosphate transporter 1, 2; SAT-2 = system A amino acid transporter 2; SGLT-1 = sodium glucose cotransporter; Xtrp-3 = X-transporter protein 3; Znt-1 = Zinc transporter 1.

contribute to the drug resistance observed in neuropathological conditions (e.g., cancer, epilepsy and infections) (Jodoin et al. 2003, Sawchuk et al. 2000, Schinkel et al. 1996).

ABC transporters are diverse and some of their mechanisms are yet unknown, the majority of them are promiscuous efflux carriers. The first multidrug resistance protein (MDR) from the ABC family that was described was P glycoprotein (P-gp) (Schinkel et al. 1996, Schinkel and Jonker 2003). P-gp is highly expressed in the luminal surface of brain endothelial cells, though it is also expressed in perivascular astrocyte endfeet and pericytes (Fig. 3B) (Bendayan et al. 2006, Lee and Bendayan 2004). P-gp extrudes anti-parasitic molecules (e.g., ivermectin), anti-epileptic compounds (e.g., phenytoin), chemotherapy agents (e.g., imatinib), H1 histamine receptor antagonists, anti-HIV drugs and verapamil (Spector 2010, Zlokovic 2008). In addition to P-gp, brain endothelial cells also express the MDR proteins and the breast cancer resistant protein (BCRP). P-gp and BCRP have synergistic interactions (Zhou et al. 2009), whereas MDR proteins are responsible for efflux transport of glutathione-, glucoronate- and sulfate-conjugated molecules (Zlokovic 2008). ABC transporters and their putative functions are presented in Table 3.

Table 3. ABC transporters present in brain endothelial cells.

ABC transporters	Function
ABC1 (ABC subfamily A)	
Member 2	Lipid transport and lipid metabolism
MDR TAP (ABC subfamily B)	
P-gp	Drugs and xenobiotics efflux
Member 6	Transporter of iron sulfur from mitochondria to cytosol
Member 8	Transport of phospholipids to the mitochondrial membrane
CFTR MRP (ABC subfamily C)	
SUR2	K transport channel
MRP1	Multispecific anion transporter
MRP5	Drug, putative glutathione and conjugate pump
MRP3	Organic anions, drugs and bile salt efflux
MRP4	Organic anion transporter, putative efflux pump for cAMP, cGMP
MRP6	Organic anions amphipatics efflux
MRP8	Regulator of K^+ channels
ALD (ABC subfamily D)	
PMP70	Peroxisomal import of fatty acid and fatty acyl coA
OABP (ABC subfamily E)	
Member 1	RNAse inhibitor
GCN20 (ABC subfamily F)	
Member 1	mRNA translation
WHITE (ABC subfamily G)	
BCRP1	Drugs and xenobiotics efflux
Member 1	Transfer of cellular cholesterol to HDL

Based on Enerson and Drewes 2006. Abbreviations: ABC = ATP-binding cassette, MDR = multidrug resistance, TAP = transporter associated with antigen processing, P-gp = P glycoprotein, CFTR = cystic fibrosis transmembrane conductance regulator, MRP = multidrug resistance-associated protein, SUR = sulfonylurea receptor, ALD = adrenoleukodystrophy, PMP = peroxisomal membrane protein, OABP = oligoadenylate binding protein, GCN = general control nonderepressible, BCRP = breast cancer resistance protein, cAMP = cyclic adenosine monophosphate, cGMP = cyclic guanosine monophosphate, coA = coenzyme A, mRNA = messenger ribonucleic acid, HDL = high-density lipoprotein

3.4.2.3 Vesicle-Mediated Endocytosis

Endothelial cells use a variety of endocytotic pathways for the internalization of exogenous materials, including clathrin-mediated endocytosis, macropinocytosis and caveolar endocytosis. Vesicle-mediated transport in brain endothelial cells is limited to only 5–10 vesicles/μm^2 in comparison with endothelial cells of other organs and tissues (Nag 2011). Macropinocytosis is a form of endocytosis in which bending of single surface lamellipodia gives rise to curved ruffles, followed by sealing of the aperture and vacuole formation. The large size of macropinosomes provides an efficient route for non-selective endocytosis of macromolecules (Swanson and Watts 1995). Macropinosomes have been described as the entry routes of bacteria and virus into the brain (Falcone et al. 2006).

Adsorptive endocytosis has the potential for binding and uptake of cationic molecules to the luminal surface of endothelial cells and for exocytosis at the abluminal surface. The flux rate of adsorptive endocytosis is proportional to the extracellular concentration of solutes and also the cellular ability to form vesicles. Cationic charges in macromolecules promote their electrostatic interaction with the negatively charged endothelial cell surface, leading to an enhanced cellular uptake via adsorptive endocytosis (Georgieva et al. 2011, Kumagai et al. 1987). Adsorptive endocytosis is potentially valuable as a drug delivery system for large-molecules into the brain; for example, the dynorphin-like peptide, ebiratide, is transported into the brain by this pathway. Other proteins also cross the blood-brain barrier by adsorptive endocytosis; such as albumin, pro-thrombin, plasminogen (Frank et al. 2003, Abbott et al. 2010), β-endorphin-cationized albumin complex (Kumagai et al. 1987), histones and avidin (Tamai et al. 1997).

Adsorptive endocytosis takes place in 50–100 nm vesicles, also called caveolae, that are formed by invagination of the endothelial cell membrane (Frank et al. 2003, Razani et al. 2002). The invagination of caveolae is controlled by the distinctive material present in the inner membrane surface: cholesterol, sphingolipids, glycolipids, glycosylphosphatidylinositol-linked proteins, caveolin-1, and to a lesser proportion, caveolin-2 (Fig. 3C) (Anderson 1993, Lourenco et al. 2010). Caveolae formation and transit across brain endothelial cells depends on the cellular metabolic state; tight junction disentanglement and tyrosine phosphorylation of caveolin-1 both promote caveolae formation (Georgieva et al. 2011, Kim et al. 2009, Lourenco et al. 2010). Calcium and cyclic adenosine monophosphate also increase caveolae formation by activating signaling pathways; such as those involving non-receptor tyrosine kinases and endothelial nitric oxide synthase (Anderson 1993).

On the other hand, receptor-mediated endocytosis is mediated by the clathrin cover of endocytic vesicles (Fig. 3C). The molecules which are transported by receptor-mediated endocytosis include transferrin, insulin, high- and low-density lipoprotein (Zlokovic 2008), IL-1α, IL-1β, IL-6, TNFα (Banks et al. 1995, Banks 2005), β-amyloid protein, apolipoprotein E, 2-macroglobulin, S-100 protein, IgG, epidermal growth factor (Abbott et al. 2010), leptin (Kastin and Pan 2000) and insulin-like growth factor (IGF-

1) (García-Segura et al. 1991). The transport of these molecules from the blood-to-brain interstitial fluid takes several seconds and is characterized by dissociation of the ligand-receptor complex. Thus, when the vesicle attaches to the abluminal membrane, the ligand is released into the brain interstitial fluid whereas the membrane and receptor are recycled (Abbott et al. 2010, Nag 2011). Endothelial cells express receptors depending on the requirements of the brain, it has been reported that leptin receptor expression varies as a function of circadian rhythms and the blood concentration of estrogens, triglycerides, glucose, insulin and adrenaline (Banks 2001, Banks et al. 2004, Kastin and Pan 2000, Kastin et al. 2001, Kastin and Akerstrom 2001); this produces differences in the transport rate of leptin in distinct brain regions (Banks et al. 2000).

3.5 Neuronal-Glial-Endothelial Cell Interactions Regulate Blood-Brain Barrier Permeability

For many years the blood-brain barrier was generally thought to be an impermeable physical and chemical barrier with selective transport of the energy substrates and nutrients required for proper brain function, equally impermeable at all times in all brain regions. Nevertheless, it is currently accepted that CBF varies as a function of the activity of each brain region, a phenomenon formerly called functional hyperemia (Mosso 1880, Roy and Sherrington 1890) now known as neurovascular coupling (Cauli and Hamel 2010, Filosa and Blanco 2007). It has been recently shown that there is a significant change in the transport rate of glucose and other metabolic substrates in response to increased neuronal activity in each brain capillary; thus the phenomenon is called neurobarrier coupling (Leybaert 2005, Leybaert et al. 2007). Both neurovascular and neurobarrier coupling guarantee an adequate supply of glucose, oxygen and other molecules to highly active regions (Cauli and Hamel 2010, Leybaert 2005, Leybaert et al. 2007).

3.5.1 Neurovascular Coupling

Neurovascular coupling involves changes in the vasodynamic properties of brain vessels mediated by the contraction or relaxation of perivascular smooth muscle cells and pericytes (Bleys and Cowen 2001, Hamilton et al. 2010). Brain blood vessels are regulated by extrinsic and intrinsic chemical signals. The extrinsic signals are of two types, blood-circulating signals (Huber et al. 2001, Sharma et al. 1996) and signals coming from the autonomic innervation of large extracerebral vessels (Bleys and Cowen 2001, Edvinsson et al. 1987). Sympathetic fibers release noradrenaline, neuropeptide Y and serotonin (Edvinsson et al. 1987), which function as the principal effectors; while parasympathetic fibers release nitric oxide (Lee 2000). Autonomic sympathetic mediators are generally vasoconstrictors and parasympathetic mediators are mostly vasodilators (Bleys and Cowen 2001, Edvinsson et al. 1987).

Intrinsic innervation of the large and medium blood vessels within the brain derives from the locus coeruleus, dorsal raphe, tractus solitarius and basal forebrain (Kalaria et al. 1989, Reis and Iadecola 1986, Sato and Sato 1992). Cortical and subcortical nerve terminals innervate brain capillaries; nerve terminals containing dopamine, vasoactive intestinal peptide (VIP), NO, acetylcholine and γ-amino butyric acid have been found in the vicinity of parenchymal capillaries (Arneric et al. 1988, Gragera et al. 1993, Benagiano et al. 1996, Krimer et al. 1998, Roufail et al. 1995). Histamine, noradrenaline and serotonin induce vasoconstriction of capillaries (Kelley et al. 1988, Markhotina et al. 2007); while VIP, NO, adenosine and a drop in extracellular pH or O_2 induce vasodilatation (Carmignoto and Gomez-Gonzalo 2010, Chen and Andreson 1997, Haefliger et al. 1994, Hamilton et al. 2010, Matsugi et al. 1997, Markhotina et al. 2007).

Neurovascular coupling is accomplished by the joint activity of astrocytes and pericytes. Given their perivascular position and contractile properties, pericytes are the main vascular constrictors and dilators, while astrocytes maintain CBF changes related to increased local synaptic activity. Astrocytes sense neuronal activity by activation of neurotransmitter receptors located in their perisynaptic endfeet and by detecting changes in interstitial fluid O_2 concentration (Carmignoto and Gomez-Gonzalo 2010, Yamamoto and Kato 2002). In response, astrocytes have Ca^{2+} waves that propagate through their entire cytoplasm to their perivascular endfeet at a rate of 20 µm/s in rodents and 40 µm/s in humans (Matyash and Kettenmann 2010, Zonta et al. 2003a, 2003b). Intracellular Ca^{2+} accumulation leads to release of arachidonic acid derivatives (Amruthesh et al. 1992, Zonta et al. 2003a), lactate, prostaglandin E_2 (Gordon et al. 2008), K^+ (Filosa et al. 2006) and NO from astrocyte perivascular endfeet (Gordon et al. 2008, Metea and Newman 2006). The final effect depends on the concentration of O_2 in the brain interstitial fluid; under low O_2 conditions the release of vasodilators prevails, while under high O_2 conditions, mainly vasoconstrictors are released (Gordon et al. 2008, Metea and Newman 2006). Pericytes contract the capillary wall in response to both neural- and astroglial-generated signals. Vasoconstrictors depolarize pericytes by increasing cytoplasmic Ca^{2+} concentration (Sakagami et al. 1999) and pericyte relaxation is accomplished by cell hyperpolarization (Filosa et al. 2006, Hamilton et al. 2010).

Neurovascular coupling has been reported in several *in vivo* studies using MRI and two-photon microscopy. In the rat olfactory bulb, increases in CBF have been consistently observed one to two seconds after acute exposure to a pure odorant; increased CBF is present only in capillaries irrigating the stimulated glomerulus instead of occurring at upstream arterioles irrigating nearby glomeruli (Chaigneau et al. 2003, Petzold et al. 2008). At least in one study, the observed CBF increase was concomitant with astroglial Ca^{2+} waves in the individual glomeruli (Petzold et al. 2008). In the rat somatosensory cortex, astroglial Ca^{2+} waves also occurred concurrent to CBF increases after mechanical stimulation of the limbs; the hemodynamic changes became apparent within one second after limb stimulation and the latency to the induction of astroglial Ca^{2+} waves was only 0.5 seconds (Winship et al. 2007). Other studies pinpoint even earlier hemodynamic changes associated to increased local synaptic activity. Ferezou et al. (2006) showed that the CBF increased in the rat primary somatosensory cortex during the search for potential stimulus sources, when animals were moving toward an object and presenting whisker behavior.

3.5.2 Neurobarrier Coupling

In addition to CBF changes, increases in local synaptic activity modify the transport efficiency of energy substrates and other molecules across brain endothelial cells. Although the first experiments reporting neurobarrier coupling appeared in 2000 (Cornford et al. 2000), currently little is known about the mechanisms regulating the changes in blood-brain barrier permeability to glucose and other solutes required for proper neural function. Both neural and astroglial-derived signals seem to be the major players in neurobarrier coupling, both cellular types may potentially induce changes in the expression of carriers at the endothelial cell membrane by the release of ATP and other molecules (Boado et al. 1999, Braet and Leybaert 2000).

Neurobarrier coupling constitutes the subjacent phenomenon for the blood oxygen level-dependent (BOLD) signal in the functional MRI (fMRI) (Figley and Stroman 2011, Logothetis 2002). Increases in local synaptic activity change O_2 levels within the microregion. During the first 300–500 milliseconds after neural stimulation O_2 levels decrease, followed by a rapid rise in oxygenation secondary to vasodilatation and increased CBF (Buxton 2001, Carmignoto and Gomez-Gonzalo 2010, Yamamoto and Kato 2002). Using fMRI, Devonshire et al. (2012) showed that electrical stimulation of the rat whisker-pad induced differential BOLD responses along the somatosensory pathway (trigeminal spinal interpolaris nucleus, thalamus and somato-sensory cortex). All these regions showed augmented BOLD signal as a function of increased synaptic activity, measured by local field potentials. However, high frequency stimuli produced higher BOLD signals in the subcortical nuclei (thalamus and trigeminal spinal interpolaris nucleus) than in the somatosensory cortex. In contrast, higher BOLD signals were seen in the somatosensy cortex when low frequency stimuli were applied to the whisker pad (Devonshire et al. 2012). Despite numerous reports using the BOLD signal as an indirect measure of neural activity, it has been suggested that the free diffusion of O_2 through endothelial cell membranes exceeds the requirements of surrounding neurons irrespective of their activation. Moreover, vasodilatation increases the surface area for O_2 exchange in brain endothelial cells ensuring higher transport efficiency (Fox and Raichle 1986, Heeger and Ress 2002, Logothetis 2002). Consequently, changes in glucose consumption appear to better reflect the local synaptic activity (Figley and Stroman 2011).

Cornford et al. (2000) showed that generalized increases in brain activity, induced in rats by penthylenetetrazole seizures, were accompanied by higher glucose transport efficiency in almost the whole brain within the first three minutes after seizures began. Although this was observed during abnormal brain activity (seizures), it clearly exemplifies neurobarrier coupling. Under basal physiological conditions, brain interstitial fluid glucose concentration is around 0.35–2.6 mM (Leybaert 2005). Increases in synaptic activity reduce 10 to 52 percent the local glucose concentration (Chen et al. 1993, Lowry et al. 1998) increasing the gradient difference in glucose concentration between the blood and parenchyma compartments, promoting facilitated glucose transport through brain endothelial cells (Leybaert 2005). Moreover, glucose transport in highly active brain regions also depends on changes in membrane expression of GLUT-1. In the rat, only 60 per cent of GLUT-1 protein is expressed in the membrane of brain endothelial cells, while the rest is localized within their cytoplasm (Dobrogowska and Vorbrodt 1999, Farrel and Pardridge 1991, Leybaert 2005, Leybaert et al. 2007). Increased synaptic activity promotes GLUT-1 translocation from the cytoplasm to the endothelial cell membrane, mediated by intra-cellular Ca^{2+} signaling (Nishimura and Simpson 1994).

Finally, neurobarrier coupling has also been described for growth factors. In a rodent model, Nishijima et al. (2010) showed that electrical stimulation of the cerebellar peduncles increased the rate of IGF-1 receptor-mediated endocytosis in the cerebellar cortex, when compared to non-stimulated brain regions. Mechanical stimulation of the rodent vibrissae also increased blood-to-brain IGF-1 transfer, specifically in the barrel neocortex. Furthermore, increased transport of IGF-1 across brain endothelial cells was observed in more natural conditions, such as enriched environment exposure. IGF-1 passage through hippocampal endothelial cells was higher in rats exposed to an enriched environment than in the hippocampus of rodents living under standard laboratory conditions (Nishijima et al. 2010).

3.6 Final Remarks

Brain endothelial cells are key mediators of brain homeostasis maintenance; they act as physical and chemical barriers, precluding the passage of blood-borne exogenous molecules to the brain. Brain endothelial cells also selectively transport glucose, amino acids and other molecules (e.g., lactate and neurotrophic factors) from the blood to the brain and extrude potentially neurotoxic substances from the brain interstitial fluid. However, their barrier characteristics depend on cell contacts with parenchymal cells and on molecules released by these cells. Pericytes contribute to mantaining the physical properties of brain endothelial cells by promoting entanglement and support of tight junctions and contributing to their proper cytoskeletal arrangement. Meanwhile, astroglial cells are required for the maintenance of the chemical properties of brain endothelial cells. Brain endothelial cells require the signals released by astrocytes, pericytes and neurons to allow the passage of energy substrates and other molecules at the proper time and in the proper brain region. It is clearly established that CBF varies as a function of local synaptic activity, as both increase in unison. The phenomenon, known as neurovascular coupling is induced by the coordinated action of pericytes and astroglia. Pericytes control the vasodilatation associated to increased CBF, while signals from astroglia and neurons regulate the vasodynamic properties of arterioles and capillaries. On the other hand, brain endothelial cells modify the transport rate of glucose, oxygen and growth factors from the blood-to-brain in response to increased neuronal activity. In the case of neurobarrier coupling, it seems that the main players are signals released by neurons and glia. Although much is known about brain endothelial cells specializations, more research is needed to elucidate the mechanisms governing neurovascular and neurobarrier coupling, as well as to determine the role of the vasculature in information processing.

Acknowledgements

This work was supported by a postdoctoral fellowship to Beatriz Gómez-González from PROMEP-SEP.

References

Abbott, N.J., A.A.K. Patabendige, D.E.M. Dolman, S.R. Yusof and D.J. Begley. 2010. Structure and function of the blood-brain barrier. Neurobiol. Dis. 37: 13–25.

Abbott, N.J., L. Ronnback and E. Hansson. 2006. Astrocyte-endothelial interactions at the blood-brain barrier. Nat. Rev. Neurosci. 7: 41–53.

Abbott, N.J. 2005. Dynamics of CNS barriers: evolution, differentiation, and modulation. Cell. Mol. Neurobiol. 25: 5–23.

Allt, G. and J.G. Lawrenson. 2001. Pericytes: cell biology and pathology. Cells Tissues Organs 169: 1–11.
Amruthesh, S.C., J.R. Falck and E.F. Ellis. 1992. Brain synthesis and cerebrovascular action of epoxygenase metabolites of arachidonic acid. J. Neurochem. 58: 503–510.
Anand-Apte, B. and J.G. Hollyfield. 2010. Developmental anatomy of the retinal and choroidal vasculature. pp. 9–15. *In*: D.A. Dartt, J. Besharse and R. Dana [eds.]. Encyclopedia of the Eye. Elsevier, Oxford, GB.
Anderson, R.G. 1993. Caveolae: where incoming and outgoing messengers meet. Proc. Natl. Acad. Sci. USA 90: 10909–10913.
Andrae, J., R. Gallini and C. Betsholtz. 2008. Role of platelet-derived growth factor in physiology and medicine. Genes Dev. 22: 1276–1312.
Arneric, S.P., M.A. Honig, T.A. Milner, S. Greco, C. Iadecola and D.J. Reis. 1988. Neuronal and endothelial sites of acetylcholine synthesis and release associated with microvessels in rat cerebral cortex: ultrastructural and neurochemical studies. Brain Res. 454: 11–30.
Balanov, R. and P. Dore-Duffy. 1998. Role of the CNS microvascular pericyte in the blood-brain barrier. J. Neurosci. Res. 53: 637–644.
Bandopadhyay, R., C. Orte, J.G. Lawrenson, A.R. Reid, S. De Silva and G. Allt. 2001. Contractile proteins in pericytes at the blood-brain and blood-retinal barriers. J. Neurocytol. 30: 35–44.
Banks, W.A. 2001. Enhanced leptin transport across the blood–brain barrier by α1-adrenergic agents. Brain Res. 899: 209–217.
Banks, W.A. 2005. Blood-brain barrier transport of cytokines: a mechanism for neuropathology. Curr. Pharm. Des. 11: 973–984.
Banks, W.A., C.M. Clever and C.L. Farrell. 2000. Partial saturation and regional variation in the blood to brain transport of leptin in normal weight mice. Amer. J. Physiol. 278: E1158–E1165.
Banks, W.A., A.B. Coon, S.M. Robinson, A. Moinuddin, J.M. Shultz, R. Nakaoke and J.E. Morley. 2004. Triglycerides induce leptin resistance at the blood–brain barrier. Diabetes 53: 1253–1260.
Banks, W.A., A.J. Kastin and R.D. Broadwell. 1995. Passage of cytokines across the blood-brain barrier. Neuroimmunomod. 2: 241–248.
Barrers, B.A. 2008. The mystery and magic of glia: a perspective on their roles in health and disease. Neuron 60: 430–440.
Barros, L.F. and J.W. Deitmer. 2010. Glucose and lactate supply to the synapse. Brain Res. Rev. 63: 149–159.
Bartanusz, V., D. Jezova, B. Alajajian and M. Dgicaylioglu. 2011. The blood-spinal cord barrier: morphology and clinical implications. Ann. Neurol. 70: 194–206.
Bell, M.A. and A.G. Wedell. 1984. A morphometric study of intrafascicular vessels in mammalian sciatic nerve. Muscle Nerve 7: 524–534.
Benagiano, V., D. Virgintino, E. Maiorano, A. Rizzi, S. Palombo, L. Roncali and G. Ambrosi. 1996. VIP-like immunoreactivity within neurons and perivascular neuronal processes of the human cerebral cortex. Eur. J. Histochem. 40: 53–56.
Bendayan, R., P.T. Ronaldson, D. Gingras and M. Bendayan. 2006. *In situ* localization of P-glycoprotein (ABCB1) in human and rat brain. J. Histochem. Cytochem. 54: 1159–1167.
Bergersen, L., O. Waerhaug, J. Helm, M. Thomas, P. Laake, A.J. Davies, M.C. Wilson, A.P. Halestrap and O.P. Ottersen. 2001. A novel postsynaptic density protein, the monocarboxylate transporter MCT2 is co-localized with δ-glutamate receptors in postsynaptic densities of parallel fiber-Purkinje cell synapses. Exp. Brain Res. 136: 523–534.
Black, J.A. and S.G. Waxman. 2012. Sodium channels and microglial function. Exp. Neurol. 234: 302–315.
Bleys, R.L. and T. Cowen. 2001. Innervation of cerebral blood vessels: morphology, plasticity, age-related, and Alzheimer's disease-related neurodegeneration. Microsc. Res. Tech. 53: 106–118.
Boado, R.J., D. Wu and M. Windisch. 1999. *In vivo* upregulation of the blood–brain barrier GLUT1 glucose transporter by brain-derived peptides. Neurosci. Res. 34: 217–224.

Borowsky, I.W. and R.C. Collins. 1989. Metabolic anatomy of brain: a comparison of regional capillary density, glucose metabolism, and enzyme activities. J. Comp. Neurol. 288: 401–413.

Braet, K. and L. Leybaert. 2000. Endothelial GLUT-1-mediated glucose uptake is acutely stimulated by ATP and histamine. Eur. J. Physiol. 440: R6.

Brightman, M.W. and T.S. Reese. 1969. Junctions between intimately apposed cell membranes in the vertebrate brain. J. Cell Biol. 40: 648–677.

Burkel, W.E. 1967. The histological fine structure of the perineurium. Anat. Rec. 158: 177–190.

Butt, A.M., H.C. Jones and N.J. Abbott. 1990. Electrical resistance across the blood–brain barrier in anaesthetized rats: a developmental study. J. Physiol. 429: 47–62.

Buxton, R.B. 2001. The elusive initial dip. Neuroimage 13: 953–958.

Cannon, W.B. 1932. The wisdom of the body. W.W. Norton & Company Inc., California, USA.

Carmignoto, G. and M. Gomez-Gonzalo. 2010. The contribution of astrocyte signaling to neurovascular coupling. Brain Res. Rev. 63: 138–148.

Carvey, P.M., B. Hendey and A.J. Monahan. 2009. The blood–brain barrier in neurodegenerative disease: a rhetorical perspective. J. Neurochem. 111: 291–314.

Cataldo, A.M. and R.D. Broadwell. 1986. Cytochemical identification of cerebral glycogen and glucose-6-phosphatasa activity under normal and experimental conditions. II. Choroid plexus and ependymal epithelia, endothelia and pericytes. J. Neurocytol. 15: 511–524.

Cauli, B. and E. Hamel. 2010. Revisiting the role of neurons in neurovascular coupling. Front. Neuroenerg. 2: 9.

Chaigneau, E., M. Oheim, E. Audinat and S. Charpak. 2003. Two-photon imaging of capillary blood flow in olfactory bulb glomeruli. Proc. Natl. Acad. Sci. USA 100: 13081–13086.

Chen, Q. and D.R. Anderson. 1997. Effect of CO_2 on intracellular pH and contraction of retinal capillary pericytes. Invest. Ophthalmol. Visual Sci. 38: 643–651.

Chen, W., E.J. Novotny, X.H. Zhu, D.L. Rothman and R.G. Shulman. 1993. Localized 1H NMR measurement of glucose consumption in the human brain during visual stimulation. Proc. Natl. Acad. Sci. USA 90: 9896–9900.

Chiba, H., M. Osanai, M. Murata, T. Kojima and N. Sawada. 2008. Transmembrane proteins of tight junctions. Biochim. Biophys. Acta 1778: 729–756.

Choi, Y.K. and K.-W. Kim. 2008. Blood-neural barrier: its diversity and coordinated cell-to-cell communication. BMB Reports 41: 345–352.

Clark, A.K. and M. Malcangio. 2011. Microglial signaling mechanisms: Cathepsin S and Fractalkine. Exp. Neurol. 234: 283–292.

Colegio, O.R., C.M. van Itallie, H.J. McCrea, C. Rahner and J.M. Anderson. 2002. Claudins create charge-selective channels in the paracellular pathway between epithelial cells. Am. J. Physiol. Cell Physiol. 283: 142–147.

Cornford, E.M. and S. Hyman. 2005. Localization of brain endothelial luminal and abluminal transporters with immunogold electron microscopy. NeuroRx 2: 27–43.

Cornford, E.M., E.V. Nguyen and E.M. Landlaw. 2000. Acute upregulation of blood-brain barrier glucose transporter activity in seizures. Amer. J. Physiol. Heart Circ. Physiol. 279: H1346–H1354.

Cserr, H.F. 1971. Physiology of the choroid plexus. Physiol. Rev. 51: 273–311.

De Graaf, R.A., J.W. Pan, F. Telang, J.H. Lee, P. Brown, E.J. Novotny, H.P. Hetherington and D.L. Rothman. 2001. Differentiation of glucose transport in human brain gray and white matter. J. Cereb. Blood Flow Metab. 21: 483–492.

Deli, M.A. 2009. Potential use of tight junction modulators to reversibly open membranous barriers and improve drug delivery. Biochim. Biophys. Acta 1788: 892–910.

Devonshire, I.M., N.G. Papadakis, M. Port, J. Berwick, A.J. Kennerly, J.E.W. Mayhew and P.G. Overton. 2012. Neurovascular coupling is brain region-dependent. NeuroImage 59: 1997–2006.

Dobrogowska, D.H. and A.W. Vorbrodt. 1999. Quantitative immunocytochemical study of blood–brain barrier glucose transporter (GLUT-1) in four regions of mouse brain. J. Histochem. Cytochem. 47: 1021–1030.

Dohgu, S., F. Takata, A. Yamauchi, S. Nakagawa, T. Egawa, M. Naito, T. Tsuruo, Y. Sawada, M. Niwa and Y. Kataoka. 2005. Brain pericytes contribute to the induction and up-regulation of blood–brain barrier functions through transforming growth factor-β production. Brain Res. 1038: 208–215.

Dommise, G.F. 1980. The arteries, arterioles, and capillaries of the spinal cord. Surgical guidelines in the prevention of postoperative paraplegia. Ann. Royal Coll. Surg. England 62: 369–376.

Dwyer, D.S., S.J. Vannucci and I.A. Simpson. 2002. Expression, regulation, and functional role of glucose transporters (GLUTs) in brain. Int. Rev. Neurobiol. 51: 159–188.

Edvinsson, L.M., E.T. MacKenzie, J. McCulloch and R. Uddman. 1987. Perivascular innervation and receptor mechanisms in cerebrovascular bed. pp. 145–172. *In*: J.H. Wood [ed.]. Cerebral Blood Flow, Physiologic and Clinical Aspects. McGraw-Hill Book Company, New York, USA.

Ekdahl, C.T. 2012. Microglial activation-tuning and pruning adult neurogenesis. Front. Pharmacol. 3: 41.

Emsley, J.G. and J.D. Macklis. 2006. Astroglial heterogeneity closely reflects the neuronal-defined anatomy of the adult murine CNS. Neuron Glia Biol. 2: 175–186.

Enerson, B.E. and L.R. Drewes. 2006. The rat blood–brain barrier transcriptome J. Cereb. Blood Flow Metab. 26: 959–973.

Falcone, S., E. Cocucci, P. Podini, T. Kirchhausen, E. Clementi and J. Meldolesi. 2006. Macropinocytosis: regulated coordination of endocytic and exocytic membrane traffic events. J. Cell Sci. 119: 4758–4769.

Farquhar, M.G. and G.E. Palade. 1963. Junctional complexes in various epithelia. J. Cell Biol. 17: 375–412.

Farrell, C.L. and W.M. Pardridge. 1991. Blood–brain barrier glucose transporter is asymmetrically distributed on brain capillary endothelial luminal and abluminal membranes: an electron microscopic immunogold study. Proc. Natl. Acad. Sci. USA 88: 5779–5583.

Ferezou, I., S. Bolea and C.C. Petersen. 2006. Visualizing the cortical representation of whisker touch: voltage-sensitive dye imaging in freely moving mice. Neuron 50: 617–629.

Fetler, L. and S. Amigorena. 2005. Brain under surveillance: the microglia patrol. Science 309: 392–393.

Fiedler, U. and H.G. Augustin. 2006. Angiopoietins: a link between angiogenesis and inflammation. Trends Immunol. 27: 552–558.

Figley, C.R. and P.W. Stroman. 2011. The role(s) of astrocytes and astrocyte activity in neurometabolism, neurovascular coupling, and the production of functional neuroimaging signals. Eur. J. Neurosci. 33: 577–588.

Filosa, J.A., A.D. Bonev, S.V. Straub, A.L. Meredith, M.K. Wilkerson, R.W. Aldrich and M.T. Nelson. 2006. Local potassium signaling couples neuronal activity to vasodilation in the brain. Nat. Neurosci. 9: 1397–1403.

Filosa, J.A. and V.M. Blanco. 2007. Neurovascular coupling in the mammalian brain. Exp. Physiol. 92: 641–646.

Fishman, R.A. 1990. Blood-brain barrier. pp. 43–62. *In*: R.A. Fishman [ed.]. Cerebrospinal Fluid in Diseases of the Nervous System. Saunders, Philadelphia, USA.

Fox, P.T. and M.E. Raichle. 1986. Focal physiological uncoupling of cerebral blood flow and oxidative metabolism during somatosensory stimulation in human subjects. Proc. Natl. Acad. Sci. USA 83: 1140–1144.

Frank, P.G., S.E. Woodman, D.S. Park and M.P. Lisanti. 2003. Caveolin, caveolae, and endothelial cell function. Arterioscler. Thromb. Vasc. Biol. 23: 729–736.

Frederickson, R.G. and F.N. Low. 1969. Blood vessels and tissue space associated with the brain of the rat. Amer. J. Anat. 125: 123–146.

Furuse, M., H. Sasaki and S. Tsukita. 1999. Manner of interaction of heterogeneous claudin species within and between tight junction strands. J. Cell Biol. 147: 891–903.

Furuse, M., K. Fujita, T. Hiiragi, K. Fujimoto and S. Tsukita. 1998. Claudin-1 and -2: novel integral membrane proteins localizing at tight junctions with no sequence similarity to occludin. J. Cell Biol. 141: 1539–1550.

Furuse, M., M. Itoh, T. Hirase, A. Nagafuchi, S. Yonemura, S. Tsukita and S. Tsukita. 1994. Direct association of occludin with ZO-1 and its possible involvement in the localization of occludin at tight junctions. J. Cell Biol. 127: 1617–1626.

Furuse, M., T. Hirase, M. Itoh, A. Nagafuchi, S. Yonemura and S. Tsukita. 1993. Occludin: a novel integral membrane protein localizing at tight junctions. J. Cell Biol. 123: 1777–1788.

Gao, L. and H.H. Lipowsky. 2010. Composition of the endothelial glycocalyx and its relation to its thickness and diffusion of small solutes. Microvasc. Res. 80: 394–401.

Gao, Z. and S.E. Tsirka. 2011. Animal Models of MS Reveal Multiple Roles of Microglia in Disease Pathogenesis. Neurol. Res. Int. 383087.

Garcia-Segura, L.M., J. Perez, S. Pons, M.T. Rejas and I. Torres-Aleman. 1991. Localization of insulin-like growth factor I (IGF-I)-like immunoreactivity in the developing and adult rat brain. Brain Res. 560: 167–174.

Ge, S. and J.S. Patcher. 2006. Isolation and culture of microvascular endothelial cells from murine spinal cord. J. Neuroimmunol. 177: 209–214.

Georgieva, J.V., P.C. Kalicharan, I.A. Romero, B. Weksler, D. Hoekstra and I.S. Zuhorn. 2011. Surface characteristics of nanoparticles determine their intracellular fate in and processing by human blood–brain barrier endothelial cells *in vitro*. Mol. Ther. 19: 318–325.

Gomez-Gonzalez, B., H.M. Larios and A. Escobar. 2011. Increased transvascular transport of WGA-peroxidase after chronic perinatal stress in the hippocampal microvasculature of the rat. Int. J. Dev. Neurosci. 29: 839–846.

Gordon, G.R., H.B. Choi, R.L. Rungta, G.C. Ellis-Davies and B.A. MacVicar. 2008. Brain metabolism dictates the polarity of astrocyte control over arterioles. Nature 456: 745–749.

Gragera, R.R., E. Muniz and R. Martinez-Rodriguez. 1993. Electron microscopic immunolocalization of GABA and glutamic acid decarboxylase (GAD) in cerebellar capillaries and their microenvironment. Cell. Mol. Biol. (Noisy-le-grand) 39: 809–817.

Guillemin, G.J. and B.J. Brew. 2004. Microglia, macrophages, perivascular macrophages, and pericytes: a review of function and identification. J. Leuk. Biol. 75: 388–397.

Haefliger, I.O., A. Zschauer and D.R. Anderson. 1994. Relaxation of retinal pericyte contractile tone through the nitric oxide-cyclic guanosine monophosphate pathway. Invest. Ophthal. Visual Sci. 35: 991–997.

Hamilton, N.B., D. Attwell and C.N. Hall. 2010. Pericyte-mediated regulation of capillary diameter: a component of neurovascular coupling in health and disease. Front. Neuroenerg. 2: 5.

Hawkins, R.A., R.L. O'Kane, I.A. Simpson and J.R. Vina. 2006. Structure of the blood-brain barrier and its role in the transport of amino acids. J. Nutr. 136: 218S–226S.

Heeger, D.J. and D. Ress. 2002. What does fMRI tell us about neuronal activity? Nat. Rev. Neurosci. 3: 142–151.

Hellstrom, M., H. Gerhardt, M. Kalen, X. Li, U. Eriksson, H. Wolburg and C. Betsholtz. 2001. Lack of pericytes leads to endothelial hyperplasia and abnormal vascular morphogenesis. J. Cell Biol. 153: 543–553.

Hewet, J.A. 2009. Determinants of regional and local diversity within the astroglial lineaje of the normal central nervous system. J. Neurochem. 110: 1717–1736.

Hori, S., S. Ohtsuki, K. Hosoya, E. Nakashima and T.A. Terasaki. 2004. A pericyte-derived angiopoietin-1 multimeric complex induces occludin gene expression in brain capillary endothelial cells trhough Tie-2 activation *in vitro*. J. Neurochem. 89: 503–513.

Hosaya, K. and M. Tachikawa. 2009. Inner blood-retinal barrier transporters: role of retinal drug delivery. Pharm. Res. 26: 2055–2065.

Houades, V., N. Rouach, P. Ezan, F. Kirchhoff, A. Koulakoff and C. Giaume. 2006. Shapes of astrocyte networks in the juvenile brain. Neuron Glia Biol. 2: 3–14.

Huber, J.D., R.D. Egleton and T.P. Davis. 2001. Molecular physiology and pathophysiology of tight junctions in the blood-brain barrier. Trends Neurosci. 24: 719–725.

Hurley, J.V., R.M. Anderson and P.T. Sexton. 1981. The fate of plasma protein which escapes from blood-vessels of the choroid plexus of the rat—an electro microscopy study. J. Pathol. 134: 57–70.

Huxley, V.H. and F.E. Curry. 1985. Albumin modulation of capillary permeability: test of an adsorption mechanism. Am. J. Physiol. 248: H264–273.

Inoue, K. and M. Tsuda. 2012. Purinergic systems, neuropathic pain and the role of microglia. Exp. Neurol. 234: 293–301.

Itoh, M., H. Sasaki, M. Furuse, H. Ozaki, T. Kita and S. Tsukita. 2001. Junctional adhesion molecule (JAM) binds to PAR-3: a possible mechanism for the recruitment of PAR-3 to tight junctions. J. Cell Biol. 154: 491–497.

Job, K.M., R.O. Dull and V. Hlady. 2012. Use of reflectance interference contrast microscopy to characterize the endothelial glycocalyx stiffness. Am. J. Physiol. Lung Cell. Mol. Physiol. 302: L1242–L1249.

Jodoin, J., M. Demeule, L. Fenart, R. Cecchelli, S. Farmer, K.J. Linton, C.F. Higgins and R. Beliveau. 2003. P-glycoprotein in blood-brain barrier endothelial cells: interaction and oligomerization with caveolins. J. Neurochem. 87: 1010–1023.

Johanson, C.E., E.G. Stopa and P.N. McMillan. 2011. The blood-cerebrospinal fluid barrier: structure and functional significance. pp. 101–132. *In*: S. Nag [ed.]. The Blood-Brain Barrier and Other Neural Barriers: Reviews and Protocols. Methods in Molecular Biology, Humana Press, New York, USA.

Jones, H.C., R.F. Keep and A.M. Butt. 1992. The development of ion regulation at the blood–brain barrier. Prog. Brain Res. 91: 123–131.

Juhn, S.K., B.A. Hunter and R.M. Odland. 2001. Blood-labyrinth barrier and fluid dynamics of the inner ear. Int. Tinnitus J. 7: 72–83.

Kalaria, R.N., C.A. Stockmeier and S.I. Harik. 1989. Brain micrivessels are innervated by locus coeruleus noradrenergic neurons. Neurosci. Lett. 97: 203–208.

Kamouchi, M., T. Ago and T. Kitazono. 2011. Brain pericytes: emerging concepts and functional roles in brain homeostasis. Cell. Mol. Neurobiol. 31: 175–193.

Kastin, A.J. and V. Akerstrom. 2001. Glucose and insulin increase the transport of leptin through the blood–brain barrier in normal mice but not in streptozotocin-diabetic mice. Neuroendocrinol. 73: 237–242.

Kastin, A.J. and W. Pan. 2000. Dynamic regulation of leptin entry into brain by the blood-brain barrier. Regul. Pept. 92: 37–43.

Kastin, A.J., V. Akerstrom and L.M. Maness. 2001. Chronic loss of ovarian function decreases transport of leptin into mouse brain. Neurosci. Lett. 310: 69–71.

Keep, R.F. and H.C. Jones. 1990. A morphometric study on the development of the lateral ventricle choroid plexus, choroid plexus capillaries and ventricular ependyma in the rat. Dev. Brain Res. 56: 47–53.

Kelley, C., P. D'Amore, H.B. Hechtman and D. Shepro. 1988. Vasoactive hormones and cAMP affect pericyte contraction and stress fibres *in vitro*. J. Muscle Res. Cell Mot. 9: 184–194.

Kim, M.P., S.I. Park, S. Kopetz and G.E. Gallick. 2009. Src family kinases as mediators of endothelial permeability: effects on inflammation and metastasis. Cell Tissue Res. 335: 249–259.

Kovbasnjuk, O.N., U. Szmulowicz and K.R. Spring. 1998. Regulation of the MDCK cell tight junction. J. Membr. Biol. 161: 93–104.

Krause, G., L. Winkler, S.L. Mueller, R.F. Haseloff, J. Piontek and I.E. Blasig. 2008. Structure and function of claudins. Biochim. Biophys. Acta 1778: 631–645.

Krimer, L.S., E.C. Muly III, G.V. Williams and P.S. Goldman-Rakic. 1998. Dopaminergic regulation of cerebral cortical microcirculation. Nat. Neurosci. 1: 286–289.

Kumagai, A.K., J.B. Eisenberg and W.M. Pardridge. 1987. Adsorptive-mediated endocytosis of cationized albumin and a beta-endorphin-cationized albumin chimeric peptide by isolated brain capillaries. Model system of blood-brain barrier transport. J. Biol. Chem. 262: 15214–15219.

Kumar, P., Q. Shen, C.D. Pivetti, E.S. Lee, M.H. Wu and S.Y. Yuan. 2009. Molecular mechanisms of endothelial hyperpermeability: implications in inflammation. Expert Rev. Mol. Med. 11: e19.

Lai, C.H. and K.H. Kuo. 2005. The critical component to establish *in vitro* BBB model: pericyte. Brain Res. Rev. 50: 258–265.

Lambert, D., C.A. O'Neill and P.J. Padfield. 2005. Depletion of Caco-2 cell cholesterol disrupts barrier function by altering the detergent solubility and distribution of specific tight-junction proteins. Biochem. J. 387: 553–560.

Lawrenson, J.G., J.P. Cassela, A.J. Hayes, J.A. Firth and G. Allt. 2000. Endothelial glycoconjugates: a comparative lectin study of the brain, retina and myocardium. J. Anat. 196: 55–60.

Lee, G. and R. Bendayan. 2004. Functional expression and localization of P-glycoprotein in the central nervous system: relevance to the pathogenesis and treatment of neurological disorders. Pharm. Res. 21: 1313–1330.

Lee, T.J. 2000. Nitric oxide and the cerebral vascular function. J. Biomed. Sci. 7: 16–26.

Leybaert, L. 2005. Neurobarrier coupling in the brain: a partner of neurovascular and neurometabolic coupling? J. Cereb. Blood Flow Metab. 25: 2–16.

Leybaert, L., M. De Bock, M. Van Moorhem, E. Decrock and E. De Vuyst. 2007. Neurobarrier coupling in the brain: adjusting glucose entry with demand. J. Neurosci. Res. 85: 3213–3220.

Li, N., J.Y. Sul and P.G. Haydon. 2003. A calcium-induced calcium influx factor, nitric oxide, modulates the refilling of calcium stores in astrocytes. J. Neurosci. 23: 10302–10310.

Loane, D.J. and K.R. Byrnes. 2010. Role of microglia in neurotrauma. Neurotherapeutics 7: 366–377.

Logothetis, N.K. 2002. The neural basis of the blood-oxygen-level-dependent functional resonance imaging signal. Philos. Trans. R. Soc. Lond. B Biol. Sci. 357: 1003–1037.

Lourenco, F.C., D. Brites and M.A. Brito. 2010. Looking at the blood–brain barrier: Molecular anatomy and possible investigation approaches. Brain Res. Rev. 64: 328–363.

Lowry, J.P., R.D. O'Neil, M.G. Boutelle and M. Fillenz. 1998. Continuous monitoring of extracellular glucose concentrations in the striatum of freely moving rats with an implanted glucose biosensor. J. Neurochem. 70: 391–396.

Mark, K.S. and T.P. Davis. 2002. Cerebral microvascular changes in permeability and tight junctions induced by hypoxia-reoxygenation. Am. J. Physiol. Heart Circ. Physiol. 282: H1485–H1494.

Markhotina, N., G.J. Liu and D.K. Martin. 2007. Contractility of retinal pericytes grown on silicone elastomer substrates is through a protein kinase A-mediated intracellular pathway in response to vasoactive peptides. IET Nanobiotechnol. 1: 44–51.

Martin-Padura, I., S. Lostaglio, M. Schneemann, L. Williams, M. Romano, P. Fruscella, C. Panzeri, A. Stoppacciaro, L. Ruco, A. Villa, D. Simmons and E. Dejana. 1998. Junctional adhesion molecule, a novel member of the immunoglobulin superfamily that distributes at intercellular junctions and modulates monocyte transmigration. J. Cell Biol. 142: 117–127.

Martin, D.L. 1992. Synthesis and release of neuroactive substances by glial cells. Glia 5: 81–94.

Matsugi, T., Q. Chen and D.R. Anderson. 1997. Adenosine-induced relaxation of cultured bovine retinal pericytes. Invest. Ophthalmol. Visual Sci. 38: 2695–2701.

Matyash, V. and H. Kettenmann. 2010. Heterogeneity in astrocyte morphology and physiology. Brain Res. Rev. 63: 2–10.

McCaffrey, G., W.D. Staatz, C.A. Quigley, N. Nametz, M.J. Seelbach, C.R. Campos, T.A. Brooks, R.D. Egleton and T.P. Davis. 2007. Tight junctions contain oligomeric protein assembly critical for maintaining blood–brain barrier integrity *in vivo*. J. Neurochem. 103: 2540–2555.

McCaslin, A.F.H., B.R. Chen, A.J. Radosevich, B. Cauli and E.M.C. Hillman. 2011. *In vivo* 3D morphology of astrocyte-vasculature interactions in the somatosensory cortex: implications for neurovascular coupling. J. Cereb. Blood Flow Metab. 31: 795–806.

Metea, M.R. and E.A. Newman. 2006. Glial cells both dilate and constrict blood vessels: a mechanism of neurovascular coupling. J. Neurosci. 26: 2862–2870.

Mizisin, A.P. and A. Weerasuriya. 2011. Homeostatic regulation of the endoneurial microenvironment during development, aging and in response to trauma, disease and toxic insult. Acta Neuropathol. 121: 291–312.

Mosso, A. 1880. Sulla circolazione del cervello dell'uomo. Atti R. Accad. Lincei. 5: 237–358.

Nag, S. 2011. Morphology and properties of brain endothelial cells. pp. 3–47. *In*: S. Nag [ed.]. The Blood-Brain Barrier and Other Neural Barriers: Reviews and Protocols. Methods in Molecular Biology, Humana Press, New York, USA.

Nag, S. and D.J. Begley. 2005. Blood–brain barrier, exchange of metabolites and gases. pp. 22–29. *In*: H. Kalimo [ed.]. Pathology and Genetics. Cerebrovascular Diseases. ISN Neuropathology Press, Basel, Switzerland.

Navarro, P., L. Ruco and E. Dejana. 1998. Differential localization of VE- and N-cadherins in human endothelial cells: VE-cadherin competes with N-cadherin for junctional localization. J. Cell Biol. 140: 1475–1484.

Nedergaard, M., B. Ransom and S.A. Goldman. 2003. New roles for astrocytes: redefining the functional architecture of the brain. Trends Neurosci. 26: 523–530.

Neher, J.J., U. Neniskyte and G.C. Brown. 2012. Primary phagocytosis of neurons by inflamed microglia: potential roles in neurodegeneration. Front. Pharmacol. 3: 27.

Niessen, C.M. and C.J. Gottardi. 2008. Molecular components of the adherens junction. Biochim. Biophys. Acta 1778: 562–571.

Nishijima, T., J. Piriz, S. Duflot, A.M. Fernandez, G. Gaitan, U. Gomez-Pinedo, J.M. García Verdugo, F. Leroy, A. Nuñez and I. Torres-Aleman. 2010. Neuronal activity drives localized blood-brain barrier transport of serum insulin-like growth factor-1 into the CNS. Neuron 67: 834–846.

Nishimura, H. and I.A. Simpson. 1994. Staurosporine inhibits phorbol 12-myristate 13-acetate- and insulin-stimulated translocation of GLUT1 and GLUT4 glucose transporters in rat adipose cells. Biochem. J. 302: 271–277.

Nitta, T., M. Hata, S. Gotoh, Y. Seo, H. Sasaki, N. Hashimoto, M. Furuse and S. Tsukita. 2003. Size-selective loosening of the blood–brain barrier in claudin-5-deficient mice. J. Cell Biol. 161: 653–660.

Noble, L.J., A.E.M. Mauters and J.J. Hall. 1996. Characterization of the microvascular glycocalyx in normal and injured spinal cord in the rat. J. Comp. Neurol. 376: 542–556.

Nusrat, A., G.T. Brown, J. Tom, A. Drake, T.T. Bui, C. Quan and R.J. Mrsny. 2005. Multiple protein interactions involving proposed extracellular domains of the tight junction protein occludin. Mol. Biol. Cell 16: 1725–1734.

Nusrat, A., C.A. Parkos, P. Verkade, C.S. Foley, T.W. Liang, W. Innis-Whitehouse, K.K. Eastburn and J.L. Madara. 2000. Tight junctions are membrane microdomains. J. Cell Sci. 113: 1771–1781.

Oberheim, N.A., T. Takano, X. Han, W. He, J.H.C. Lin, F. Wang, Q. Xu, J.D. Wyatt, W. Pilcher, J.G. Ojeman, B.R. Ransom, S.A. Goldman and M. Nedergaard. 2009. Uniquely hominid features of adult human astrocytes. J. Neurosci. 29: 3276–3287.

Ohsawa, K. and S. Kohsaka. 2011. Dynamic motility of microglia: purinergic modulation of microglial movement in the normal and pathological brain. Glia 59: 1793–1799.

Palte, M.J. and R.T. Raines. 2012. Interaction of nucleic acids with the glycocalyx. J. Amer. Chem. Soc. 134: 6218–6223.

Pan, W., W.A. Banks and A.J. Kastin. 1997. Permeability of the blood-brain barrier and blood-spinal cord barriers to interferons. J. Neuroimmunol. 76: 105–111.

Pardridge, W.M. 1983. Brain metabolism: a perspective from the blood-brain barrier. Physiol. Rev. 63: 1481–1535.

Pardridge, W.M. 2005. The blood-brain barrier: bottleneck in brain drug development. NeuroRx 2: 3–14.

Pardridge, W.M. 2007. Blood-brain barrier delivery. Drug Discov. Today 12: 54–61.

Pardridge, W.M., R.J. Boado and C.R. Farrell. 1990. Brain-type glucose transporter (GLUT1) is selectively localized to the blood-brain barrier. J. Biol. Chem. 285: 18035–18040.

Parpura, V., M.T. Heneka, V. Montana, S.H.R. Oliet, A. Schousboe, P.G. Haydon, R.F. Staut Jr., D.C. Spray, A. Reichenbach, T. Pannicke, M. Pekna, R. Zorec and A.Verkhratsky. 2012. Glial cells in (phato) physiology. J. Neurochem. 121: 4–27.

Patel, U. 1983. Non-random distribution of blood vessels in the posterior region of the rat somatosensory cortex. Brain Res. 289: 65–70.

Peppiatt, C.M., C. Howarth, P. Mobbs and D. Attwell. 2006. Bidirectional control of CNS capillary diameter by pericytes. Nature 443: 700–704.

Persidsky, Y., S.H. Ramirez, J. Haorah and G.D. Kanmogne. 2006. Blood-brain barrier: structural components and function under physiologic and pathologic conditions. J. Neuroimmune Pharmacol. 1: 223–236.

Peters, A., S.L. Palay and H.F. Webster. 1991. Blood vessels. pp. 344–355. *In*: A. Peters, S.L. Palay and H.F. Webster [eds.]. The Fine Structure of the Nervous System. Neurons and their Supporting Cells. Oxford University Press, New York, USA.

Petzold, G.C., D.F. Albeanu, T.F. Sato and V.N. Murthy. 2008. Coupling of neural activity to blood flow in olfactory glomeruli is mediated by astrocytic pathways. Neuron 58: 897–910.

Prockop, L.D., K.A. Naidu, J.E. Binard and J. Ransohoff. 1995. Selective permeability of [3H]-D-mannitol and [14C]-carboxyl-inulin across the blood-brain barrier and blood-spinal cord barrier in the rabbit. J. Spinal Cord Med. 18: 221–226.

Qiu, M.G. and X.M. Zhu. 2004. Aging changes of the angioarchitecture and arterial morphology of the spinal cord in rats. Gerontology 50: 360–365.

Queiroz, G., D.K. Meyer, A. Meyer, K. Starke and I. von Kügelgen. 1999. A study of the mechanism of the release of ATP from rat cortical astroglial cells evoked by activation of glutamate receptors. Neurosci. 91: 1171–1181.

Ransohoff, R.M. and V.H. Perry. 2009. Microglial physiology: unique stimuli, specialized responses. Annu. Rev. Immunol. 27: 119–145.

Razani, B., S.E. Woodman and M.P. Lisanti. 2002. Caveolae: from cell biology to animal physiology. Pharmacol. Rev. 54: 431–467.

Rehder, D., S. Iden, I. Nasdala, J. Wegener, M.K. Brickwedde, D. Vestweber and K. Ebnet. 2006. Junctional adhesion molecule-A participates in the formation of apico-basal polarity through different domains. Exp. Cell Res. 312: 3389–3403.

Reis, D.J. and C. Iadecola. 1986. Regulation by the brain of its blood flow and metabolism: role of intrinsic neuronal networks and circulating catecholamines. pp. 129–145. *In*: C. Owman and J. Hardebo [eds.]. Neural Regulation of Brain Circulation. Elsevier Science Publishers, Amsterdam, Netherlands.

Reitsma, S., D.W. Slaaf, H. Vink, M.A. van Zandvoort and M.G. oude Egbrink. 2007. The endothelial glycocalyx: composition, functions, and visualization. Eur. J. Physiol. 454: 345–359.

Riera, J.J., A. Schousboe, H.S. Waagepetersen, C. Howart and F. Hyder. 2008. The micro-architecture of the cerebral cortex: functional neuroimaging models and metabolism. Neuroimage 40: 1436–1459.

Risau, W. 1991. Induction of blood-brain barrier endothelial cell differentiation. Ann. New York Acad. Sci. 633: 405–419.

Roufail, E., M. Stringer and S. Rees. 1995. Nitric oxide synthase immunoreactivity and NADPH diaphorase staining are colocalised in neurons closely associated with the vasculature in rat and human retina. Brain Res. 684: 36–46.

Roy, C.S. and C. Sherrington. 1890. On the regulation of the blood supply of the brain. J. Physiol. 11: 85–108.

Sakagami, K., D.M. Wu and D.G. Puro. 1999. Physiology of rat retinal pericytes: modulation of ion channel activity by serum-derived molecules. J. Physiol. 521: 637–650.

Sato, A. and Y. Sato. 1992. Regulation of regional cerebral blood flow by cholinergic fibers originating in the basal forebrain. Neurosci. Res. 14: 242–274.

Sawchuk, R.J. and W.F. Elmquist. 2000. Microdialysis in the study of drug transporters in the CNS. Adv. Drug Deliv. Rev. 45: 295–307.

Schinkel, A.H. and J.W. Jonker. 2003. Mammalian drug efflux transporters of the ATP binding cassette (ABC) family: an overview. Adv. Drug Deliv. Rev. 55: 3–29.

Schinkel, A.H., E. Wagenaar, C. Mol and L. Van Deemter. 1996. P-glycoprotein in the blood-brain barrier of mice influences the brain penetration and pharmacological activity of many drugs. J. Clin. Invest. 97: 2517–2524.

Schneeberger, E.E. and R.D. Lynch. 2004. The tight junction: a multifunctional complex. Am. J. Physiol. Cell Physiol. 286: C1213–C1228.

Schomberg, D. and J.K. Olson. 2012. Immune responses of microglia in the spinal cord: contribution to pain states. Exp. Neurol. 234: 262–270.

Shanthaveerappa, T.R. and G.H. Bourne. 1962. The 'perineurial epithelium', a metabolically active, continuous, protoplasmic cell barrier surrounding peripheral nerve fasciculi. J. Anat. 96: 527–537.

Sharma, H.S., J. Westman, J. Cervós, P.K. Dey and F. Nyberg. 1996. Probable involvement of serotonin in the increased permeability of the blood-brain barrier by forced swimming. An experimental study using Evans blue and ^{131}I-sodium tracers in the rat. Behav. Brain Res. 72: 189–196.

Sharma, H.S. 2005. Pathophysiology of blood-spinal cord barrier in traumatic injury and repair. Curr. Pharm. Des. 11: 1353–1389.

Shepro, D. and N.M.L. Morel. 1993. Pericyte physiology. FASEB Journal 7: 1031–1038.

Sherwood, C.C., C.D. Stimpson, M.A. Raghanti, D.E. Wildman, M. Uddin, L.I. Grossman, M. Goodman, J.C. Redmond, C.J. Bonar, J.M. Erwin and P.R. Hof. 2006. Evolution of increased glia-neuron ratios in the human frontal cortex. Proc. Natl. Acad. Sci. USA 103: 13606–13611.

Shimizu, F., Y. Sano, T. Maeda, M.A. Abe, H. Nakayama, R. Takahashi, M. Ueda, S. Ohtsuki, T. Terasaki, M. Obinata and T. Kanda. 2008. Peripheral nerve pericytes originating from the blood–nerve barrier expresses tight junctional molecules and transporters as barrier-forming cells. J. Cell. Physiol. 217: 388–399.

Simpson, I.A., A. Carruthers and S.J. Vannucci. 2007. Supply and demand in cerebral energy metabolism: the role of nutrient transporters. J. Cereb. Blood Flow Metab. 27: 1766–1791.

Sims, D.E. 1986. The pericyte—a review. Tissue Cell 18: 153–174.

Skaper, S.D., P. Giusti and L. Facci. 2012. Microglia and mast cells: two tracks on the road to neuroinflammation. FASEB J. 26: 3103–3117.

Smith, J.A., A. Das, S.K. Ray and N.L. Banik. 2012. Role of pro-inflammatory cytokines released from microglia in neurodegenerative diseases. Brain Res. Bull. 87: 10–20.

Sofroniev, M.V. and H.V. Vinters. 2010. Astrocytes: biology and phatology. Acta Neuropathol. 119: 7–35.

Spector, R. 2010. Nature and consequences of mammalian brain and CSF efflux transporters: four decades of progress. J. Neurochem. 112: 13–23.

Stamatovic, S.M., O.B. Dimitrijevic, R.F. Keep and A.V. Andjelkovic. 2006. Protein kinase C-α: Rhoa cross talk in CCL2-induced alterations in brain endothelial permeability. J. Biol. Chem. 281: 8379–8388.

Stevenson, B.R. and D.A. Begg. 1994. Concentration-dependent effects of cytochalasin D on tight junctions and actin filaments in MDCK epithelial cells. J. Cell Sci. 107: 367–375.

Stockton, R.A., E. Schaefer and M.A. Schwartz. 2004. p21-activated kinase regulates endothelial permeability through modulation of contractility. J. Biol. Chem. 279: 46621–46630.

Streit, W.J., S.A. Walter and N.A. Pennell. 1999. Reactive microgliosis. Prog. Neurobiol. 57: 563–581.

Stuart, R.O., A. Sun, M. Panichas, S.C. Hebert, B.M. Brenner and S.K. Nigam. 1994. Critical role for intracellular calcium in tight junction biogenesis. J. Cell Physiol. 159: 423–433.

Swanson, J.A. and C. Watts. 1995. Macropinocytosis. Trends Cell Biol. 5: 424–428.

Taber, K.H. and R.A. Hurley. 2008. Astroglia: not just glue. J. Neuropsychiatry Clin. Neurosci. 20: iv–129.

Tamai, I., Y. Sai, H. Kobayashi, M. Kamata, T. Wakamiya and A. Tsuji. 1997. Structure-internalization relationship for adsorptive-mediated endocytosis of basic peptides at the blood-brain barrier. J. Pharmacol. Exp. Ther. 280: 410–415.

Tang, V.W. and D.A. Goodenough. 2003. Paracellular ion channel at the tight junction. Biophys. J. 84: 1660–1673.

Thomas, W.E. 1999. Brain macrophages: on the role of pericytes and perivascular cells. Brain Res. Rev. 31: 42–57.

Trang, T., S. Beggs and M.W. Salter. 2012. ATP receptors gate microglia signaling in neuropathic pain. Exp. Neurol. 234: 354–361.

Tsuji, A. 2005. Small molecular drug transfer across the blood-brain barrier via carrier-mediated transport systems. NeuroRx 2: 54–62.

Ueno, M. 2007. Molecular anatomy of the brain endothelial barrier: an overview of the distributional features. Curr. Med. Chem. 14: 1199–1206.

Ueno, M. 2009. Mechanisms of the penetration of blood-borne substances into the brain. Curr. Neuropharmacol. 7: 142–149.

Vasioukhin, V., C. Bauer, M. Yin and E. Fuchs. 2000. Directed actin polymerization is the driving force for epithelial cell–cell adhesion. Cell 100: 209–219.

Weerasuriya, A. and A.P. Mizisin. 2011. The blood-nerve barrier: structure and functional significance. pp. 149–173. *In*: S. Nag [ed.]. The Blood-Brain Barrier and Other Neural Barriers: Reviews and Protocols. Methods in Molecular Biology, Humana Press, New York, USA.

Weerasuriya, A., S.I. Rapoport and R.E. Taylor. 1979. Modification of permeability of frog perineurium to (^{14}C) sucrose by stretch and hypertonicity. Brain Res. 173: 503–512.

Weerasuriya, A., S.I. Rapoport and R.E. Taylor. 1980. Ionic permeabilities of the frog perineurium. Brain Res. 191: 405–415.

Weiss, N., F. Miller, S. Cazaubon and P.O. Couraud. 2009. The blood–brain barrier in brain homeostasis and neurological diseases. Biochim. Biophys. Acta 1788: 842–857.

Winship, I.R., N. Plaa and T.H. Murphy. 2007. Rapid astrocyte calcium signals correlate with neuronal activity and onset of the hemodynamic response *in vivo*. J. Neurosci. 27: 6268–6272.

Wolburg, H. and A. Lippoldt. 2002. Tight junctions of the blood-brain barrier: development, composition and regulation. Vascul. Pharmacol. 38: 323–337.

Wolburg, H., S. Noell, A. Mack, K. Wolburg-Buchholz and P. Fallier-Becker. 2009. Brain endothelial cells and the glio–vascular complex. Cell Tissue Res. 335: 75–96.

Wong, W.T., M. Wang and W. Li. 2011. Regulation of microglia by ionotropic glutamatergic and GABAergic neurotransmission. Neuron Glia Biol. 14: 1–6.

Woolsey, T.A., C.M. Rovainen, S.B. Cox, M.H. Henegar, G.E. Liang, D. Liu, Y.E. Moskalenko, J. Sui and L. Wei. 1996. Neuronal units linked to microvascular modules in cerebral cortex: response elements for imaging the brain. Cereb. Cortex 6: 647–660.

Yamamoto, T. and T. Kato. 2002. Paradoxical correlation between signal in functional magnetic resonance imaging and deoxygenated haemoglobin content in capillaries: a new theoretical explanation. Physics Med. Biol. 47: 1121–1141.

Yang, I., S.J. Han, G. Kaur, C. Crane and A.T. Parsa. 2010. The role of microglia in central nervous system immunity and glioma immunology. J. Clin. Neurosci. 17: 6–10.

Yen, W., B. Cai, M. Zeng, J.M. Tarbell and B.M. Fu. 2012. Quantification of the endothelial surface glycocalyx on rat and mouse blood vessels. Microvasc. Res. 83: 337–346.

Zheng, W. and A.D. Monnot. 2012. Regulation of brain iron and copper homeostasis by brain barrier systems: implication in neurodegenerative diseases. Pharmacol. Therap. 133: 177–188.

Zhou, L., K. Schmidt, F.R. Nelson, V. Zelesky, M.D. Troutman and B. Feng. 2009. The effect of breast cancer resistance protein and P-glycoprotein on the brain penetration of Flavopiridol, Imatinib Mesylate (Gleevec), Prazosin and 2-methoxy-3-(4-(2-(5-methyl-2-phenyloxazol-4-yl)ethoxy)phenyl)propanoic acid (PF-407288) in mice. Drug Metab. Disp. 37: 946–955.

Zlokovic, B.V. 2008. The blood-brain barrier in health and chronic neurodegenerative disorders. Neuron 57: 178–201.

Zonta, M., A. Sebelin, S. Gobbo, T. Fellin, T. Pozzan and G. Carmignoto. 2003a. Glutamate-mediated cytosolic calcium oscillations regulate a pulsatile prostaglandin release from cultured rat astrocytes. J. Physiol. 553: 407–414.

Zonta, M., M.C. Angulo, S. Gobbo, B. Rosengarten, K.-A. Hossmann, T. Pozzan and G. Carmignoto. 2003b. Neuron-to-astrocyte signaling is central to the dynamic control of brain microcirculation. Nat. Neurosci. 6: 43–50.

CHAPTER 4

Dynamic Control of Neural Reproductive Centers by Endothelial Cells

Charlotte Vanacker, Ariane Sharif and *Vincent Prévot**

4.1 Introduction

It is commonly accepted that neurons, glia and brain capillaries are organized into well-structured neuro-glio-vascular units, in which individual astroglial cells support the function of specific neuronal populations and territories, and communicate with associated segments of the microvasculature (Iadecola and Nedergaard 2007, Barres 2008, Halassa and Haydon 2010, Prevot et al. 2010b). These micro-functional domains are likely to play an important role in maintaining a precisely regulated microenvironment for reliable neuronal signaling in an ever-changing physiological context. Gaining new insights into how cellular events that involve neurons, astroglia and vascular cells are orchestrated is therefore fundamental to an improved understanding of brain function.

Blood vessels and axons employ similar mechanisms and follow common guidance cues for growth and navigation during embryonic development (Carmeliet and Tessier-Lavigne 2005, Larrivee et al. 2009). Blood vessels influence axonal trajectories to reach their appropriate destinations (Makita et al. 2008). In the adult brain, blood vessels communicate with neurons and

Inserm, Laboratory of Development and Plasticity of the Postnatal Brain, Jean-Pierre Aubert Research Centre, U837, University of Lille 2, France.
* Corresponding author: vincent.prevot@inserm.fr

glia to meet physiological demands (Iadecola 2004). Endothelial cells are positioned to sense peripheral inputs and ideally suited to convey signals that could influence neuronal structure and synaptic plasticity. However, the question of whether these cells are capable of influencing the function of the mature central nervous system remains largely understudied.

The median eminence of the hypothalamus, which constitutes the ventral border of the third ventricle, provides an excellent model in which to investigate the complex relationship between neural secretion, functional morphological plasticity involving neuronal-glial-endothelial interactions and the expression of key physiological functions. It is one of the seven so-called circumventricular organs (CVO) (Duvernoy and Risold 2007, Langlet et al. 2013a) and thus constitutes a window of exchanges between the hypothalamus and the periphery that is facilitated by the presence of permeable brain capillaries featuring fenestrated endothelium (Ciofi et al. 2009, Mullier et al. 2010, Ciofi 2011, Schaeffer et al. 2013). In contrast to other CVOs, it seems that the most important function associated with the lack of a blood-brain barrier in the median eminence is that it permits the release of neurohormones, produced by neuroendocrine cells, from terminals into the pituitary portal circulation. It is also important to acknowledge that the cellular processes through which neuroendocrine terminals release their neuropeptides into the circulation could be subjected to the direct modulatory influence of blood-borne factors acting on this region. The peculiar cytoarchitecture of the median eminence is mainly conferred by tanycytes, which are specialized non-ciliated ependymoglial cells that form a belt lining the floor of the third ventricle (Page 2006). A dominant feature of tanycytes is their marked polarization; although their cell bodies line the border of the third ventricle, they also send processes to the vascular walls, where they make contacts through "endfeet" specializations (Sharif et al. 2013). In addition, tanycytes were recently shown to express efficient tight junction complexes at their apex that bestow them with properties of the blood-brain barrier (Mullier et al. 2010, Balland et al. 2014). These tight junctions are highly plastic, they depend on the metabolic status of the individual and are tightly regulated by communication between tanycytes and vascular endothelial cells (Langlet et al. 2013b, Myers 2013, Prevot et al. 2013). Although tanycytes are the dominant cell type, astrocytes also reside within the internal zone of the median eminence (Fig. 1).

Reproductive function requires the coordinated and timely activation of gonadotropin-releasing hormone (GnRH) neurons (Herbison 2006). These neurons, which in rodents are located in the preoptic region of the hypothalamus, extend their neurosecretory axons to the median eminence where they make contact with the basal lamina and open into the pericapillary space of the primary hypophyseal portal plexus. Upon reaching the pituitary portal system, GnRH travels to the pituitary to

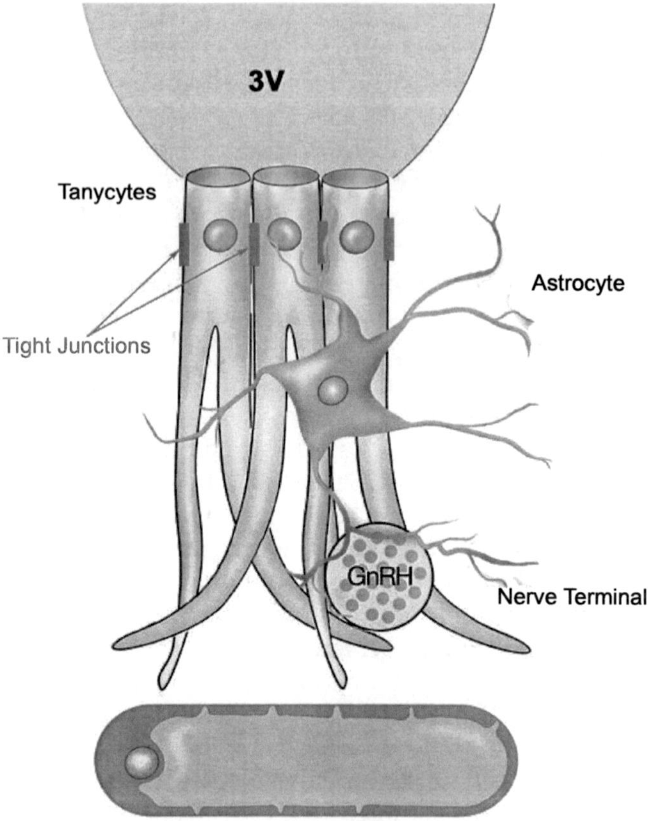

Figure 1. Schematic representation of the cell types (tanycytes, astrocytes and endothelial cells) and neuronal elements (neuroendocrine terminals) that reside within the hypothalamic median eminence. The median eminence of the hypothalamus is the brain structure that establishes the floor of the third ventricle (3V). It is one of the circumventricular organs of the brain, capable of conveying information from the brain to the periphery via the release of neurohormones into the circulation and conversely sensing information reaching the brain via the bloodstream. GnRH= gonadotropin-releasing hormone. From (Prevot et al. 2010a) with permission.

Color image of this figure appears in the color plate section at the end of the book.

stimulate the synthesis and secretion of pituitary gonadotropins: luteinizing hormone (LH) and follicle stimulating hormone (FSH). Blood-borne LH and FSH act on target cells in the gonads (the ovaries) to direct production of gametes, as well as the secretion of steroid hormones. Within the brain, gonadal steroids influence GnRH secretion via neuroendocrine feedback loops (Fig. 2).

This chapter will review the findings that have unveiled some of the cell-cell communication processes involving non-neuronal cells—such as

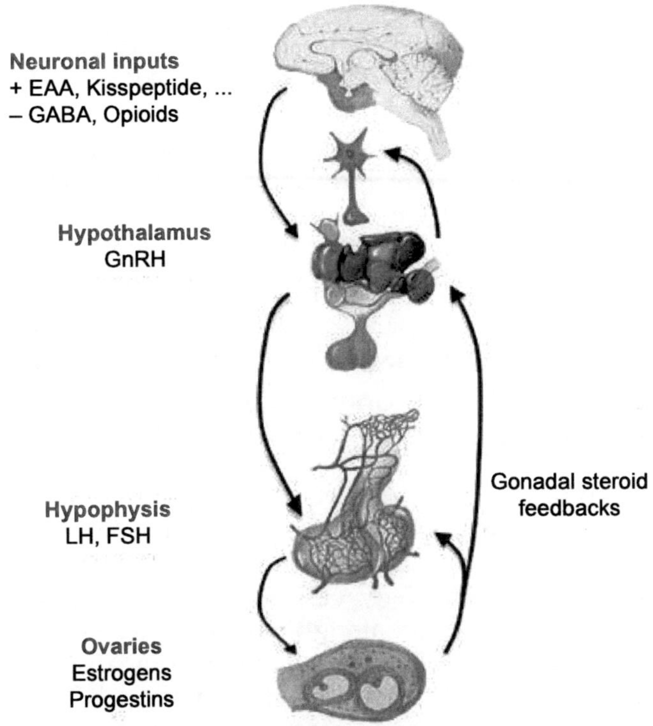

Figure 2. Neuronal control of the pituitary gonadal axis. Hypothalamic gonadotropin-releasing hormone (GnRH) neurons are the final common pathway for central control of gonatropin secretion, they are subjected to a complex array of excitatory and inhibitory trans-synaptic inputs that modulate their activity. GnRH neuroendocrine neurons project to the median eminence, where they make contact with the basal lamina and open into the pericapillary space of the primary hypophyseal portal plexus. Upon reaching the pituitary portal system, GnRH goes to the pituitary to stimulate the synthesis and secretion of pituitary gonadotropins: luteinizing hormone (LH) and follicle stimulating homone (FSH). Blood-borne LH and FSH act on target cells in the gonads (here the ovaries) to direct production of gametes, as well as the secretion of steroid hormones. Within the brain, gonadal steroids influence GnRH secretion via neuroendocrine feedback loops. EAA: excitatory amino acids, GABA: gamma-aminobutyric acid. From (Prevot 2010, Prevot et al. 2010b) with permission.

Color image of this figure appears in the color plate section at the end of the book.

vascular endothelial cells, astrocytes and tanycytes—which locally regulate GnRH neurohaemal junction formation and GnRH release within the median eminence. These modulatory mechanisms use signaling molecules that might represent some of the synchronizing cues that coordinate GnRH release from the scattered GnRH neuroendocrine terminals, which may allow for the occurrence of functionally meaningful episodes of GnRH secretion and thus contribute to fertility. More specifically, endothelial-cell-derived signals appear to regulate GnRH neurosecretion via two alternative, but not mutually exclusive pathways: i) promoting retraction of the tanycytic distal processes that sheath GnRH nerve terminals and ii) regulating outgrowth of GnRH neuroendocrine axons. This chapter will first detail the structural changes regulating the establishment of the GnRH neurovascular junction during the ovarian cycle. Then it will discuss the molecular mechanisms underlying the morphological alteration of the tanycitic sheath controlling the direct access of GnRH neurons to the pericapillary space. Finally, it will review the evidence suggesting that vascular endothelial cells play a dynamic role in the control of these morphological and functional changes.

4.2 Neuronal-Glial-Vascular Interactions in the Median Eminence Modulate Neurohaemal Junction Formation and Neuro-secretion at the GnRH Nerve Terminals

During the pre-ovulatory surge of GnRH and LH, morphological remodeling in the external zone of the median eminence results in the formation of direct neurovascular contacts with GnRH neurons (Prevot et al. 1998, Prevot et al. 1999a). Tanycytes release the engulfed terminals, retract from their initial position along the basal lamina and bring the pericapillary space to the GnRH neuroendocrine terminals (Fig. 3). Some GnRH nerve terminals sprout toward the basal lamina of the endothelial wall and eventually make direct contact with it. These phenomena seem to occur concurrently on the day of proestrus and allow the GnRH nerve terminals to directly contact the pericapillary space, so that GnRH can reach the fenestrated capillaries of the pituitary portal vasculature more efficiently and thus trigger the pre-ovulatory surge of GnRH and LH. During the last decade, significant efforts have been directed toward deciphering the cell-cell mechanisms and the underlying signaling pathways responsible for these structural changes that allow neuroendocrine GnRH neurons to establish evident neurovascular junctions every four days during the rat estrous cycle. It is becoming increasingly clear that these plastic changes require the cooperative actions of multiple signals whose cellular sources and targets are alternatively tanycytes, astrocytes, vascular endothelial cells and the GnRH neurons themselves.

4.3 Molecular Mechanisms Involved in the Retraction of Tanyctic Processes During the Ovarian Cycle

Tanycytes together with cerebellar Bergman glia and retinal Müller cells are the persistent radial glial cells of the mature central nervous system (Reichenbach and Wolburg 2005). Tanycytes are bipolar ependymoglial cells that extend long processes throughout the tissue. Remarkably, tanycytes

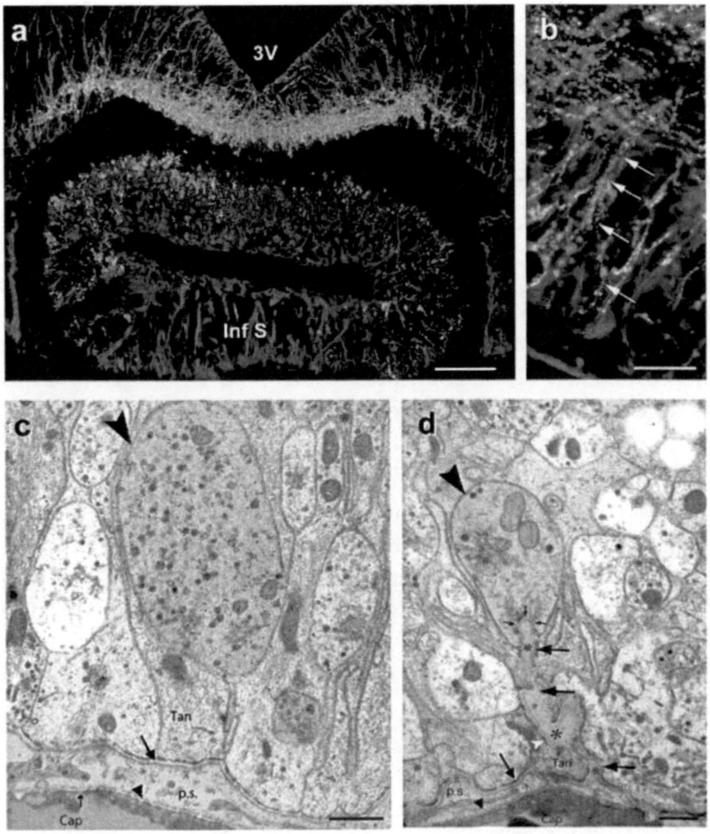

Figure 3. Close association between gonadotropin-releasing hormone (GnRH) nerve terminals and tanycytic processes in the median eminence of the hypothalamus. a) Fluorescent micrograph of a frontal section showing dense GnRH-immunoreactivity (green) and tanycytic processes visualized by their immunoreactivity for vimentin (red) in the external zone of the median eminence. 3V: third ventricle, Inf S: infundibular stem. b) High magnification image showing GnRH axon fibers (green) alongside tanycytic processes (red) projecting to the capillary zone of the median eminence (arrows). c & d) Electron micrographs which illustrate the dynamic changes occurring in the external zone of the median eminence that control the direct access of GnRH nerve terminals to the pericapillary space during the estrous cycle in the rat. Cap: capillaries, Tan: tanacyte endfeet, PS: pericapillary space. c) Electron micrograph of GnRH-immunoreactive terminals (large arrowhead, green) in the external zone of the median eminence in close proximity to fenestrated capillaries (red) of the portal vasculature. During most stages of the reproductive cycle, GnRH nerve terminals (labeled with 15-nm gold particles) are entirely embedded in tanycytic endfeet (yellow), which prevent them from contacting the pericapillary space (pink) delineated by the parenchymatous basal lamina (arrow). Arrowhead: basal lamina of the endothelium, short arrows: endothelial fenestration. Scale bar: 0.5 µm. d) On the day of proestrus, the time when the pre-ovulatory surge of GnRH and LH occurs, a significant fraction of GnRH nerve endings (large arrowhead, green) directly contact the pericapillary space (pink). Scale bar: 0.5 µm. From (Prevot et al. 1998, Prevot et al. 1999a) with permission.

Color image of this figure appears in the color plate section at the end of the book.

share some properties with embryonic radial glia and neural stem cells (Kriegstein and Alvarez-Buylla 2009, Sharif et al. 2014); they express immature glial cell markers such as vimentin (Chauvet et al. 1995, Chauvet et al. 1998, Prieto et al. 2000, Mullier et al. 2010) and nestin (Xu et al. 2005, Baroncini et al. 2007) *in vivo*, they give rise to neurons when they divide (Xu et al. 2005, Lee et al. 2012, Haan et al. 2013, Robins et al. 2013) and they are capable of supporting the migration of neuronal elements (Chauvet et al. 1995, Chauvet et al. 1996, Chauvet et al. 1998, Prieto et al. 2000). These highly plastic features of tanycytes may be conferred by the peculiar microvascular bed of the median eminence, which could provide a permissive environment for neural stem cell expansion, neuronal differentiation and parenchymal migration (Louissaint et al. 2002, Goldman and Chen 2011) by releasing regulatory signals such as nitric oxide (NO) (Prevot et al. 2000a) and brain derived neurotrophic factor (Givalois et al. 2004).

Insights into cell-cell signaling mechanisms underlying tanycyte plasticity in the reproductive neuroendocrine brain were gained recently from studies demonstrating that, in addition to being involved in dynamic communication between astrocytes and GnRH neurons (Clasadonte et al. 2011b), glial erbB (epidermal growth factor receptor or EGFR) signaling may also account for part of the plastic remodeling taking place in the external zone of the median eminence during the estrous cycle (Ojeda et al. 2008, Clasadonte et al. 2011a). The use of primary cultures of tanycytes from the median eminence as a model system showed that these cells express functional erbB-1 and erbB-2 receptors but lack erbB-4 receptors (Prevot et al. 2003a), an expression repertoire identical to *in vivo* tanycytes (Ma et al. 1992, Ma et al. 1999). Within the median eminence, tanycytes also express transforming growth factor (TGF)α (Ma et al. 1992) which, like all other EGF-related peptides, is synthesized as a membrane-bound precursor protein that has to be cleaved by zinc-dependent metalloproteinases to release a mature form that can activate erbB-1 receptors in an autocrine or paracrine manner (Junier 2000). Experiments using phorbolmyristate treatment, which promotes the processing of erbB ligand precursors via a cellular signaling pathway involving protein kinase C (Pandiella and Massague 1991, Prenzel et al. 1999), demonstrated that cultured tanycytes also expressed EGF-like peptides capable of activating endogenous erbB-1 receptors (Prevot et al. 2003a). Like hypothalamic astrocytes, tanycytes respond to TGFα by releasing prostaglandin E2 (PGE_2); however, in contrast to hypothalamic astrocytes, tanycytes also release $TGF\beta_1$. $TGF\beta_1$ is a glial growth factor known to stimulate GnRH mRNA expression in both native and immortalized GnRH cells (Galbiati et al. 2001, Bouret et al. 2004). Although $TGF\beta_1$ is known to stimulate GnRH release from GnRH-secreting cell lines (Melcangi et al. 1995, Buchanan et al. 2000), it does not stimulate GnRH release from median eminence GnRH nerve terminals (Ojeda et al. 1990), suggesting that this growth factor facilitates GnRH secretion via a direct action exerted on GnRH neuron cell bodies. The intracellular mechanisms underlying the stimulatory effect of TGFα on $TGF\beta_1$ release in tanycytes involve the formation of PGE_2, since increases in the release of both PGE_2 and $TGF\beta_1$ after the initial exposure to TGFα require cyclooxygenase (COX) activity (Prevot et al. 2003a). Quite unexpectedly, morphometric studies revealed that TGFα and $TGF\beta_1$ had dramatically opposite effects on tanycyte plasticity *in vitro* (Prevot et al. 2003a). After a 12-hr treatment $TGF\beta_1$ induced the retraction of tanycytic processes, whereas TGFα promoted tanycytic outgrowth. Noticeably, longer exposure of the tanycytes to TGFα caused tanycytic retraction in 60 per cent of the cultures; this effect was abolished by immunoneutralization of $TGF\beta_1$, indicating that the retraction is attributable to TGFα-induced $TGF\beta_1$ formation (Prevot et al. 2003a). These studies also showed that $TGF\beta_1$-promoted retraction of tanycytes requires the activity of matrix metalloproteinases (Prevot et al. 2003a), which were shown to be expressed and differentially active during the estrous cycle in the median eminence (Estrella et al. 2004). In summary, stimulation of erbB-1 receptors in tanycytes results in biphasic plastic changes, characterized by an initial outgrowth phase and a secondary retraction phase (Fig. 4). Although the initial outgrowth is independent of the TGFβ system, the successive retraction requires PGE_2 synthesis and a PGE_2-dependent increase in the production of $TGF\beta_1$. Importantly, tanycytes were shown to express TGFβ serine threonine receptors *in vivo* (Prevot et al. 2000b, Bouret et al. 2004). Taken all together, these data provide strong support for the concept that activation of erbB-1 signaling may be a key component of the cell-to-cell communication pathways used by tanycytes to facilitate GnRH release *in vivo*.

Because circulating levels of estrogens rise dramatically at the onset of the pre-ovulatory surge of GnRH and LH (Smith et al. 1975) and because median eminence tanycytes were shown to express estrogen receptors (Langub and Watson 1992), estrogens are likely to be key humoral factors involved in the orchestration of glia-to-neuron communication processes that allow GnRH neurons to directly contact the pituitary portal blood vessels at proestrus. *In vivo*, estrogens have been shown to activate erbB-mediated signaling events by stimulating the expression of both TGFα (Ma et al. 1992, Ma et al. 1994a) and erbB-1 (Ma et al. 1994b). Estrogens have also been shown to enhance the effect of PGE_2 on GnRH release by increasing the synthesis of PGE_2 receptors in hypothalamic GnRH neurons (Rage et al. 1997). Recent data demonstrate that the activity of tumor necrosis factor α converting enzyme, the principal proteolytic enzyme responsible for shedding TGFα from its trans-membrane precursor (Peschon et al. 1998, Sahin et al. 2004), increases selectively in the median eminence at the onset of the pre-ovulatory surge of GnRH and LH (Lomniczi et al. 2006). This

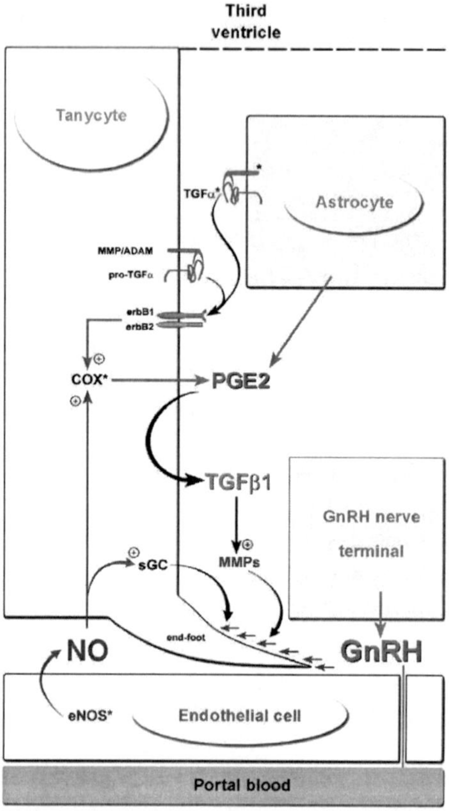

Figure 4. Morphological plasticity of glial coverage of gonadotropin-releasing hormone (GnRH) neurons in the median eminence. Under conditions of low gonadotropin output such as in diestrus, tanycytes envelop neuro-secretory GnRH nerve terminals and, through their endfeet, prevent them from directly contacting the perivascular space and releasing GnRH into the vascular compartment. Transforming growth factor alpha (TGFα), which is produced by tanycytes and astrocytes, activates epidermal growth factor receptor (erbB) 1 and 2 heterodimers to stimulate prostaglandin E2 (PGE_2) production and the subsequent PGE_2-dependent release of TGFβ1 from tanycytes. Astrocytes also constitute a source of PGE_2 in response to epidermal growth factor (EGF) peptides. Tanycyte-derived TGFβ1 acts in an autocrine manner to induce retraction of tanycytic processes through metalloproteinase (MMP)-mediated degradation of the extracellular matrix. Endothelial cells also regulate the morphological plasticity of tanycytes via the release of nitric oxide (NO), which diffuses towards tanycytes to stimulate the enzymatic activity of cyclooxygenase (COX) and the NO-sensitive soluble guanylyl cyclase (sGC). Notably, the mechanism by which the activation of sGC promotes actin cytoskeleton remodeling in tanycytes remains unknown. The retraction of tanycytic endfeet enables GnRH nerve endings to directly contact the pericapillary space (i.e., the space delineated by the parenchymatous basal lamina on one side and by the endothelial basal lamina on the other side; not shown). The released GnRH then enters the portal blood via the fenestrations of endothelial cells. Asterisks indicate the molecular components shown to be positively regulated by gonadal steroids in primary cultures of astrocytes, tanycytes and endothelial cells (see main text for details). From (Sharif et al. 2013) with permission.

Color image of this figure appears in the color plate section at the end of the book.

evidence suggests that the bioavailability of the mature form of TGFα is maximal at this stage of the estrous cycle. Along the same lines, estrogens were shown to up-regulate $TGF\beta_1$ mRNA levels in the hypothalamus (Galbiati et al. 2001), to stimulate $TGF\beta_1$ release in cultured hypothalamic astrocytes (Buchanan et al. 2000) and to modulate the expression of hypothalamic TGFβ receptors (Bouret et al. 2002).

In addition, tanycytes are also a target for other factors, including peptide hormones, that might influence structural plasticity at the GnRH nerve terminals during the estrous cycle (Garcia-Segura et al. 1996). For example, insulin-like growth factor 1, which is taken up by tanycytes from the portal blood in a gonadal steroid-controlled manner (Duenas et al. 1994, Garcia-Segura et al. 2008) and which has a well-established stimulatory effect on GnRH release (Hiney et al. 1996) may also contribute to the modulation of these plastic changes.

Intriguingly, in seasonal birds, subtractive hybridization analysis performed in the tuberal region of the quail recently identified a gene encoding TGFα as being differentially regulated during the photoperiodic control of reproduction (Takagi et al. 2007). *In situ* hybridization experiments showed that TGFα mRNA expression was markedly and rapidly induced in tanycytes of the median eminence during long days (Takagi et al. 2007), which is when GnRH secretion is induced and GnRH nerve terminals are seen in close proximity to the basal lamina in the quail median eminence (Yamamura et al. 2004), as described previously. Notably, intracerebroventricular infusion of physiological doses of TGFα induced LH secretion and gonadal growth under short-day conditions, when birds are sexually inactive (Takagi et al. 2007). These results raise the exciting possibility that the activation of TGFα-erbB signaling may be a key component of the cell-to-cell communication pathways used by the neuroendocrine brain to regulate structural changes between tanycytes and GnRH nerve terminals during the photoperiodic control of reproduction in seasonal breeders.

4.4 Endothelial-Cell-Promoted Neuronal-Glial Structural Plasticity

Using immunopanning techniques to purify the vascular endothelial cells of the median eminence and co-culture systems with isolated tanycytes, studies performed a decade ago have shown that median eminence endothelial cells can convey signals to tanycytes to promote acute remodeling of their actin cytoskeleton (de Seranno et al. 2004). These changes in tanycyte cytoarchitecture prompted by endothelial cells appear to involve the highly labile and diffusible factor NO (Garthwaite 2008). Experiments showed that pre-treatment of endothelial cells with L-NAME, an inhibitor of NO synthase (NOS) activity (de Seranno et al. 2004), or infection of endothelial cells with an adenovirus expressing a dominant negative form of endothelial NOS (eNOS) (de Seranno et al. 2010) blunted endothelial cell-promoted actin cytoskeleton remodeling in tanycytes, whereas tanycyte treatment with physiological doses of NO donors mimicked the co-culture effects (de Seranno et al. 2004). NO, which travels readily across cellular membranes, mediates most of its effects by binding to the prosthetic heme group of the NO-sensitive guanylyl cyclase (sGC) enzyme, resulting in increased production of cyclic guanosine monophosphate (cGMP) (Koesling and Friebe 1999, Bredt 2003, Garthwaite 2008). NO can also regulate the activity of COX-1 and COX-2 (other heme containing enzymes) and thus elicits prostaglandin release (Salvemini et al. 1993). The findings that the inhibition of either sGC or COX abrogates NO-mediated actin cytoskeleton remodeling in tanycytes (de Seranno et al. 2004) indicated that the intracellular mechanisms underlying the stimulatory effect of NO on tanycyte plasticity involve the participation of both cGMP and prostaglandins-thromboxanes. Additional co-culture experiments showed that increased NO production by endothelial cells, elicited by the NO precursor L-arginine, triggers tanycyte retraction (de Seranno et al. 2010). These *in vitro* results suggest that endothelial NO may induce neuronal-glial remodeling in the median eminence. Using electron microscopy of isolated median eminence explants, it was found that activation of endogenous NO production via L-arginine treatment induces rapid structural changes, allowing direct access of GnRH neuro-secretory axons to the portal vasculature, likely by promoting tanycyte endfoot retraction (de Seranno et al. 2004).

Recent studies revealed that estradiol acts on purified endothelial cells and isolated tanycytes that express estrogen receptors *in vitro* (de Seranno et al. 2010) to trigger communication between endothelial cells and glia, that leads to a sudden and massive retraction of tanycyte processes in an NO-dependent manner (de Seranno et al. 2010). Several lines of evidence support the conclusion that the estradiol-induced acute tanycyte retraction mediated by NO depends mainly on the action of COX enzyme byproducts. First, inhibition of COX activity abrogates NO-mediated actin cytoskeleton remodeling in tanycytes (de Seranno et al. 2004). Second, isolated tanycytes secrete PGE_2 both under control and under stimulated conditions (Prevot et al. 2003a). Third, estradiol up-regulates COX-1 and COX-2 expression in tanycytes, whereas the expression of sGC remains unchanged (de Seranno et al. 2010). Finally, in simple tanycyte cultures, PGE_2 mimics the estrogen-induced acute cellular retraction of tanycytes seen in co-cultures with endothelial cells, while membrane-permeable cGMP analogues do not (de Seranno et al. 2010). These results provide evidence for a major role of COX products, such as PGE_2, in the estradiol-induced tanycyte retraction mediated by endothelial cells; although additional factors released by tanycytes and/or endothelial cells may also contribute to this process (Fig. 4).

Direct evidence for the ability of PGE_2 to control neuronal-glial plasticity at the neurovascular junction was obtained from experiments in which PGE_2 was applied directly to median eminence explants at concentrations known to stimulate GnRH release and structural remodeling was observed in a matter of minutes (Ojeda and Negro-Vilar 1985, Ojeda et al. 1990, Prevot et al. 2003b). PGE_2 treatment caused the advancement of GnRH neuro-secretory terminals toward the pericapillary space, a phenomenon that probably results from the retraction of tanycyte endfeet (de Seranno et al. 2010). Intriguingly, PGE_2 failed to promote direct neurovascular contacts between the GnRH axons and the vascular wall. This is in contrast to the effects of L-arginine treatment mentioned earlier and suggests that additional downstream signaling pathways, such as those involving cGMP, are required for NO to fully exert its effects on neuronal-glial plasticity. The recent demonstration of an involvement of NO-cGMP signaling in axonal elongation and/or growth cone orientation (Song et al. 1998, Seidel and Bicker 2000, Nishiyama et al. 2003, de Seranno et al. 2004) supports this interpretation.

Results obtained *in vivo* using adult rats corroborate the role of PGE_2 in the mechanism that controls neuronal-glial plasticity at the neurovascular junction. Intracerebral infusion of the same COX antagonist that blocks NO-mediated actin cytoskeleton remodeling in primary cultures (indomethacin) (de Seranno et al. 2004) into the median eminence resulted in a discernible impairment of the ovarian cycle (de Seranno et al. 2010), which requires coordinated delivery of GnRH into the hypothalamo-hypophyseal portal system. Local inhibition of prostaglandin synthesis arrested the ovarian cycle at either the diestrus or estrus phase (de Seranno et al. 2010), when GnRH release is low (Levine and Ramirez 1982) and GnRH neuroendocrine terminals are enclosed by tanycyte endfeet (Prevot et al. 1998, Prevot et al. 1999a). These results highlight the physiological importance of eicosanoids in the cell to cell communication processes regulating GnRH release. These results are in full agreement with pioneering data obtained more than 30 years ago, demonstrating that PGE_2 synthesis increases within the hypothalamic region containing the median eminence during the onset of the first pre-ovulatory surge at puberty (Ojeda and Campbell 1982). These results are also consistent with the findings that intracerebral infusion of PGE_2 promotes LH release (Harms et al. 1973) and, in contrast, inhibitors of prostaglandin synthesis alter the onset of the steroid-induced LH surge in ovariectomized female rats (Ojeda et al. 1975).

Altogether, these results raise the possibility that endothelial NO is a molecule involved in the process by which estrogen stimulates GnRH secretion at proestrus (Fig. 4). However, is NO really produced in the median eminence? Several experiments using either L-arginine, NO donors or NOS inhibitors have demonstrated that this gaseous transmitter can stimulate GnRH release from GnRH neuroendocrine terminals in the median eminence (Rettori et al. 1992, Bonavera et al. 1993, Moretto et al. 1993, Rettori et al. 1993, Bonavera et al. 1994, Seilicovich et al. 1995, Bonavera et al. 1996, Canteros et al. 1996, Kohsaka et al. 1999). The use of amperometric measurements to measure NO production in isolated median eminence explants revealed that estradiol promotes acute increases in the release of both NO and GnRH (Prevot et al. 1999b). Amperometric methods also showed that NO is released spontaneously in isolated median eminence explants throughout the estrous cycle, with levels increasing dramatically at the onset of the pre-ovulatory surge (Knauf et al. 2001). Critically, blockade of NO production with L-NIO, a selective eNOS inhibitor, prevented estradiol- and proestrus-stimulated release of both GnRH and NO (Prevot et al. 1999b, Knauf et al. 2001) (Fig. 5). It was also found that increases in median eminence NO release can be induced in ovariectomized rats by injection of estrogens (Knauf et al. 2001). Additionally, it was shown that intracerebroventricular injection of antisense oligodeoxynucleotides to eNOS inhibits the gonadal-steroid-induced surge of GnRH and LH in ovariectomized rats (Aguan et al. 1996). Finally, it was reported that inhibition of NO synthesis in the median eminence with a local infusion of NOS inhibitors blunts estrous cyclicity (de Seranno et al. 2004).

Taken together, these findings provide strong evidence that vascular endothelial cells are crucial targets for the action of estrogen on the structural remodeling that takes place at the projection site of GnRH neurons and the action of estrogen on GnRH neurosecretion during the estrous cycle. Estrogen may induce endothelial cells and tanycytes in the median eminence to produce NO and up-regulate cyclooxygenase expression, respectively. Endothelial NO diffuses into neural tissue and acts on tanycytes to stimulate PGE_2 production, which in turn acts with additional unidentified factors to promote the acute retraction of tanycyte endfeet that allows GnRH nerve terminals to form direct neurovascular junctions.

Do endothelial cells talk directly to GnRH neurons themselves? Electron microscopic data suggest that on the day of proestrus some GnRH axons sprout new terminals that appear to be attracted by the vascular wall (Prevot et al. 1999a). Interestingly, mature GnRH neurons were shown to express growth-associated-43 protein (GAP-43), an intrinsic determinant of axonal growth (Aigner et al. 1995, Benowitz and Routtenberg 1997) and the proportion of GnRH perikarya expressing GAP-43 mRNA reached a maximum on the day of proestrus (Prevot et al. 2000c). Conceivably, an accumulation of GAP-43 in GnRH nerve terminals at proestrus could enhance the responsiveness of individual GnRH nerve endings to growth-

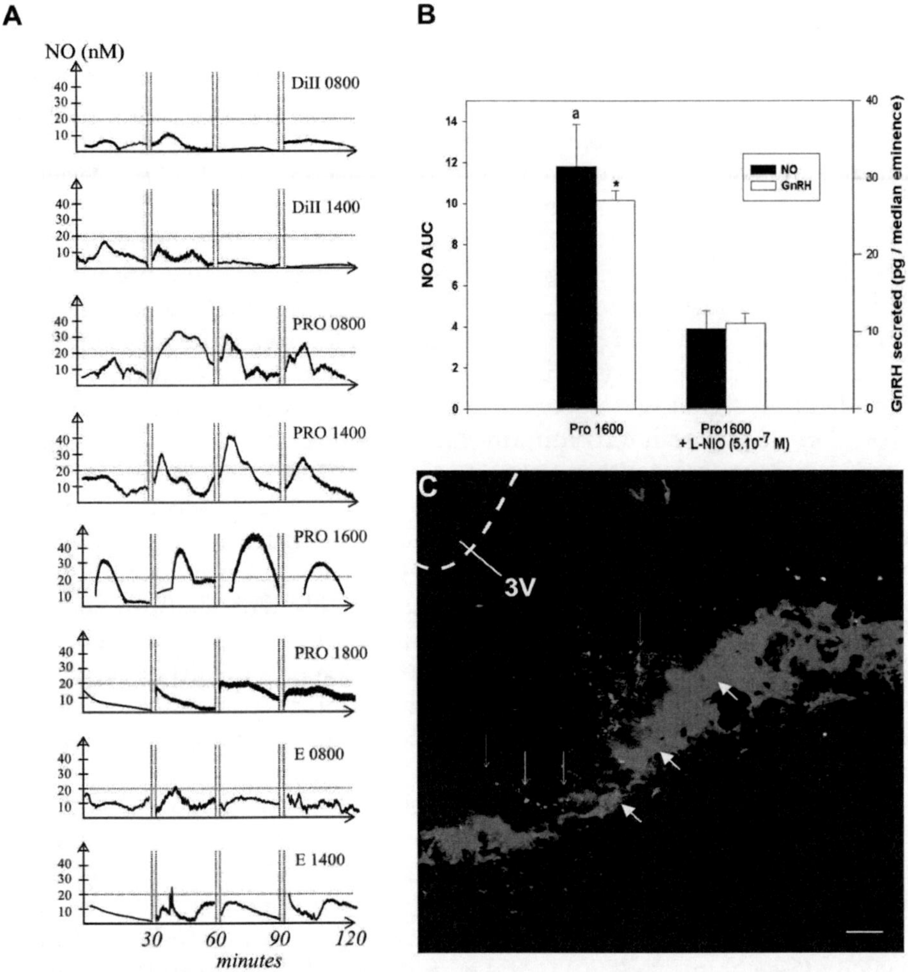

Figure 5. In the median eminence of the hypothalamus, endothelial nitric oxide (NO) secretion may represent one of the synchronizing cues that coordinates gonadotropin-releasing hormone (GnRH) release from GnRH neuroendocrine terminals, that are distributed over two mm within the median eminence, allowing the occurrence of functionally meaningful episodes of GnRH secretion. A) Real time amperometric measurement of spontaneous NO release from median eminence explants at different stages of the rat estrous cycle. DiII: diestrus II, PRO: proestrus, E: estrus. B) On the afternoon of proestrus, the pre-ovulatory GnRH-NO release is blocked with L-NIO, an NO synthase (NOS) inhibitor selective for endothelial NOS (eNOS) at 0.5 μM. * and a: significantly different from treated samples, p<0:05. AUC: area under the curve during a 30 min period. C) Photomicrograph showing GnRH axonal fibers in the external zone of the median eminence (green fluorescence, arrows) in close apposition to eNOS-immunoreactive portal vasculature (red fluorescence, arrowheads). 3V: third ventricle. The dotted lines outline the third ventricle. Scale bar: 75 μm. From (Prevot et al. 2000a, Knauf et al. 2001) with permission.

Color image of this figure appears in the color plate section at the end of the book.

promoting factors (Benowitz and Routtenberg 1997). Screening for the molecular cues known to play a role in the control of axon guidance indicated Semaphorin3A (Sema3A) as a likely candidate (Campagne et al. 2008, Giacobini and Prevot 2013). Semaphorins are chemotropic factors recently shown to intervene in the control of GnRH neuronal migration during embryogenesis (Cariboni et al. 2007, Giacobini et al. 2008, Messina et al. 2011). Sema3A, by binding to the trans-membrane protein neuropilin-1 (Nrp1), has been shown to be essential for the patterning of the axonal scaffold used by GnRH neurons in their migration from the nose to the brain during embryonic life (Cariboni et al. 2011, Hanchate et al. 2012). *In situ* hybridization within the adult brain, revealed robust Sema3A mRNA expression in the capillary zone of the median eminence, while only scant hybridization signal was observed in the nervous tissue (Giacobini et al. 2014). On the other hand, immunohistochemical staining showed that Sema3A immunoreactivity is restricted to the portal blood capillaries of the median eminence (Giacobini et al. 2014). Western blot analyses investigating whether Sema3A protein expression changes during the reproductive cycle, found that the expression of the 65-kDa Sema3A isoform increased significantly on the afternoon of proestrus (Giacobini et al. 2014). Further experiments showed that GnRH nerve terminals might be capable of sensing Sema3A; expression of Nrp1 receptor was observed in GnRH neurons at both the mRNA and the protein level and the proportion of GnRH perikarya expressing Nrp1 mRNA appears to be maximal at proestrus, a phase of the estrous cycle when GnRH nerve terminals have direct access to the portal vasculature (Giacobini et al. 2014). The use of electron microscopy further established that activation of Sema3A and Nrp1 signaling in the median eminence promotes sprouting of GnRH axons towards the pericapillary space of the pituitary portal blood vessels (Giacobini et al. 2014). Finally, local *in vivo* neutralization of Nrp1 activity was shown to disrupt estrous cyclicity (Giacobini et al. 2014). Taken together all these results raise the exciting possibility that endothelial cells of the median eminence may use chemotropic factors of the semaphorin family to regulate GnRH axonal plasticity.

4.5 Conclusion

The evidence reviewed in this chapter demonstrates the vital participation of vascular endothelial cells, astrocytes and tanycytes in cellular mechanisms that regulate function-related structural plasticity of the GnRH system. As such, their roles need to be integrated into current models of the central control of fertility, in which neurons are not the sole contributors to the neuroendocrine release of GnRH. Rather, neurons, astrocytes and vascular endothelial cells are components of a basic unit of functional organization, which we can conceptualize as a neuro-glio-vascular unit (Nedergaard et al. 2003) and must coordinate their actions to maintain homeostasis of the reproductive neuroendocrine brain. In a broader perspective, the demonstration that non-neural cells (vascular endothelial cells) participate in morphological and functional plasticity of neurons and glial cells in the adult central nervous system opens new avenues of investigation in neuroscience.

Acknowledgments

This research was supported by the Agence National pour la Recherche (ANR, France) grants ANR-07-NEURO-026-03 and ANR-09-BLAN-0267, the Fondation pour la RechercheMédicale (EquipeFRM 2005 & DEQ20130326524, France) and the European INTERREG program TC2N.

References

Aguan, K., V.B. Mahesh, L. Ping, G. Bhat and D.W. Brann. 1996. Evidence for a physiological role for nitric oxide in the regulation of the LH surge: effect of central administration of antisense oligonucleotides to nitric oxide synthase. Neuroendocrinology 64: 449–455.

Aigner, L., S. Arber, J.P. Kapfhammer, T. Laux, C. Schneider, F. Botteri, H.R. Brenner and P. Caroni. 1995. Overexpression of the neural growth-associated protein GAP-43 induces nerve sprouting in the adult nervous system of transgenic mice. Cell 83: 269–278.

Balland, E., J. Dam, F. Langlet, E. Caron, S. Steculorum, A. Messina, S. Rasika, A. Falluel-Morel, Y. Anouar, B. Dehouck, E. Trinquet, R. Jockers, S.G. Bouret and V. Prevot. 2014 Hypothalamic tanycytes are an ERK-gated conduit for leptin into the brain. Cell Metab. 19: 293–301.

Baroncini, M., C. Allet, D. Leroy, J.C. Beauvillain, J.P. Francke and V. Prevot. 2007. Morphological evidence for direct interaction between gonadotrophin-releasing hormone neurones and astroglial cells in the human hypothalamus. J. Neuroendocrinol. 19: 691–702.

Barres, B.A. 2008. The mystery and magic of glia: a perspective on their roles in health and disease. Neuron 60: 430–440.

Benowitz, L.I. and A. Routtenberg. 1997. GAP-43: an intrinsic determinant of neuronal development and plasticity. Trends Neurosci. 20: 84–91.

Bonavera, J.J., A. Sahu, P.S. Kalra and S.P. Kalra. 1993. Evidence that nitric oxide may mediate the ovarian steroid-induced luteinizing hormone surge: involvement of excitatory amino acids. Endocrinology 133: 2481–2487.

Bonavera, J.J., A. Sahu, P.S. Kalra and S.P. Kalra. 1994. Evidence in support of nitric oxide (NO) involvement in the cyclic release of prolactin and LH surges. Brain Res. 660: 175–179.

Bonavera, J.J., P.S. Kalra and S.P. Kalra. 1996. L-arginine/nitric oxide amplifies the magnitude and duration of the luteinizing hormone surge induced by estrogen: involvement of neuropeptide Y. Endocrinology 137: 1956–1962.

Bouret, S., S. De Seranno, J.C. Beauvillain and V. Prevot. 2004. Transforming growth factor beta1 may directly influence gonadotropin-releasing hormone gene expression in the rat hypothalamus. Endocrinology 145: 1794–1801.

Bouret, S., V. Prevot, T. Takumi, J.C. Beauvillain and V. Mitchell. 2002. Regulation by gonadal steroids of the mRNA encoding for a type I receptor for TGF-beta in the female rat hypothalamus. Neuroendocrinology 76: 1–7.

Bredt, D.S. 2003. Nitric oxide signaling specificity—the heart of the problem. J. Cell Sci. 116: 9–15.

Buchanan, C.D., V.B. Mahesh and D.W. Brann. 2000. Estrogen-astrocyte-luteinizing hormone-releasing hormone signaling: a role for transforming growth factor-beta(1). Biol. Reprod. 62: 1710–1721.

Campagne, C., S.G. Bouret, D. Leroy, J.C. Beauvillain and V. Prevot. 2008. Semaphorin3A may be a chemotropic factor used by endothelial cells of the median eminence to regulate GnRH axon plasticity during the rat estrous cycle. Soc. Neuroci. Abstr. 618.613.

Canteros, G., V. Rettori, A. Genaro, A. Suburo, M. Gimeno and S.M. McCann. 1996. Nitric oxide synthase content of hypothalamic explants: increase by norepinephrine and inactivated by NO and cGMP. Proc. Natl. Acad. Sci. USA 93: 4246–4250.

Cariboni, A., J. Hickok, S. Rakic, W. Andrews, R. Maggi, S. Tischkau and J.G. Parnavelas. 2007. Neuropilins and their ligands are important in the migration of gonadotropin-releasing hormone neurons. J. Neurosci. 27: 2387–2395.

Cariboni, A., K. Davidson, S. Rakic, R. Maggi, J.G. Parnavelas and C. Ruhrberg. 2011. Defective gonadotropin-releasing hormone neuron migration in mice lacking SEMA3A signalling through NRP1 and NRP2: implications for the aetiology of hypogonadotropic hypogonadism. Hum. Mol. Genet. 20: 336–344.

Carmeliet, P. and M. Tessier-Lavigne. 2005. Common mechanisms of nerve and blood vessel wiring. Nature 436: 193–200.

Chauvet, N., A. Privat and G. Alonso. 1996. Aged median eminence glial cell cultures promote survival and neurite outgrowth of cocultured neurons. Glia 18: 211–223.

Chauvet, N., M. Prieto and G. Alonso. 1998. Tanycytes present in the adult rat mediobasal hypothalamus support the regeneration of monoaminergic axons. Exp. Neurol. 151: 1–13.

Chauvet, N., M.L. Parmentier and G. Alonso. 1995. Transected axons of adult hypothalamo-neurohypophysial neurons regenerate along tanycytic processes. J. Neurosci. Res. 41: 129–144.

Ciofi, P. 2011. The arcuate nucleus as a circumventricular organ in the mouse. Neurosci. Lett. 487: 187–190.

Ciofi, P., M. Garret, O. Lapirot, P. Lafon, A. Loyens, V. Prevot and J.E. Levine. 2009. Brain-endocrine interactions: a microvascular route in the mediobasal hypothalamus. Endocrinology 150: 5509–5519.

Clasadonte, J., A. Sharif, M. Baroncini and V. Prevot. 2011a. Gliotransmission by prostaglandin E2: a prerequisite for GnRH neuronal function. Front. Endocrinology 2: 1–12.

Clasadonte, J., P. Poulain, N.K. Hanchate, G. Corfas, S.R. Ojeda and V. Prevot. 2011b. Prostaglandin E2 release from astrocytes triggers gonadotropin-releasing hormone (GnRH) neuron firing via EP2 receptor activation. Proc. Natl. Acad. Sci. USA 108: 16104–16109.

de Seranno, S., C. Estrella, A. Loyens, A. Cornea, S.R. Ojeda, J.C. Beauvillain and V. Prevot. 2004. Vascular endothelial cells promote acute plasticity in ependymoglial cells of the neuroendocrine brain. J. Neurosci. 24: 10353–10363.

de Seranno, S., X. d'Anglemont de Tassigny, C. Estrella, A. Loyens, S. Kasparov, D. Leroy, S.R. Ojeda, J.C. Beauvillain and V. Prevot. 2010. Role of estradiol in the dynamic control of tanycyte plasticity mediated by vascular endothelial cells in the median eminence. Endocrinology 151: 1760–1772.

Duenas, M., S. Luquin, J.A. Chowen, I. Torres-Aleman, F. Naftolin and L.M. Garcia-Segura. 1994. Gonadal hormone regulation of insulin-like growth factor-I-like immunoreactivity in hypothalamic astroglia of developing and adult rats. Neuroendocrinology 59: 528–538.

Duvernoy, H.M. and P.Y. Risold. 2007. The circumventricular organs: an atlas of comparative anatomy and vascularization. Brain Res. Rev. 56: 119–147.

Estrella, C., S. De Seranno, E. Caron, X. d'Anglemont de Tassigny, V. Mitchell, J.C. Beauvillain and V. Prevot. 2004. Matrix metalloproteinases are expressed in the median eminence of the hypothalamus and their activities vary during the rat estrous cycle. Endocrine Soc. Abstr. P3–266.

Galbiati, M., V. Magnaghi, L. Martini and R.C. Melcangi. 2001. Hypothalamic transforming growth factor beta1 and basic fibroblast growth factor mRNA expression is modified during the rat oestrous cycle. J. Neuroendocrinol. 13: 483–489.

Garcia-Segura, L.M., B. Lorenz and L.L. DonCarlos. 2008. The role of glia in the hypothalamus: implications for gonadal steroid feedback and reproductive neuroendocrine output. Reproduction 135: 419–429.

Garcia-Segura, L.M., J.A. Chowen and F. Naftolin. 1996. Endocrine glia: roles of glial cells in the brain actions of steroid and thyroid hormones and in the regulation of hormone secretion. Front. Neuroendocrinol. 17: 180–211.

Garthwaite, J. 2008. Concepts of neural nitric oxide-mediated transmission. Eur. J. Neurosci. 27: 2783–2802.

Giacobini, P., J. Parkash, C. Campagne, A. Messina, F. Casoni, C. Vanacker, F. Langlet, B. Hobo, G. Cagnoni, S. Gallet, N.K. Hanchate, D. Mazur, M. Tanaguchi, M. Mazzone, J. Verhaagen, P. Ciofi, S.G. Bouret, L. Tamagnone and V. Prevot. 2014. Brain Endothelial Cells Control Fertility through Ovarian-Steroid-Dependent Release of Semaphorin 3A PLoS Biol. 12: e1001808.

Giacobini, P. and V. Prevot. 2013. Semaphorins in the development, homeostasis and disease of hormone systems. Semin. Cell Dev. Biol. 24: 190–198.

Giacobini, P., A. Messina, F. Morello, N. Ferraris, S. Corso, J. Penachioni, S. Giordano, L. Tamagnone and A. Fasolo. 2008. Semaphorin 4D regulates gonadotropin hormone-releasing hormone-1 neuronal migration through PlexinB1-Met complex. J. Cell Biol. 183: 555–566.

Givalois, L., S. Arancibia, G. Alonso and L. Tapia-Arancibia. 2004. Expression of brain-derived neurotrophic factor and its receptors in the median eminence cells with sensitivity to stress. Endocrinology 145: 4737–4747.

Goldman, S.A. and Z. Chen. 2011. Perivascular instruction of cell genesis and fate in the adult brain. Nat. Neurosci. 14: 1382–1389.

Haan, N., T. Goodman, A. Najdi-Samiei, C.M. Stratford, R. Rice, E. El Agha, S. Bellusci and M.K. Hajihosseini. 2013. Fgf10-expressing tanycytes add new neurons to the appetite/energy-balance regulating centers of the postnatal and adult hypothalamus. J. Neurosci. 33: 6170–6180.

Halassa, M.M. and P.G. Haydon. 2010. Integrated brain circuits: astrocytic networks modulate neuronal activity and behavior. Annu. Rev. Physiol. 72: 335–355.

Hanchate, N.K., P. Giacobini, P. Lhuillier, J. Parkash, C. Espy, C. Fouveaut, C. Leroy, S. Baron, C. Campagne, C. Vanacker, F. Collier, C. Cruaud, V. Meyer, A. García-Piñero, D. Dewailly, C. Cortet-Rudelli, K. Gersak, C. Metz, G. Chabrier, M. Pugeat, J. Young, J.P. Hardelin, V. Prevot and C. Dodé. 2012. SEMA3A, a Gene Involved in axonal pathfinding, is mutated in patients with Kallmann syndrome. PLoS Genet 8: e1002896.

Harms, P.G., S.R. Ojeda and S.M. McCann. 1973. Prostaglandin involvement in hypothalamic control of gonadotropin and prolactin release. Science 181: 760–761.

Herbison, A.E. 2006. Physiology of the gonadotropin-releasing hormone neuronal network. pp. 1415–1482. *In*: E. Knobil and J.D. Neill [eds.]. Knobil and Neill's Physiology of Reproduction. Elsevier, New York, USA.

Hiney, J.K., V. Srivastava, C.L. Nyberg, S.R. Ojeda and W.L. Dees. 1996. Insulin-like growth factor I of peripheral origin acts centrally to accelerate the initiation of female puberty. Endocrinology 137: 3717–3728.

Iadecola, C. 2004. Neurovascular regulation in the normal brain and in Alzheimer's disease. Nat. Rev. Neurosci. 5: 347–360.

Iadecola, C. and M. Nedergaard. 2007. Glial regulation of the cerebral microvasculature. Nat. Neurosci. 10: 1369–1376.

Junier, M.P. 2000. What role(s) for TGFalpha in the central nervous system? Prog. Neurobiol. 62: 443–473.

Knauf, C., V. Prevot, G.B. Stefano, G. Mortreux, J.C. Beauvillain and D. Croix. 2001. Evidence for a spontaneous nitric oxide release from the rat median eminence: influence on gonadotropin-releasing hormone release. Endocrinology 142: 2343–2350.

Koesling, D. and A. Friebe. 1999. Soluble guanylyl cyclase: structure and regulation. Rev. Physiol. Biochem. Pharmacol. 135: 41–65.

Kohsaka, A., H. Watanobe, Y. Kakizaki and T. Suda. 1999. A comparative study of the effects of nitric oxide and carbon monoxide on the *in vivo* release of gonadotropin-releasing hormone and neuropeptide Y from rat hypothalamus during the estradiol-induced luteinizing hormone surge: estimation by push-pull perfusion. Neuroendocrinology 69: 245–253.

Kriegstein, A. and A. Alvarez-Buylla. 2009. The glial nature of embryonic and adult neural stem cells. Annu. Rev. Neurosci. 32: 149–184.

Langlet, F., A. Mullier, S.G. Bouret, V. Prevot and B. Dehouck. 2013a. Tanycyte-like cells form a blood-cerebrospinal fluid barrier in the circumventricular organs of the mouse brain. J. Comp. Neurol. 521: 3389–3405.

Langlet, F., B.E. Levin, S. Luquet, M. Mazzone, A. Messina, A.A. Dunn-Meynell, E. Balland, A. Lacombe, D. Mazur, P. Carmeliet, S.G. Bouret, V. Prevot and B. Dehouck. 2013b. Tanycytic VEGF-A boosts blood-hypothalamus barrier plasticity and access of metabolic signals to the arcuate nucleus in response to fasting. Cell Metab. 17: 607–617.

Langub, M.C. Jr. and R.E. Watson Jr. 1992. Estrogen receptor-immunoreactive glia, endothelia, and ependyma in guinea pig preoptic area and median eminence: electron microscopy. Endocrinology 130: 364–372.

Larrivee, B., C. Freitas, S. Suchting, I. Brunet and A. Eichmann. 2009. Guidance of vascular development: lessons from the nervous system. Circ. Res. 104: 428–441.

Lee, D.A., J.L. Bedont, T. Pak, H. Wang, J. Song, A. Miranda-Angulo, V. Takiar, V. Charubhumi, F. Balordi, H. Takebayashi, S. Aja, E. Ford, G. Fishell and S. Blackshaw. 2012. Tanycytes of the hypothalamic median eminence form a diet-responsive neurogenic niche. Nat. Neurosci. 15: 700–702.

Levine, J.E. and V.D. Ramirez. 1982. Luteinizing hormone-releasing hormone release during the rat estrous cycle and after ovariectomy, as estimated with push-pull cannulae. Endocrinology 111: 1439–1448.

Lomniczi, A., A. Cornea, M.E. Costa and S.R. Ojeda. 2006. Hypothalamic tumor necrosis factor-alpha converting enzyme mediates excitatory amino acid-dependent neuron-to-glia signaling in the neuroendocrine brain. J. Neurosci. 26: 51–62.

Louissaint, A. Jr., S. Rao, C. Leventhal and S.A. Goldman. 2002. Coordinated interaction of neurogenesis and angiogenesis in the adult songbird brain. Neuron 34: 945–960.

Ma, Y.J., D.F. Hill, K.E. Creswick, M.E. Costa, A. Cornea, M.N. Lioubin, G.D. Plowman and S.R. Ojeda. 1999. Neuregulins signaling via a glial erbB-2-erbB-4 receptor complex contribute to the neuroendocrine control of mammalian sexual development. J. Neurosci. 19: 9913–9927.

Ma, Y.J., D.F. Hill, M.P. Junier, M.E. Costa, S.E. Felder and S.R. Ojeda. 1994b. Expression of epidermal growth factor receptor changes in the hypothalamus during the onset of female puberty. Mol. Cell. Neurosci. 5: 246–262.

Ma, Y.J., K. Berg-von der Emde, M. Moholt-Siebert, D.F. Hill and S.R. Ojeda. 1994a. Region-specific regulation of transforming growth factor alpha (TGF alpha) gene expression in astrocytes of the neuroendocrine brain. J. Neurosci. 14: 5644–5651.

Ma, Y.J., M.P. Junier, M.E. Costa and S.R. Ojeda. 1992. Transforming growth factor-alpha gene expression in the hypothalamus is developmentally regulated and linked to sexual maturation. Neuron 9: 657–670.

Makita, T., H.M. Sucov, C.E. Gariepy, M. Yanagisawa and D.D. Ginty. 2008. Endothelins are vascular-derived axonal guidance cues for developing sympathetic neurons. Nature 452: 759–763.

Melcangi, R.C., M. Galbiati, E. Messi, F. Piva, L. Martini and M. Motta. 1995. Type 1 astrocytes influence luteinizing hormone-releasing hormone release from the hypothalamic cell line GT1-1: is transforming growth factor-beta the principle involved? Endocrinology 136: 679–686.

Messina, A., N. Ferraris, S. Wray, G. Cagnoni, D.E. Donohue, F. Casoni, P.R. Kramer, A.A. Derijck, Y. Adolfs, A. Fasolo, R.J. Pasterkamp and P. Giacobini. 2011. Dysregulation of Semaphorin7A/beta1-integrin signaling leads to defective GnRH-1 cell migration, abnormal gonadal development and altered fertility. Hum. Mol. Genet. 20: 4759–4774.

Moretto, M., F.J. Lopez and A. Negro-Vilar. 1993. Nitric oxide regulates luteinizing hormone-releasing hormone secretion. Endocrinology 133: 2399–2402.

Mullier, A., S.G. Bouret, V. Prevot and B. Dehouck. 2010. Differential distribution of tight junction proteins suggests a role for tanycytes in blood-hypothalamus barrier regulation in the adult mouse brain. J. Comp. Neurol. 518: 943–962.

Myers, M.G. Jr. 2013. How is the hungry brain like a sieve? Cell Metab. 17: 467–468.

Nedergaard, M., B. Ransom and S.A. Goldman. 2003. New roles for astrocytes: redefining the functional architecture of the brain. Trends Neurosci. 26: 523–530.

Nishiyama, M., A. Hoshino, L. Tsai, J.R. Henley, Y. Goshima, M. Tessier-Lavigne, M.M. Poo and K. Hong. 2003. Cyclic AMP/GMP-dependent modulation of Ca2+ channels sets the polarity of nerve growth-cone turning. Nature 424: 990–995.

Ojeda, S.R. and A. Negro-Vilar. 1985. Prostaglandin E2-induced luteinizing hormone-releasing hormone release involves mobilization of intracellular Ca+2. Endocrinology 116: 1763–1770.

Ojeda, S.R. and W.B. Campbell. 1982. An increase in hypothalamic capacity to synthesize prostaglandin E2 precedes the first preovulatory surge of gonadotropins. Endocrinology 111: 1031–1037.

Ojeda, S.R., A. Lomniczi and U.S. Sandau. 2008. Glial-gonadotrophin hormone (GnRH) neurone interactions in the median eminence and the control of GnRH secretion. J. Neuroendocrinol. 20: 732–742.

Ojeda, S.R., H.F. Urbanski, M.E. Costa, D.F. Hill and M. Moholt-Siebert. 1990. Involvement of transforming growth factor alpha in the release of luteinizing hormone-releasing hormone from the developing female hypothalamus. Proc. Natl. Acad. Sci. USA 87: 9698–9702.

Ojeda, S.R., P.G. Harms and S.M. McCann. 1975. Effect of inhibitors of prostaglandin synthesis on gonadotropin release in the rat. Endocrinology 97: 843–854.

Page, R.B. 2006. The anatomy of the hypothalamo-hypophysial complex. pp. 1309–1413. *In*: E. Knobil and J.D. Neill [eds.]. Knobil and Neill's Physiology of Reproduction. Elsevier, New York, USA.

Pandiella, A. and J. Massague. 1991. Multiple signals activate cleavage of the membrane transforming growth factor-alpha precursor. J. Biol. Chem. 266: 5769–5773.

Peschon, J.J., J.L. Slack, P. Reddy, K.L. Stocking, S.W. Sunnarborg, D.C. Lee, W.E. Russell, B.J. Castner, R.S. Johnson, J.N. Fitzner, R.W. Boyce, N. Nelson, C.J. Kozlosky, M.F. Wolfson, C.T. Rauch, D.P. Cerretti, R.J. Paxton, C.J. March and R.A. Black. 1998. An essential role for ectodomain shedding in mammalian development. Science 282: 1281–1284.

Prenzel, N., E. Zwick, H. Daub, M. Leserer, R. Abraham, C. Wallasch and A. Ullrich. 1999. EGF receptor transactivation by G-protein-coupled receptors requires metalloproteinase cleavage of proHB-EGF. Nature 402: 884–888.

Prevot, V. 2010. Plasticity of neuroendocrine systems. Eur. J. Neurosci. 32: 1987–1988.

Prevot, V., A. Cornea, A. Mungenast, G. Smiley and S.R. Ojeda. 2003a. Activation of erbB-1 signaling in tanycytes of the median eminence stimulates transforming growth factor beta1 release via prostaglandin E2 production and induces cell plasticity. J. Neurosci. 23: 10622–10632.

Prevot, V., C. Rio, G.J. Cho, A. Lomniczi, S. Heger, C.M. Neville, N.A. Rosenthal, S.R. Ojeda and G. Corfas. 2003b. Normal female sexual development requires neuregulin-erbB receptor signaling in hypothalamic astrocytes. J. Neurosci. 23: 230–239.

Prevot, V., D. Croix, C.M. Rialas, P. Poulain, G.L. Fricchione, G.B. Stefano and J.C. Beauvillain. 1999b. Estradiol coupling to endothelial nitric oxide stimulates gonadotropin-releasing hormone release from rat median eminence via a membrane receptor. Endocrinology 140: 652–659.

Prevot, V., D. Croix, S. Bouret, S. Dutoit, G. Tramu, G.B. Stefano and J.C. Beauvillain. 1999a. Definitive evidence for the existence of morphological plasticity in the external zone of the median eminence during the rat estrous cycle: implication of neuro-glio-endothelial interactions in gonadotropin-releasing hormone release. Neuroscience 94: 809–819.

Prevot, V., F. Langlet and B. Dehouck. 2013. Flipping the tanycyte switch: how circulating signals gain direct access to the metabolic brain. Aging (Albany NY) 5: 332–334.

Prevot, V., N. Bellefontaine, M. Baroncini, A. Sharif, N.K. Hanchate, J. Parkash, C. Campagne and S. de Seranno. 2010a. GnRH nerve terminals, tanycytes and neurohaemal junction remodeling in the adult median eminence: functional consequences for reproduction and dynamic role of vascular endothelial cells. J. Neuroendocrinol. 22: 639–649.

Prevot, V., N.K. Hanchate, N. Bellefontaine, A. Sharif, J. Parkash, C. Estrella, C. Allet, S. de Seranno, C. Campagne, X. de Tassigny and M. Baroncini. 2010b. Function-related structural plasticity of the GnRH system: a role for neuronal-glial-endothelial interactions. Front. Neuroendocrinol. 31: 241–258.

Prevot, V., S. Bouret, D. Croix, G. Alonso, L. Jennes, V. Mitchell, A. Routtenberg and J.C. Beauvillain. 2000c. Growth-associated protein-43 messenger ribonucleic acid expression in gonadotropin-releasing hormone neurons during the rat estrous cycle. Endocrinology 141: 1648–1657.

Prevot, V., S. Bouret, D. Croix, T. Takumi, L. Jennes, V. Mitchell and J.C. Beauvillain. 2000b. Evidence that members of the TGFbeta superfamily play a role in regulation of the GnRH neuroendocrine axis: expression of a type I serine-threonine kinase receptor for TGRbeta and activin in GnRH neurones and hypothalamic areas of the female rat. J. Neuroendocrinol. 12: 665–670.

Prevot, V., S. Bouret, G.B. Stefano and J. Beauvillain. 2000a. Median eminence nitric oxide signaling. Brain Res. Rev. 34: 27–41.

Prevot, V., S. Dutoit, D. Croix, G. Tramu and J.C. Beauvillain. 1998. Semi-quantitative ultrastructural analysis of the localization and neuropeptide content of gonadotropin releasing hormone nerve terminals in the median eminence throughout the estrous cycle of the rat. Neuroscience 84: 177–191.

Prieto, M., N. Chauvet and G. Alonso. 2000. Tanycytes transplanted into the adult rat spinal cord support the regeneration of lesioned axons. Exp. Neurol. 161: 27–37.

Rage, F., B.J. Lee, Y.J. Ma and S.R. Ojeda. 1997. Estradiol enhances prostaglandin E2 receptor gene expression in luteinizing hormone-releasing hormone (LHRH) neurons and facilitates the LHRH response to PGE2 by activating a glia-to-neuron signaling pathway. J. Neurosci. 17: 9145–9156.

Reichenbach, A. and H. Wolburg. 2005. Astrocytes and ependymal glia. pp. 19–36. *In*: A. Reichenbach and H. Wolburg [eds.]. Neuroglia. Oxford University Press, New York, USA.

Rettori, V., M. Gimeno, K. Lyson and S.M. McCann. 1992. Nitric oxide mediates norepinephrine-induced prostaglandin E2 release from the hypothalamus. Proc. Natl. Acad. Sci. USA 89: 11543–11546.

Rettori, V., N. Belova, W.L. Dees, C.L. Nyberg, M. Gimeno and S.M. McCann. 1993. Role of nitric oxide in the control of luteinizing hormone-releasing hormone release *in vivo* and *in vitro*. Proc. Natl. Acad. Sci. USA 90: 10130–10134.

Robins, S.C., I. Stewart, D.E. McNay, V. Taylor, C. Giachino, M. Goetz, J. Ninkovic, N. Briancon, E. Maratos-Flier, J.S. Flier, M.V. Kokoeva and M. Placzek. 2013. α-Tanycytes of the adult hypothalamic third ventricle include distinct populations of FGF-responsive neural progenitors. Nat. Commun. 4: 2049.

Sahin, U., G. Weskamp, K. Kelly, H.M. Zhou, S. Higashiyama, J. Peschon, D. Hartmann, P. Saftig and C.P. Blobel. 2004. Distinct roles for ADAM10 and ADAM17 in ectodomain shedding of six EGFR ligands. J. Cell Biol. 164: 769–779.

Salvemini, D., T.P. Misko, J.L. Masferrer, K. Seibert, M.G. Currie and P. Needleman. 1993. Nitric oxide activates cyclooxygenase enzymes. Proc. Natl. Acad. Sci. USA 90: 7240–7244.

Schaeffer, M., F. Langlet, C. Lafont, F. Molino, D.J. Hodson, T. Roux, L. Lamarque, P. Verdie, E. Bourrier, B. Dehouck, J.L. Baneres, J. Martinez, P.F. Mery, J. Marie, E. Trinquet, J.A. Fehrentz, V. Prevot and P. Mollard. 2013. Rapid sensing of circulating ghrelin by hypothalamic appetite-modifying neurons. Proc. Natl. Acad. Sci. USA 110: 1512–1517.

Seidel, C. and G. Bicker. 2000. Nitric oxide and cGMP influence axonogenesis of antennal pioneer neurons. Development 127: 4541–4549.

Seilicovich, A., B.H. Duvilanski, D. Pisera, S. Theas, M. Gimeno, V. Rettori and S.M. McCann. 1995. Nitric oxide inhibits hypothalamic luteinizing hormone-releasing hormone release by releasing gamma-aminobutyric acid. Proc. Natl. Acad. Sci. USA 92: 3421–3424.

Sharif, A., M. Baroncini and V. Prevot. 2013. Role of glia in the regulation of gonadotropin-releasing hormone neuronal activity and secretion. Neuroendocrinology 98: 1–15.

Sharif, A., S.R. Ojeda and V. Prevot. 2014 Neurogenesis and gliogenesis in the postnatal hypothalamus: a new level of plasticity for the regulation of hypothalamic function. *In*: M.P. Junier and S.G. Kernie (eds.). Endogenous Stem Cell-Based Brain Remodeling in Mammals, Stem Cell Biology and Regenerative Medicine. New York: Springer Science+Business Media.

Smith, M.S., M.E. Freeman and J.D. Neill. 1975. The control of progesterone secretion during the estrous cycle and early pseudopregnancy in the rat: prolactin, gonadotropin and steroid levels associated with rescue of the corpus luteum of pseudopregnancy. Endocrinology 96: 219–226.

Song, H., G. Ming, Z. He, M. Lehmann, L. McKerracher, M. Tessier-Lavigne and M. Poo. 1998. Conversion of neuronal growth cone responses from repulsion to attraction by cyclic nucleotides. Science 281: 1515–1518.

Takagi, T., T. Yamamura, T. Anraku, S. Yasuo, N. Nakao, M. Watanabe, M. Iigo, S. Ebihara and T. Yoshimura. 2007. Involvement of transforming growth factor alpha in the photoperiodic regulation of reproduction in birds. Endocrinology 148: 2788–2792.

Xu, Y., N. Tamamaki, T. Noda, K. Kimura, Y. Itokazu, N. Matsumoto, M. Dezawa and C. Ide. 2005. Neurogenesis in the ependymal layer of the adult rat 3rd ventricle. Exp. Neurol. 192: 251–264.

Yamamura, T., K. Hirunagi, S. Ebihara and T. Yoshimura. 2004. Seasonal morphological changes in the neuro-glial interaction between gonadotropin-releasing hormone nerve terminals and glial endfeet in Japanese quail. Endocrinology 145: 4264–4267.

CHAPTER 5

Alterations of Cerebral Endothelial Cells in Drug-Resistant Epilepsy: Drug Transporters

Cecilia Zavala-Tecuapetla[1] and *Liliana Carmona-Aparicio*[2]

Recent biochemical and molecular advances regarding the involvement of transporter systems in drug-resistant epilepsy therapeutics are considerable and controversial. This chapter focuses on the role and alterations of the transporter systems expressed in the cerebral endothelial cells that conform the blood brain barrier, in epileptogenic brain areas from humans and experimental models. We conclude with the relevance of transporter systems in drug-resistant epilepsy and their therapeutic implications.

5.1 Introduction

Cerebral endothelial cells are the principal cell types that comprise the blood-brain barrier. The blood-brain barrier is an active interface between the blood circulation and the central nervous system. The blood-brain barrier

[1] Laboratorio de Nanotecnología, Instituto Nacional de Neurología y Neurocirugía, Insurgentes Sur 3877, Col. La Fama, México D.F., 14269, México.
Email: cztecua@yahoo.com.mx

[2] Laboratorio de Neuroquímica; Instituto Nacional de Pediatría, Insurgentes Sur 3700-C, Col. Insurgentes Cuicuilco, Del. Coyoacán, México D.F. 04530, México.
Email: c_apariccio@ yahoo.com.mx

is integrated by brain microvascular endothelial cells, which restrict the free movement of different substances between the two compartments and play a crucial role in the maintenance of central nervous system homeostasis (Persidsky et al. 2006, Wilhelm et al. 2011). Although the main function of the blood-brain barrier is to protect the brain from toxic substances in the blood and to restrict the transport of potentially toxic or harmful substances from the bloodstream, it also functions as a carrier and is responsible for the transport of nutrients to the brain and the removal of metabolites (Persidsky et al. 2006, Dyer and Patterson 2010, Wilhelm et al. 2011).

The cerebral endothelial cells of the blood-brain barrier express a large number of carriers (ABC, ATP-binding cassette transporters), such as ABCB1 (P-gp, P-glycoprotein), ABCC1, ABCC4 and ABCG2, with the function of transporting nutrients to the brain and removing metabolites. Whereas small lipid-soluble molecules and the blood gases oxygen and carbon dioxide diffuse passively across the blood-brain barrier, essential polar nutrients such as glucose and amino acids require specific transport proteins (solute carriers, SLC) to cross the blood-brain barrier (Wilhelm et al. 2011). However, the blood-brain barrier is also a limiting factor for the delivery of therapeutic agents into the central nervous system (Wilhelm et al. 2011).

Alterations of the blood-brain barrier, such as increased impermeability or alterations of the transport systems, play an important role in the pathogenesis and therapeutics of central nervous system diseases, for example in Alzheimer's disease (Cirrito et al. 2005), Parkinson's disease (Kortekaas et al. 2005), ischemia (Spudich et al. 2006) and epilepsy (Dombrowski et al. 2001, Marroni et al. 2003, Volk et al. 2005, Kwan et al. 2010b).

Epilepsy is the most common chronic neurological disorder (Aiken and Brown 2000) and is characterized by recurrent seizures (Berg et al. 2010). Between 30 and 40 per cent of epileptic patients do not respond to pharmacotherapy (Regesta and Tanganelli 1999, Kwan and Brodie 2000, Brodie 2005) regardless of different antiepileptic drugs, with dissimilar pharmacokinetics and mechanisms of action (Zhang et al. 2012), administered at the proper dosing regimen (Kwan et al 2010a).

Drug-resistant epilepsy is one of the greatest challenges in epilepsy therapeutics. Studies carried out in human epileptogenic areas and in experimental models of drug-resistant epilepsy have shown overexpression of the blood-brain barrier transporter systems, for instance multidrug resistance-related proteins (MRP; Dombrowski et al. 2001, Aronica et al. 2004, Marchi et al. 2010), P-gp (Tishler et al. 1995, Rizzi et al. 2002, Marroni et al. 2003, Weiss et al. 2003, Aronica et al. 2004, Potschka et al. 2004, Volk et al. 2005, Kwan et al. 2010b), the multidrug transporter MRP-1 (Aronica et al. 2004) and MRP-2 (Dombrowski et al. 2001, Hoffmann et al. 2006) and the monocarboxylate transporter 1 (Lauritzen et al. 2011). However, it is unknown if alterations in the expression of transporter systems described in drug-resistant epilepsy are due to chronic exposure to anti-seizure drugs or if these changes participate in the pathological mechanisms that underlie drug-resistant epilepsy (Marroni et al. 2003, Volk et al. 2005, Marchi et al. 2010).

5.2 Endothelial Cells as a Part of the Blood-Brain Barrier

The blood-brain barrier is composed by the functional neurovascular unit, comprised by the microvascular endothelium, astrocytes, basement membrane, pericytes and neurons that are in physical proximity to the endothelium (Fig. 1) (Persidsky et al. 2006, Dyer and Patterson 2010).

Cerebral endothelial cells form a continuous sheet covering the inner surface of capillaries and are interconnected by tight junctions, which form a belt-like structure at the apical region of the cells. These have both endothelial-like features (i.e., expression of von Willebrand factor, uptake of acetylated low-density lipoprotein, high activity of alkaline phosphatase and gamma-glutamyl transpeptidase) and epithelial-like features (i.e., a continuous line of tight junctions, a low level of pinocytosis and a high trans-endothelial electrical resistance; Wilhelm et al. 2011).

Cerebral endothelial cells shield the brain from toxic substances in the blood, supply the brain with nutrients and filter harmful compounds from the brain to the bloodstream (Persidsky et al. 2006, Dyer and Patterson

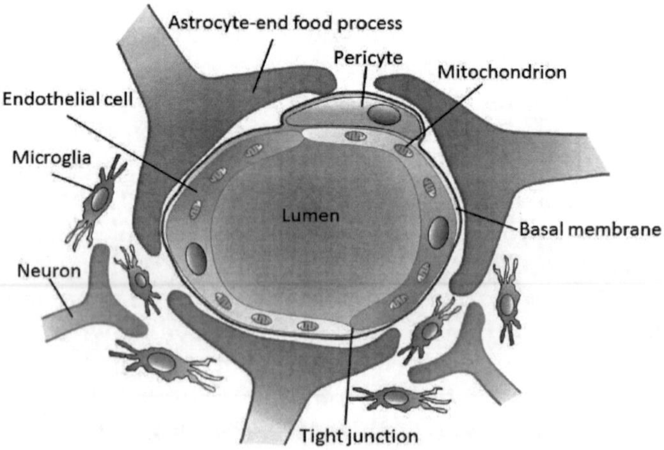

Figure 1. Scheme of the neurovascular unit. This unit is integrated basically by cerebral endothelial cells, the basal membrane, neurons, astrocytes and pericytes.

2010). The close interaction between cerebral endothelial cells and other components of the neurovascular unit ensures the proper functioning of the central nervous system. Transport across the blood-brain barrier is strictly limited through both physical (tight junctions) and metabolic barriers (enzymes, diverse transport systems; Persidsky et al. 2006, Dyer and Patterson 2010).

5.2.1 Distribution and Function of Endothelial Cells

The cerebral endothelial cells are situated at the interface between the blood and the brain and their principal biological functions are: (1) to act as a barrier; (2) transport of micronutrients and macronutrients; (3) receptor-mediated signaling; (4) leukocyte trafficking; (5) osmoregulation and (6) homeostasis (Doolittle et al. 2005, Persidsky et al. 2006). Cerebral endothelial cells are known to modulate neurogenesis, through secretion of soluble factors including nerve growth factor, brain-derived neurotrophic factor and vascular endothelial growth factor, suggesting a complex and dynamic cross-talk between endothelial cells and neuronal cells, including neural stem cells (Kim et al. 2004, Ward and Lamanna 2004, Li et al. 2006).

The structural components of cerebral endothelial cells responsible for their properties include the following: (1) tight junctions composed of proteins, for instance occludin, claudins, cingulin and zonula occludens; (2) adherens junctions composed of proteins, such as cadherins, catenins, vinculin and actinin; as well as (3) junctional adhesion molecules (Doolittle et al. 2005). Cerebral endothelial cells have more mitochondria than the endothelium of other organs because more energy is necessary for the active transport of nutrients to the brain. Indeed, it has been estimated that cerebral capillaries have five to six times more mitochondria per capillary section than rat skeletal muscle capillaries (Oldendorf et al. 1977).

On the other hand, the enzymatic barrier of the cerebral endothelium is capable of metabolizing drugs and nutrients. The principal enzymes involved are: (1) gamma-glutamyl transpeptidase; (2) alkaline phosphatase and (3) aromatic acid decarboxylase (Pardridge 2005).

5.3 Therapeutic Relevance of Transporter Systems on the Blood-Brain Barrier

The blood-brain barrier has a dual function, as a barrier as well as carrier (Wilhelm et al. 2011). The barrier function means that the blood-brain barrier restricts the transport of potentially toxic or harmful substances from the blood to the brain. The carrier function means that the blood-brain barrier conveys nutrients to the brain and removes metabolites. Endothelial cells of the blood-brain barrier express transport proteins that facilitate the passage of nutrients down their concentration gradients (Zlokovic 2008). Drug transport through the blood-brain barrier is the result of a complex and concerted action of different efflux and influx pumps. The principal blood-brain barrier transport systems include the superfamily of ABC and SLC proteins that mediate cellular extrusion of many therapeutic drugs with diverse structures and clinical applications (Schinkel and Jonker 2003, Girardin 2006).

5.3.1 Adenosine Triphosphate Binding Cassette Superfamily

ABC transporters are a large family of trans-membrane proteins that use the energy of ATP hydrolysis to transport a wide range of molecules across cellular membranes (Loscher and Potschka 2005b). Cerebral endothelial cells express a great variety of ABC transporters. These proteins target a large number of drugs, preventing them from reaching their central nervous system destination in adequate concentrations. Increasing evidence supports the role of ABC proteins in various drug-resistant brain disorders (Loscher and Potschka 2005c) and much attention has been focused on transporters, such as P-gp, which can transport a large variety of lipophilic drugs out of the cerebral endothelial cells (Demeule et al. 2002, Begley 2004).

In addition, other members of the ABC transporter superfamily may also significantly contribute to the prevention of drug transport across the blood-brain barrier. Cerebral endothelial cells have been shown to express other members of the MRP family of proteins (according to the new nomenclature the ABCC family), such as MRP1, MRP4, MRP5 and MRP6. Expression of breast cancer resistance protein (ABCG2) is also well documented (Loscher and Potschka 2005c, Shen and Zhang 2010). The expression profile of these transporters largely determines the permeability properties and functionality of the blood-brain barrier.

5.3.2 Solute Carrier Superfamily

The SLC superfamily of membrane transport proteins includes more than 300 members organized in 48 subfamilies. In general, they include organic cation and anion transporters (Kusuhara and Sugiyama 2005). The organic cation transport (OCT) system has two subtypes: the H^+ gradient-dependent transporters: OCTN1 (organic cation transporter, novel) plus OCTN2; and the voltage-sensitive transporters: OCT1, OCT2 and OCT3 (Koepsell et al. 2007). OCTs recognize a wide variety of endogenous and exogenous organic cations as substrates, including neurotoxins and monoamine neurotransmitters (Koepsell et al. 2007). Organic anions are transported by several carriers, such as the organic anion transporter polypeptide family (SLC21) and the organic anion transporter family (SLC22) (Lee et al. 2001, Mizuno et al. 2003).

Brain endothelial cells express high levels of other transporters, including the glucose transporter 1 (GLUT1, SLC2 family) which is of special importance because glucose is a key energy source for the brain; cationic amino acid transporters (SLC7 family members), monocarboxylate transporters (MCT or SLC16 family members, which transport lactate) and members of the proton oligopeptide transporters such as PEPT1 (peptide transporter 1), PEPT2, PHT1 (phosphate transporter 1) and PHT2 (SLC15) (Carl et al. 2010).

5.4 Role of Brain Endothelial Cells in Neurodegenerative Diseases: Transport Systems

Blood-brain barrier breakdown, neurovascular unit impairment or alterations in the expression of the transport systems play an important role in the pathogenesis of central nervous system diseases; such as human immunodeficiency virus encephalitis, epilepsy, ischemia, tumors, multiple sclerosis, Alzheimer's and Parkinson's (Marroni et al. 2003, Cirrito et al. 2005, Kortekaas et al. 2005, Volk et al. 2005, Persidsky et al. 2006, Spudich et al. 2006). Therefore, understanding the regulation of tight junctions and factors affecting the transport systems should allow the development of therapies to improve blood-brain barrier function (Persidsky et al. 2006).

Experimental evidence has associated neurological diseases to altered transporter expression. For example, P-gp expression has been inversely correlated to deposition of β-amyloid in Alzheimer's disease (Cirrito et al. 2005); it is up-regulated in brain capillary endothelial cells after focal cerebral ischemia (Spudich et al. 2006) and its functional activity is decreased in human brain tissue derived from Parkinson's patients (Kortekaas et al. 2005). GLUT1 is also altered in different neuronal diseases, such as brain trauma, ischemia, Alzheimer's disease and epilepsy (Guo et al. 2005).

5.5 Principal Alterations of Transport Systems in Brain Endothelial Cells Related to Epilepsy

5.5.1 Epilepsy: General Characteristics

Epilepsy is the most common serious chronic neurological disorder, affecting more than 50 million people worldwide (Aiken and Brown 2000). It is characterized by recurrent seizures and is broadly classified into two types:

focal or generalized epilepsy. Focal epilepsy causes seizures originating within networks limited to one hemisphere; while generalized epilepsy causes seizures originating at some point within, and rapidly engaging, bilaterally distributed networks (Berg et al. 2010). Despite the availability of many antiepileptic drugs 30–40 per cent of patients do not respond to pharmacotherapy (Regesta and Tanganelli 1999, Kwan and Brodie 2000); although antiepileptic drugs differ from each other in pharmacokinetics and mechanisms of action (Zhang et al. 2012).

Drug-resistant epilepsy has been defined as the failure to achieve sustained seizure freedom despite the use of two well tolerated and suitably chosen antiepileptic drug schedules; whether used as monotherapies or in combination (Kwan et al. 2010a). Antiepileptic drug resistance is recognized as one of the greatest challenges in epilepsy treatment today. Hippocampal sclerosis (a characteristic pattern of neuronal loss and gliosis), malformations of cortical development (such as focal cortical dysplasia) and brain tumors (particularly glial and glioneuronal tumors) are usually associated to medically intractable epilepsy and have a poorer pharmacological outcome compared to other symptomatic epilepsies (French 2007, Van Breemen et al. 2007).

Temporal lobe epilepsy is characterized by recurrent episodes of partial seizures that may or may not be secondarily generalized. Seizures in temporal lobe epilepsy involve a network of temporal lobe and limbic structures (Spencer 2002). Approximately 40 per cent of patients suffering from temporal lobe epilepsy cannot control their seizures with the current antiepileptic drugs. The development of more effective therapies against temporal lobe epilepsy is required. A better understanding of the underlying mechanisms of drug resistant temporal lobe epilepsy is likely to facilitate the discovery of novel therapeutic targets for this disease (Lauritzen et al. 2011). Since antiepileptic drugs must traverse the blood-brain barrier to enter the brain and exert their desired effects, the overexpression of drug efflux pumps in blood-brain barrier endothelial cells probably contributes to pharmacoresistance (Loscher and Potschka 2005a). It is important to note that cerebral endothelial cells are the most important cell types of the blood-brain barrier from the point of view of its permeability (Wilhelm et al. 2011).

5.5.2 Overexpression of Drug Transporters in Drug-Resistant Epilepsy

In recent years, attention has been focused on the role of drug transporters in the regulation of drug penetration into the brain, including the epileptic brain (Loscher and Potschka 2005a, Loscher 2007, Marchi et al. 2010). One possible mechanism that could account for resistance to pharmacological treatment with a broad range of antiepileptic drugs is inadequate drug concentrations in the epileptogenic areas (Aronica et al. 2012). Accordingly, overexpression of drug transporter proteins in the blood-brain barrier has been proposed as the mechanism responsible for multidrug resistance in epilepsy (Loscher and Potschka 2005a, Loscher 2007, Marchi et al. 2010). However, it is still unclear whether this overexpression is a result of chronic exposure to anti-seizure drugs or is part of the pathogenic mechanism that is responsible for drug-resistant epilepsy (Marroni et al. 2003, Volk et al. 2005) (Fig. 2).

The hypothesis that overexpression of ABC transporters is responsible for pharmacotherapeutic resistance is well established. In contrast, the current data provides no evidence of an association between ABCB1 variant expression and anticonvulsant drug responsiveness. This discrepancy may be explained by two phenomena: (1) the impact of ABCB1 variants on drug pharmacokinetics is weak, and no change of ABCB1 expression in brain tissue with respect to ABCB1 variants could be demonstrated, and (2) there is increasing evidence that most anticonvulsants are not or are only weak substrates of P-gp *in vitro*, although a few *in vivo* studies report an inverse correlation between some anticonvulsants and P-gp expression in the brain. Based on these findings, the hypothesis that P-gp expression participates in drug-resistant epilepsy requires revision. The role of genetic variants and the interplay between the various drug transporters require further elucidation. In particular, the role of epigenetic factors possibly superimposed on any

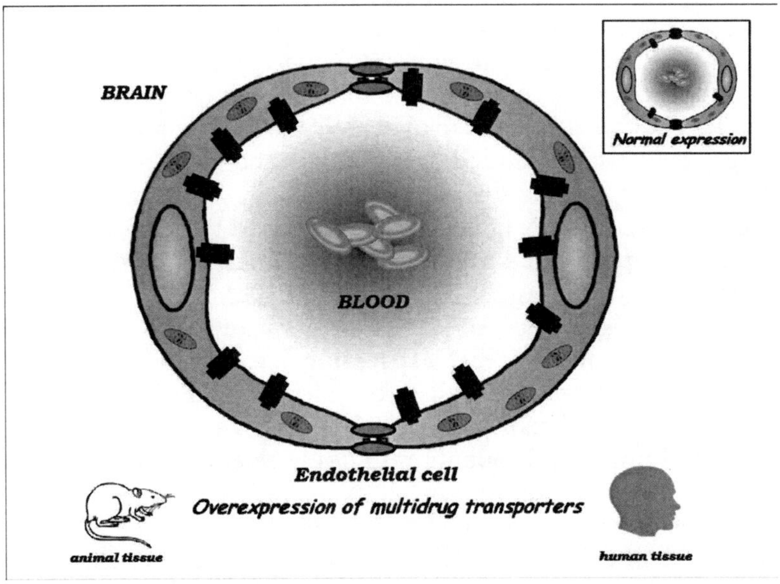

Figure 2. Schematic representation of drug transporter expression in normal (panel) and epileptic brains (observed in brain tissue from animal models and patients with drug-resistant epilepsy).

genetic effects must be taken into account when evaluating the participation of pharmacogenetics in drug-resistant epilepsy (Cascorbi 2010).

5.5.3 P-glycoprotein Transporter in Drug-Resistant Epilepsy

Most blood-brain barrier transporters are ABC proteins that mediate cellular extrusion of many therapeutic drugs with diverse structures and clinical applications. One of the most intensively studied multidrug transporters is P-gp, which plays a central role in drug absorption and distribution in many organisms (Zhang et al. 2012). Cerebral endothelial cells of the blood-brain barrier express high P-gp levels and its expression is increased in epileptic patients (Aronica et al. 2004, Loscher and Potschka 2005a, Loscher 2007). In addition, P-gp expression is higher in drug-resistant patients than in drug responsive patients (Tishler et al. 1995, Dombrowski et al. 2001). Some studies have indicated that several antiepileptic drugs are substrates or inhibitors of P-gp, implying that it plays an important role in drug-resistant epilepsy (Weiss et al. 2003, Luna-Tortos et al. 2008).

Increased expression of P-gp mRNA and protein was detected in capillary endothelial cells from surgically resected brain tissue belonging to patients with drug-resistant temporal lobe epilepsy; suggesting that P-gp expression may limit antiepileptic drug uptake, contributing to drug resistance (Tishler et al. 1995). Different models of temporal lobe epilepsy (status epilepticus induced by pilocarpine, kainate or electrical stimulation of the amygdala) also show over-expression of P-gp in brain endothelial cells, perivascular astrocytes, parenchymal astrocytes and neurons; supporting the complex role of P-gp, which involves different cell types (Rizzi et al. 2002, Potschka et al. 2004, Volk et al. 2004a,b). Even though surgery is recommended as a first line therapy for drug-resistant temporal lobe epilepsy, seizures can recur post-surgery in about one-third of the patients. It has been suggested that increased P-gp expression in the epileptogenic temporal lobe might serve as a marker for drug-resistant seizure recurrence post-surgery (Kwan et al. 2010b).

In addition to P-gp, other ABC transporters are also induced by seizure activity. Up-regulation of MRP2 expression has also been observed in endothelial, glial and neuronal cells after pilocarpine-induced convulsive status-epilepticus (Hoffmann et al. 2006). Moreover, cellular expression studies in both human and rodent brain tissue clearly show that different efflux transporters are expressed in endothelial cells and other parenchymal cells, such as astrocytes, microglia and neurons (Aronica et al. 2012).

Both clinical and experimental data show that the over-expression of multidrug transporter proteins is a common feature of different pathologies associated to drug-resistant epilepsy, including focal malformations of cortical development and temporal lobe epilepsy with hippocampal sclerosis (Aronica et al. 2012). Thus, increased knowledge about the complex regulation of drug transporters in different cell types raises new challenges for pharmacological studies aimed at preventing (or reducing) up-regulation of drug transporters in the blood-brain barrier (Aronica et al. 2012).

5.5.4 Monocarboxylate Transporter in Drug-Resistant Epilepsy

MCT1 is preferentially localized on the plasma membrane of endothelial cells. It is a key transporter of blood-derived mono-carboxylates (i.e., pyruvate, lactate and the ketone bodies: acetoacetate and beta-hydroxybutyrate) into the brain (Terasaki et al. 1991, Gerhart et al. 1997, Koehler-Stec et al. 1998, Bergersen et al. 1999, Leino et al. 1999, Pierre et al. 2000). Moreover, several studies have shown that MCT1 transports significant amounts of acidic antiepileptic drugs, such as valproic acid, across the blood-brain barrier (Kang et al. 1990, Utoguchi and Audus 2000, Fischer et al. 2008). In this regard, it seems that resistance to valproic acid in patients with mesial temporal lobe epilepsy may be due to a loss of microvessel-associated MCT1 in the epileptogenic hippocampus (Lauritzen et al. 2011). MCT1 is severely deficient in the endothelium of microvessels from the hippocampal formation in mesial temporal lobe epilepsy, compared to non-mesial temporal lobe epilepsy and neurologically normal autopsy samples. The observation that the loss of MCT1 is more pronounced in mesial temporal lobe epilepsy than in non-mesial temporal lobe epilepsy also raises the possibility that the mechanism of drug-resistance is different between temporal lobe epilepsy with and without hippocampal sclerosis (Lauritzen et al. 2011).

5.5.5 Non-ABC Transporter in Drug-Resistant Epilepsy

Recently, attention has been also focused on a non-ABC transporter, the stress-response protein, which functions as a multi-specific transporter of chemotherapeutic agents and glutathione conjugates (Awasthi et al. 2009, Singhal et al. 2009, Aronica et al. 2012). This transporter has been suggested to mediate resistance to antiepileptic drugs in patients with drug-resistant epilepsy (Awasthi et al. 2005). However, controversy exists about the expression of this protein at the level of the blood-brain barrier in the human brain (Janigro et al. 2007, Soranzo et al. 2007).

5.6 Conclusions

Drug resistance remains an unmet challenge in a variety of neurological disorders, but epilepsy is probably the refractory disease that has received the most experimental, preclinical and therapeutic attention. The introduction of many new antiepileptic drugs in the last two decades does not appear to have reduced substantially the proportion of people who are drug-resistant and continue to have seizures. Alterations of transporter systems expressed in the blood-brain barrier are relevant to understand the mechanisms that underlie drug-resistant epilepsy, although it is unknown if these changes are due to chronic exposure to antiepileptic drugs or are a part of the pathological mechanisms that underlie this type of epilepsy. Thus, it is necessary to investigate these alterations in order to clarify the origin of these changes. Increased knowledge about the complex regulation of drug transporters in different cell types raises new challenges for pharmacogenomic studies, aiming to study genetic predisposition to drug-resistance, as well as to pharmacological studies meant to prevent (or reduce) up-regulation of drug transporters in the blood-brain barrier.

However, altered expression of transporter systems, whatever its origin, does not fully explain drug-resistance. Thus, other factors may need to be considered. In this regard, new hypotheses have been proposed. One theory suggests that impaired mitochondrial energy production is the basis of drug-resistance in epilepsy (Yuen and Sander 2011). Another novel conceptual strategy that may be used for the control of drug-resistant seizures is the modulation of the homeostatic bioenergetic network, affecting whole neuronal networks synergistically by mobilizing multiple endogenous biochemical and receptor-dependent molecular pathways (Boisson et al. 2011).

Acknowledgments

We thank Aristides III Sampieri-Hernández, from the National Autonomous University of Mexico, for his technical assistance in the preparation of this manuscript.

References

Aiken, S.P. and W.M. Brown. 2000. Treatment of epilepsy: existing therapies and future developments. Front. Biosci. 5: E124–E152.

Aronica, E., J.A. Gorter, M. Ramkema, S. Redeker, F. Ozbas-Gerceker, E.A. van Vliet, G.L. Scheffer, R.J. Scheper, P. van der Valk, J.C. Baayen and D. Troost. 2004. Expression and cellular distribution of multidrug resistance-related proteins in the hippocampus of patients with mesial temporal lobe epilepsy. Epilepsia 45: 441–451.

Aronica, E., S.M. Sisodiya and J.A. Gorter. 2012. Cerebral expression of drug transporters in epilepsy. Adv. Drug Deliv. Rev. 64: 919–929.
Awasthi, S., K.L. Hallene, V. Fazio, S.S. Singhal, L. Cucullo, Y.C. Awasthi, G. Dini and D. Janigro. 2005. RLIP76, a non-ABC transporter, and drug resistance in epilepsy. BMC Neurosci. 6: 61.
Awasthi, Y.C., P. Chaudhary, R. Vatsyayan, A. Sharma, S. Awasthi and R. Sharma. 2009. Physiological and pharmacological significance of glutathione-conjugate transport. J. Toxicol. Environ. Health B Crit. Rev. 12: 540–551.
Begley, D.J. 2004. ABC transporters and the blood-brain barrier. Curr. Pharm. Des. 10: 1295–1312.
Berg, A.T., S.F. Berkovic, M.J. Brodie, J. Buchhalter, J.H. Cross, W. van Emde Boas, J. Engel, J. French, T.A. Glauser, G.W. Mathern, S.L. Moshe, D. Nordli, P. Plouin and I.E. Scheffer. 2010. Revised terminology and concepts for organization of seizures and epilepsies: report of the ILAE Commission on Classification and Terminology, 2005–2009. Epilepsia 51: 676–685.
Bergersen, L., E. Johannsson, M.L. Veruki, E.A. Nagelhus, A. Halestrap, O.M. Sejersted and O.P. Ottersen. 1999. Cellular and subcellular expression of monocarboxylate transporters in the pigment epithelium and retina of the rat. Neuroscience 90: 319–331.
Boison, D., S.A. Masino and J.D. Geiger. 2011. Homeostatic bioenergetic network regulation —a novel concept to avoid pharmacoresistance in epilepsy. Expert Opin. Drug Discov. 6: 713–724.
Brodie, M.J. 2005. Diagnosing and predicting refractory epilepsy. Acta Neurol. Scand. Suppl. 181: 36–39.
Carl, S.M., D.J. Lindley, P.O. Couraud, B.B. Weksler, I. Romero, S.A. Mowery and G.T. Knipp. 2010. ABC and SLC transporter expression and pot substrate characterization across the human CMEC/D3 blood-brain barrier cell line. Mol. Pharm. 7: 1057–1068.
Cascorbi, I. 2010. ABC transporters in drug-refractory epilepsy: limited clinical significance of pharmacogenetics? Clin. Pharmacol. Ther. 87: 15–18.
Cirrito, J.R., R. Deane, A.M. Fagan, M.L. Spinner, M. Parsadanian, M.B. Finn, H. Jiang, J.L. Prior, A. Sagare, K.R. Bales, S.M. Paul, B.V. Zlokovic, D. Piwnica-Worms and D.M. Holtzman. 2005. P-glycoprtein deficiency at the blood–brain barrier increases amyloid-β deposition in an Alzheimer disease mouse model. J. Clin. Invest. 115: 3285–3290.
Demeule, M., A. Regina, J. Jodoin, A. Laplante, C. Dagenais, F. Berthelet, A. Moghrabi and R. Beliveau. 2002. Drug transport to the brain: key roles for the efflux pump P-glycoprotein in the blood-brain barrier. Vascul. Pharmacol. 38: 339–348.
Dombrowski, S.M., S.Y. Desai, M. Marroni, L. Cucullo, K. Goodrich, W. Bingaman, M.R. Mayberg, L. Bengez and D. Janigro. 2001. Overexpression of multiple drug resistance genes in endothelial cells from patients with refractory epilepsy. Epilepsia 4: 1501–1506.
Doolittle, N.D., L.E. Abrey, W.A. Bleyer, S. Brem, T.P. Davis, P. Dore-Duffy, L.R. Drewes, W.A. Hall, J.M. Hoffman, A. Korfel, R. Martuza, L. Muldoon, D. Peereboom, D.R. Peterson, S.D. Rabkin, Q. Smith, G.H. Stevens and E.A. Neuwelt. 2005. New frontiers in translational research in neuro-oncology and the blood–brain barrier: report of the tenth annual Blood–Brain Barrier Disruption Consortium Meeting. Clin. Cancer Res. 11: 421–428.
Dyer, L.A. and C. Patterson. 2010. Development of the endothelium: an emphasis on heterogeneity. Semin. Thromb. Hemost. 36: 227–235.
Fischer, W., K. Praetor, L. Metzner, R.H. Neubert and M. Brandsch. 2008. Transport of valproate at intestinal epithelial (Caco-2) and brain endothelial (RBE4) cells: mechanism and substrate specificity. Eur. J. Pharm. Biopharm. 70: 486–492.
French, J.A. 2007. Refractory epilepsy: clinical overview. Epilepsia 48: 3–7.
Gerhart, D.Z., B.E. Enerson, O.Y. Zhdankina, R.L. Leino and L.R. Drewes. 1997. Expression of monocarboxylate transporter MCT1 by brain endothelium and glia in adult and suckling rats. Am. J. Physiol. 273: E207–E213.
Girardin, F. 2006. Membrane transporter proteins: a challenge for CNS drug development. Dialogues Clin. Neurosci. 8: 311–321.

Guo, X., M. Geng and G. Du. 2005. Glucose transporter 1, distribution in the brain and in neural disorders: its relationship with transport of neuroactive drugs through the blood–brain barrier. Biochem. Genet. 43: 175–187.

Hoffmann, K., A.M. Gastens, H.A. Volk and W. Loscher. 2006. Expression of the multidrug transporter MRP2 in the blood–brain barrier after pilocarpine-induced seizures in rats. Epilepsy Res. 69: 1–14.

Janigro, D., S. Awasthi, Y.C. Awasthi, R. Sharma, S. Yadav, S.S. Singhal and K. Hallene. 2007. RLIP76 in AED drug resistance. Epilepsia 48: 1218–1219.

Kang, Y.S., T. Terasaki and A. Tsuji. 1990. Acidic drug transport *in vivo* through the blood-brain barrier. A role of the transport carrier for monocarboxylic acids. J. Pharmacobiodyn. 13: 158–163.

Kim, H., Q. Li, B.L. Hempstead and J.A. Madri. 2004. Paracrine and autocrine functions of brain-derived neurotrophic factor (BDNF) and nerve growth factor (NGF) in brain-derived endothelial cells. J. Biol. Chem. 279: 33538–33546.

Koehler-Stec, E.M., I.A. Simpson, S.J. Vannucci, K.T. Landschulz and W.H. Landschulz. 1998. Monocarboxylate transporter expression in mouse brain. Am. J. Physiol. 275: E516–E524.

Koepsell, H., K. Lips and C. Volk. 2007. Polyspecific organic cation transporters: structure, function, physiological roles, and biopharmaceutical implications. Pharm. Res. 24: 1227–1251.

Kortekaas, R., K.L. Leenders, J.C. van Oostrom, W. Vaalburg, J. Bart, A.T. Willemsen and N.H. Hendrikse. 2005. Blood–brain barrier dysfunction in parkinsonian midbrain *in vivo*. Ann. Neurol. 57: 176–179.

Kusuhara, H. and Y. Sugiyama. 2005. Active efflux across the blood–brain barrier: role of the solute carrier family. NeuroRx 2: 73–85.

Kwan, P. and M.J. Brodie. 2000. Epilepsy after the first drug fails: substitution or add-on? Seizure 9: 464–468.

Kwan, P., A. Arzimanoglou, A.T. Berg, M.J. Brodie, W. Allen Hauser, G. Mathern, S.L. Moshe, E. Perucca, S. Wiebe and J. French. 2010a. Definition of drug resistant epilepsy: consensus proposal by the ad hoc Task Force of the ILAE Commission on Therapeutic Strategies. Epilepsia 51: 1069–1077.

Kwan, P., H.M. Li, E. Al-Jufairi, R. Abdulla, M. Gonzales, A.H. Kaye, C. Szoeke, H.K. Ng, K.S. Wong and T.J. O'Brien. 2010b. Association between temporal lobe P-glycoprotein expression and seizure recurrence after surgery for pharmacoresistant temporal lobe epilepsy. Neurobiol. Dis. 39: 192–197.

Lauritzen, F., N.C. de Lanerolle, T.S. Lee, D.D. Spencer, J.H. Kim, L.H. Bergersen and T. Eid. 2011. Monocarboxylate transporter 1 is deficient on microvessels in the human epileptogenic hippocampus. Neurobiol. Dis. 41: 577–584.

Lee, G., S. Dallas, M. Hong and R. Bendayan. 2001. Drug transporters in the central nervous system: brain barriers and brain parenchyma considerations. Pharmacol. Rev. 53: 569–596.

Leino, R.L., D.Z. Gerhart and L.R. Drewes. 1999. Monocarboxylate transporter (MCT1) abundance in brains of suckling and adult rats: a quantitative electron microscopic immunogold study. Brain Res. Dev. Brain Res. 113: 47–54.

Li, Q., M.C. Ford, E.B. Lavik and J.A. Madri. 2006. Modeling the neurovascular niche: VEGF- and BDNF-mediated cross-talk between neural stem cells and endothelial cells: An *in vitro* study. J. Neurosci. Res. 84: 1656–1668.

Loscher, W. 2007. Drug transporters in the epileptic brain. Epilepsia 48: 8–13.

Loscher, W. and H. Potschka. 2005a. Drug resistance in brain diseases and the role of drug efflux transporters. Nat. Rev. Neurosci. 6: 591–602.

Löscher, W. and H. Potschka. 2005b. Role of drug efflux transporters in the brain for drug disposition and treatment of brain diseases. Prog. Neurobiol. 76: 22–76.

Loscher, W. and H. Potschka. 2005c. Blood–brain barrier active efflux transporters: ATP binding cassette gene family. NeuroRx 2: 86–98.

Luna-Tortos, C., M. Fedrowitz and W. Loscher. 2008. Several major antiepileptic drugs are substrates for human P-glycoprotein. Neuropharmacology 55: 1364–1375.

Marchi, N., J. Gonzalez-Martinez, M.T. Nguyen, T. Granata and D. Janigro. 2010. Transporters in drug-refractory epilepsy: clinical significance. Clin. Pharmacol. Ther. 87: 13–15.

Marroni, M., N. Marchi, L. Cucullo, N.J. Abbott, K. Signorelli and D. Janigro. 2003. Vascular and parenchymal mechanisms in multiple drug resistance: a lesson from human epilepsy. Curr. Drug Targets 4: 297–304.

Mizuno, N., T. Niwa, Y. Yotsumoto and Y. Sugiyama. 2003. Impact of drug transporter studies on drug discovery and development. Pharmacol. Rev. 55: 425–461.

Oldendorf, W.H., M.E. Cornford and W.J. Brown. 1977. The large apparent work capability of the blood–brain barrier: a study of the mitochondrial content of capillary endothelial cells in brain and other tissues of the rat. Ann. Neurol. 1: 409–417.

Pardridge, W.M. 2005. Molecular biology of the blood–brain barrier. Mol. Biotechnol. 30: 57–70.

Persidsky, Y., S.H. Ramirez, J. Haorah and G.D. Kanmogne. 2006. Blood–brain barrier: structural components and function under physiologic and pathologic conditions. J. Neuroimmune Pharmacol. 1: 223–236.

Pierre, K., L. Pellerin, R. Debernardi, B.M. Riederer and P.J. Magistretti. 2000. Cell-specific localization of monocarboxylate transporters, MCT1 and MCT2, in the adult mouse brain revealed by double immunohistochemical labeling and confocal microscopy. Neuroscience 100: 617–627.

Potschka, H., H.A. Volk and W. Loscher. 2004. Pharmacoresistance and expression of multidrug transporter P-glycoprotein in kindled rats. Neuroreport 15: 1657–1661.

Regesta, G. and P. Tanganelli. 1999. Clinical aspects and biological bases of drug-resistant epilepsies. Epilepsy Res. 34: 109–122.

Rizzi, M., S. Caccia, G. Guiso, C. Richichi, J.A. Gorter, E. Aronica, M. Aliprandi, R. Bagnati, R. Fanelli, M. D'Incalci, R. Samanin and A. Vezzani. 2002. Limbic seizures induce P-glycoprotein in rodent brain: functional implications for pharmacoresistance. J. Neurosci. 22: 5833–5839.

Schinkel, A.H. and J.W. Jonker. 2003. Mammalian drug efflux transporters of the ATP binding cassette (ABC) family: an overview. Adv. Drug Deliv. Rev. 55: 3–29.

Shen, S. and W. Zhang. 2010. ABC transporters and drug efflux at the blood-brain barrier. Rev. Neurosci. 21: 29–53.

Singhal, S.S., S. Yadav, C. Roth and J. Singhal. 2009. RLIP76: a novel glutathione-conjugate and multi-drug transporter. Biochem. Pharmacol. 77: 761–769.

Soranzo, N., L. Kelly, L. Martinian, M.W. Burley, M. Thom, A. Sali, D.L. Kroetz, D.B. Goldstein and S.M. Sisodiya. 2007. Lack of support for a role for RLIP76 (RALBP1) in response to treatment or predisposition to epilepsy. Epilepsia 48: 674–683.

Spencer, S.S. 2002. Neural networks in human epilepsy: evidence of and implications for treatment. Epilepsia 43: 219–227.

Spudich. A., E. Kilic, H. Xing, U. Kilic, K.M. Rentsch, H. Wunderli-Allenspach, C.L. Bassetti and D.M. Hermann. 2006. Inhibition of multidrug resistance transporter-1 facilitates neuroprotective therapies after focal cerebral ischemia. Nat. Neurosci. 9: 487–488.

Terasaki, T., S. Takakuwa, S. Moritani and A. Tsuji. 1991. Transport of monocarboxylic acids at the blood-brain barrier: studies with monolayers of primary cultured bovine brain capillary endothelial cells. J. Pharmacol. Exp. Ther. 258: 932–937.

Tishler, D.M., K.I. Weinberg, D.R. Hinton, N. Barbaro, G.M. Annett and C. Raffel. 1995. MDR1 gene expression in brain of patients with medically intractable epilepsy. Epilepsia 36: 1–6.

Utoguchi, N. and K.L. Audus. 2000. Carrier-mediated transport of valproic acid in BeWo cells, a human trophoblast cell line. Int. J. Pharm. 195: 115–124.

Van Breemen, M.S., E.B. Wilms and C.J. Vecht. 2007. Epilepsy in patients with brain tumours: epidemiology, mechanisms, and management. Lancet Neurol. 6: 421–430.

Volk, H.A., K. Burkhardt, H. Potschka, J. Chen, A. Becker and W. Loscher. 2004a. Neuronal expression of the drug efflux transporter P-glycoprotein in the rat hippocampus after limbic seizures. Neuroscience 123: 751–759.

Volk, H.A., H. Potschka and W. Loscher. 2004b. Increased expression of the multidrug transporter P-glycoprotein in limbic brain regions after amygdala-kindled seizures in rats. Epilepsy Res. 58: 67–79.

Volk, H., H. Potschka and W. Loscher. 2005. Immunohistochemical localization of P-glycoprotein in rat brain and detection of its increased expression by seizures are sensitive to fixation and staining variables. J. Histochem. Cytochem. 53: 517–531.

Ward, N.L. and J.C. Lamanna. 2004. The neurovascular unit and its growth factors: coordinated response in the vascular and nervous systems. Neurol. Res. 26: 870–883.

Weiss, J., C.J. Kerpen, H. Lindenmaier, S.M. Dormann and W.E. Haefeli. 2003. Interaction of antiepileptic drugs with human P-glycoprotein *in vitro*. J. Pharmacol. Exp. Ther. 307: 262–267.

Wilhelm, I., C. Fazakas and I.A. Krizbai. 2011. *In vitro* models of the blood-brain barrier. Acta Neurobiol. Exp. 71: 113–128.

Yuen, A.W. and J.W. Sander. 2011. Impaired mitochondrial energy production: the basis of pharmacoresistance in epilepsy. Med. Hypotheses 77: 536–540.

Zhang, C., P. Kwan, Z. Zuo and L. Baum. 2012. The transport of antiepileptic drugs by P-glycoprotein. Adv. Drug Deliv. Rev. 64: 930–942.

Zlokovic, B.V. 2008. The blood–brain barrier in health and chronic neurodegenerative disorders. Neuron 57: 178–201.

CHAPTER 6

Angiogenesis During Tumor Proliferation and Cell Invasion

América Vanoye Carlo,[1] *Gisela Gómez Lira,*[2] *Julieta G. Mendoza Torreblanca*[3] and *Óscar A. Pérez González*[4,*]

6.1 Introduction

Recent data suggest that tumor-associated neovascularization is a central process for the growth, invasion and metastasis of brain malignancies. Neovascularization involves multiple steps and pathways dependent on the local balance between positive and negative regulatory factors, as well as interactions between the tumor, its vasculature and the surrounding extracellular tissue matrix. When a tumor remains in its dormant state the cellular proliferation rate is balanced by the apoptotic rate; thus, it is unable to grow in size beyond a few millimeters in the absence of the acquired angiogenic phenotype. The mechanism by which tumors switch to the angiogenic phenotype has not been completely elucidated.

[1] Laboratorio de Bioquímica Genética; Instituto Nacional de Pediatría, Insurgentes Sur 3700-C, Col. Insurgentes Cuicuilco, Del. Coyoacán, México D.F. 04530, México.
Email: america_vc@yahoo.com.mx

[2] Depto. de Farmacobiología, CINVESTAV, Calzada de los Tenorios 235, Col. Granjas Coapa, México D.F., 14330, México.
Email: glira@cinvestav.mx

[3] Laboratorio de Neuroquímica; Instituto Nacional de Pediatría, Insurgentes Sur 3700-C, Col. Insurgentes Cuicuilco, Del. Coyoacán, México D.F. 04530, México.
Email: julietamt14@hotmail.com

[4] Laboratorio de Oncología Experimental; Instituto Nacional de Pediatría, Insurgentes Sur 3700-C, Col. Insurgentes Cuicuilco, Del. Coyoacán, México D.F. 04530, México.
Email: dr.oscarperez@gmail.com

* Corresponding author

Nevertheless, angiogenesis is markedly involved in tumor growth and dissemination. Angiogenesis requires not only malignant and endothelial cells, but also other cell types, as well as matrix components. This process involves several molecules that may have pro-angiogenic or anti-angiogenic actions, interacting with several growth factors and cytokines secreted in the brain tissue. Angiogenesis is also a very important mechanism for the establishment and growth of secondary tumors. This chapter contains information about the molecular biology underlying expansion and invasion of primary brain tumors. It also describes tumor angiogenesis in a detailed manner, emphasizing the participation of the main pro-angiogenic and anti-angiogenic factors in each step of neovascularization, as well as tumoral cell interactions with the extracellular matrix (ECM) and endothelial cells. In addition, we approach the implications of recent advances in brain tumor angiogenesis and the development of new therapeutics tools. In this context, therapeutic agents and strategies are being devised either to interrupt or inhibit one or more of the pathogenic steps involved in tumor neovascularization or to directly target and destroy tumor vasculature.

6.2 Primary and Secondary Central Nervous System Tumors

Central nervous system tumors (CNST) have important clinical impact due to their high morbidity and mortality. Treatment protocols for these tumors use a multimodal scheme that includes surgery, radiation therapy and, in some cases, chemotherapy depending on the tumoral behavior. Unfortunately, the response rate has not improved considerably. Even non-malignant tumors that do not contain cancer cells and grow slowly may cause complications due to their size or their implantation site and might represent a clinical challenge (Rivera Luna 2006, American Cancer Society; ACS 2012, National Cancer Institute; NCI 2012). Conversely, malignant tumors have cancer cells that grow rapidly within nearby tissues by invasion and some spread far from their site of origin. Sometimes, their characteristics complicate complete surgical resection and contribute to therapeutic failure and poor patient prognosis (ACS 2012, NCI 2012).

Tumors originating from neurons and glial cells are considered primary tumors; while tumors originating from other organs such as the lung, breast or melanoma are considered secondary or metastatic tumors when they invade the central nervous system (CNS). The frequency of CNST is age-related; pediatric patients predominantly have primary tumors, adult patients, in contrast, mostly present secondary or metastatic tumors (NCI 2012).

The World Health Organization (WHO) classifies primary CNST based on their cellular origin, histological appearance and degree of malignancy (Table 1, Louis et al. 2007). Their cellular origin determines their name;

Table 1. World Health Organization Grading of Central Nervous System Tumors.

	I	II	III	IV		I	II	III	IV
1) Astrocytic tumors					7) Neuronal and mixed neuronal-glial tumors				
Subependymal giant cell astrocytoma	•				Central neurocytoma		•		
Pilocytic astrocytoma	•				Extraventricular neurocytoma		•		
Pilomyxoid astrocytoma		•			Cerebellar liponeurocytoma		•		
Diffuse astrocytoma		•			Paraganglioma of the spinal cord	•			
Pleomorphic xanthoastrocytoma		•			Papillary glioneuronal tumor	•			
Anaplastic astrocytoma			•		Rosette-forming glioneuronal tumor of the 4th ventricle	•			
Glioblastoma				•					
Giant cell glioblastoma				•	8) Pineal tumors				
Gliosarcoma				•	Pineocytoma	•			
2) Oligodendroglial tumors					Pineal parenchymal tumor of intermediate differentiation		•	•	
Oligodendroglioma		•			Pineoblastoma				•
Anaplastic oligodendroglioma			•		Papillary tumor of the pineal region		•	•	
Oligoastrocytic tumors					9) Embryonic tumors				
Oligoastrocytoma		•			Medulloblastoma				•
Anaplastic oligoastrocytoma			•		Primitive neuroectodermal tumor				•
3) Ependymal tumors					Atypical teratoid or rhabdoid				•
Subependymoma	•				10) Tumors of the cranial and paraspinal nerves				
Myxopapillary ependymoma	•				Schwannoma	•			
Ependymoma		•			Neurofibroma	•			
Anaplastic ependymoma			•		Perineurioma	•	•	•	
4) Choroid plexus tumors					Malignant peripheral nerve sheath tumor		•	•	•
Choroid plexus papilloma	•				11) Meningeal tumors				
Atypical choroid plexus papilloma		•			Meningioma	•			

Table 1. contd....

Table 1. contd....

	I	II	III	IV		I	II	III	IV
Choroid plexus carcinoma			•		Atypical meningioma		•		
5) Other neuroepithelial tumors					Anaplastic or malignant meningioma			•	
Angiocentric glioma	•				Haemangiopericytoma		•		
Chordoid glioma of the third ventricle		•			Anaplastic haemangiopericytoma			•	
6) Neuronal and mixed neuronal-glial tumors					Haemangioblastoma	•			
Gangliocytoma	•				12) Tumors of the sellar region				
Ganglioglioma	•				Craniopharyngioma	•			
Anaplastic ganglioglioma			•		Granular cell tumor of the neurohypophysis	•			
Desmoplastic infantile astrocytoma and ganglioglioma	•				Pituicytoma	•			
Dysembryoplastic neuroepithelial tumor	•				Spindle cell oncocytoma of the adenohypophysis	•			

thus, tumors developed from astrocytes are astrocytomas or glioblastomas, those that are generated from oligodendrocytes are oligodendrogliomas, those that originate from the ependymal cells are ependymomas, etc. The histological classification predicts the biological behavior of a neoplasia according to a malignancy grading system. In the clinical setting, this is a key factor for treatment protocol assignment and is defined as follows: Grade I are low proliferative lesions with the possibility of cure after surgical resection; Grade II tumors are infiltrative in nature and often recur, despite their low-level of proliferative activity (some type II tumors tend to progress to higher levels of cancer); Grade III tumors have histological evidence of malignancy, including nuclear atypia and brisk mitotic activity; and Grade IV tumors are cytologically malignant, mitotically active, necrosis-prone neoplasms typically associated with rapid disease progression and fatal outcomes (Louis et al. 2007, Chandana et al. 2008). Regarding metastatic tumors, they always get the name of the place where they originated no matter where the cancer cell came from. For example, a lung tumor that has spread to the brain is named a metastatic lung tumor (ACS 2012) and its prognosis is related to the primary tumor site.

6.3 Epidemiology

The frequency of CNST changes considerably according to the age of patients; whereas in adults primary CNST represent less than two per cent of all cancers, they are the second most common cancer during childhood and adolescence (Rivera Luna 2006). In addition, compared to other types of cancer, they have a very high impact on the quality of life (Burnet et al. 2005). The WHO International Agency for Research on Cancer has reported one of most important and complete records regarding cancer incidence, prevalence and mortality (GLOBOCAN 2008; shown partly in Table 2). CNST incidence is highest in Europe and the Unites States of America. Mortality rates show the same tendency; however, there is a narrow difference in the mortality rates between developed nations and developing countries.

CNST also show others important epidemiological differences between children and adults. The frequency and distribution of CNST, according to location and histology, differs between adults and pediatric patients as reported by the Central Brain Tumor Registry of the United States (CBTRUS; Table 3). The most common location for CNST in adults is the meninges, followed by the cerebral lobes, whereas in children CNST is most frequent in the cerebral lobes followed by the cerebellum. Histologically, the most general type of tumor in adults is non-malignant meningioma followed by glioblastoma (also called glioblastoma multiforme, GBM). On the other hand, during childhood the most widespread type of tumor by histology is pilocytic astrocytoma followed by embryonic or medulloblastoma tumors. In adults, gliomas represent 30 per cent of all tumors and 80 per cent of malignant tumors. Meanwhile, in pediatric patients they represent 46 per cent and 71 per cent respectively (CBTRUS 2012). CNST incidence also

Table 2. Estimated Incidence, Mortality and Five-Year Prevalence of Brain Cancer and Other Cancers of the Nervous System (per 100,000 individuals; GLOBOCAN 2008).

	Incidence (for all ages)		Mortality (for all ages)		Five Year Prevalence (adult population)	
	Men	Women	Men	Women	Men	Women
Worldwide	3.8	3.1	2.9	2.2	1.3	0.9
Europe	6.2	4.6	4.6	3.1	0.8	0.6
Africa	1.7	1.1	1.3	0.9	1.4	0.7
Asia	3.3	3.0	2.6	2.0	1.8	1.3
Latin America Caribbean	4.1	3.1	3.5	2.7	1.7	1.1
United States	6.3	4.9	3.6	2.3	0.8	0.6

Table 3. Distribution of Primary (Malignant and Non-malignant) Central Nervous System Tumors at Different Ages, by Site and Histology (CBTRUS 2012) (CNS = central nervous system).

Site	All ages (%)	Ages 0–19 (%)	Histology	All ages (%)	Ages 0–19 (%)
Meninges	34.3	2.8	Meningioma, non-malignant	34.7	3.1
All lobes	21.7	17.1	Glioblastoma	16.3	2.9
Pituitary	14.7	12.2	Pituitary	13.5	12.6
Craneal nerves	6.9	6.2	Nerve sheath	8.5	5.5
Spinal cord and cauda equina	3.2	4.8	Astrocytoma	6.8	23.2
Cerebellum	2.9	16.7	Other neuroepitelial	5.0	-
Cerebrum	2.0	5.6	Lymphoma	2.3	0.4
Brain stem	1.6	10.7	Oligodendrogliomas	1.9	1.8
Ventricle	1.2	5.8	Ependymomas	1.8	5.1
Pineal	0.5	2.9	Embryonal or medulloblastoma	1.1	9.7
Nasal Cavity	0.2	0.1	Craniopharyngioma	0.9	3.5
Other brain	10.2	13.6	Germ cell	0.5	4.4
Other CNS	0.6	1.6	All other	6.7	9.4

varies by age group, being highest among individuals over 85 years (71.06 per 100,000) and lowest among children aged 0–19 years (5.05 per 100,000) (Bondy et al. 2008, ACS 2012, CBTRUS 2012).

Finally, the relative survival rates for malignant brain tumors at five and 10 years are 33.7 per cent and 27.9 per cent respectively. However, there is an important variation in survival estimates depending on tumor histology. For example, five-year survival rates are 94 per cent for pilocytic astrocytomas but less than 5 per cent for glioblastomas. CBTRUS reported an age-related decrease in overall survival rates, since children and young adults have better survival rates for most histologies (Bondy et al. 2008, ACS 2012, CBTRUS 2012).

On the other hand, secondary or metastatic tumors are most frequent in adult patients. In the United States, their incidence is estimated at about 170,000 cases per year (Nathoo et al. 2005), around 10 times higher than primary tumors (Nathoo et al. 2005, Preusser et al. 2012, Siegel 2012). Brain metastasis is the most common diagnosis in patients with cancer who have had neurological complications and is a major cause of morbidity and mortality. It is estimated that 17 per cent of patients with systemic cancer develop brain metastases, although this is probably an under-estimate as a consequence of low-resolution diagnostic tools, which are useless in detecting micro-metastasis (Schouten et al. 2002, Nayak et al. 2012). In adults, several studies report that lung cancer (41–56 percent), breast cancer (13–30 per cent), melanoma (6–10 per cent), gastrointestinal cancer (6–9 per cent) and kidney cancer (2–6 per cent) are the most frequent cancers that develop brain metastases, accounting for 67–80 per cent of all cancers. Other cancers, such as prostate or colorectal cancer rarely metastasize to the CNS. In children, brain metastases are very rare (Barnholtz -Sloan et al. 2004, Nayak et al. 2012, see also Preusser et al. 2012).

6.4 Brain Tumor Growth Patterns

Brain tumor growth patterns are useful features, in addition to histological criteria, to determine the malignant grade of these neoplasms. Scherer in 1940, and later Daumas-Duport in 1982, described forms of tumor growth and discussed how these tumors interact with the surrounding parenchyma in different ways. Scherer designated three main patterns of growth—expansive, mixed and infiltrative—each with different prognostic significance (Scherer 1940). Daumas-Duport described a similar classification scheme that considered a tumor tissue nidus without surrounding parenchymal tumor cell infiltration, tumors including a nidus with parenchymal tumor cell infiltration and tumors having diffuse infiltration without a nidus. These types of tumors can be equivalent to expansive, mixed and infiltrative tumors, respectively (Moulding et al. 2008).

Pilocytic astrocytomas (WHO grade I) are representative examples of expansive tumor growth; presenting as circumscribed slowly growing lesions which very rarely undergo progression to anaplasia (Kleihues et al. 1995, Wesseling et al. 1997). In contrast, irrespective of grade, diffuse infiltrative astrocytomas show a propensity to invade the surrounding brain parenchyma and spread along blood vessels (Wesseling et al. 1997, Walker et al. 2011). These neoplasms can be divided arbitrarily into three groups of increasing malignancy: astrocytoma (WHO grade II), anaplastic astrocytoma (WHO grade III), and GBM (WHO grade IV) (Kleihues et al. 1995, Walker et al. 2011). Pathological examination demonstrates that low-grade astrocytomas begin as non-angiogenic tumors that have the ability to co-opt their blood supply from the existing vasculature. While in GBM angiogenesis induces micro-vascular proliferation (MVP) and begins the phase of malignant tumor growth (Kaur et al. 2005).

Florid MVP and necrosis emerge in gliomas of high-grade malignancy. The most prominent form of MVP is 'glomeruloid' MVP; named due to the formation of coiled, glomerulus-like capillary loops (Wesseling et al. 1997, Claes et al. 2007). Glomeruloid MVP is classically associated with GBM. In the current histopathological grading of diffuse astrocytic neoplasms, florid

MVP is an important criterion to distinguish high-grade tumors from their low-grade counterparts (Claes et al. 2007). Florid MVP is a sign of malignant progression in most gliomas, except in pilocytic astrocytomas where it does not have prognostic significance. Other brain tumors may show particular micro-vascular architectures as well. Due to the presence of florid MVP, it has been suggested that GBM depends on extensive neovascularization for continued growth; while 'endothelial-poor' tumors do not (Wesseling et al. 1997).

As mentioned above, GBM is one of the main primary brain tumors, one of the most malignant and is the most biologically representative model of glioma (Burger et al. 1988, Pytel and Lucas 2009). Rapid spread along compact myelinated fiber tracts is typical of GBM neoplasms. Infiltration through the corpus callosum with extension into the contralateral hemisphere, resulting in a bilateral, symmetric lesion ("butterfly glioma") is particularly common (McComb and Burger 1985). GBM is a rapidly evolving tumor with neoplastic infiltration of adjacent normal brain tissue and solid proliferation at the periphery. Rapid growth results in spontaneous necrosis with pseudo-palisading of tumor cells, which quickly develops into a necrotic core (Fig. 1). The essential diagnostic feature that distinguishes GBM from lower grade gliomas is abundant endothelial proliferation forming multilayered vessels and or necrosis (Walker et al. 2011).

The molecular events that lead to the transformation from low-grade gliomas (LGG) to GBM have been extensively examined. Now it is well known that the molecular pathways involved in the growth and invasion capacity of brain tumors can also induce the development of new blood vessels (angiogenesis), which allows continuous tumor growth. Several studies and more recent large-scale genomic analysis have shown that three core signaling pathways are altered in glioblastomas: (1) The receptor tyrosine kinase, rat sarcoma (Ras), Phosphatidylinositol 3-kinase (PI3K) pathway, where epidermal growth factor receptor

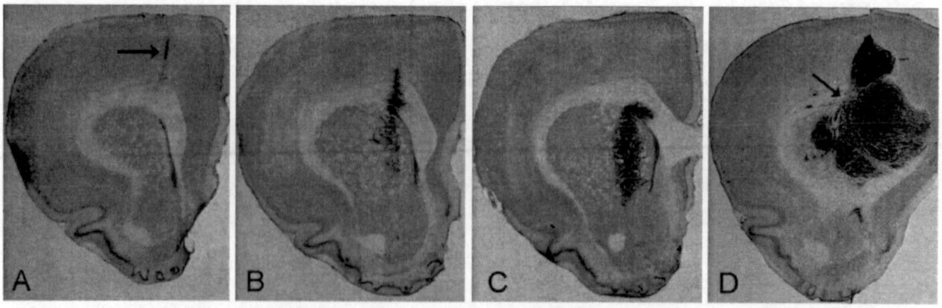

Figure 1. Glioma growth pattern in the rat brain. Implantation of ethylnitrosourea-induced RG2 (rat glioma 2) cells in the rat brain reproduces the characteristic high grade glioblastoma tumor infiltrative growth pattern (D, arrow). RG2 cell implantation site (A, arrow) and colonization at day six (B), 12 (C) and 20 (D) are shown.

(EGFR) amplification, platelet derived growth factor receptor (PDGFR)-A overexpression or amplification and phosphatase and tensin homologue inactivation are the most common genetic alterations; (2) the p53 pathway, where inactivation of p53 and p14Arf are the main genetic alterations; (3) the RB pathway, with inactivation of retinoblastoma (Rb1) and p16Ink4a (Arnold et al. 1999, Maher et al. 2001, Verhaak et al. 2010, Nazarenko et al. 2012).

The most important signaling pathway implicated in angiogenesis during tumor development is the PI3K pathway; which regulates proliferation, differentiation, migration, trafficking and neovascularization by a great diversity of lipid kinases. PI3K deregulation is a common feature of gliobastomas, as a consequence of the alterations in EGFR and PDGFR. EGFR gene amplification is a striking feature of GBM but is rare in LGG, suggesting it plays a causal role in the pathogenesis of glioblastomas. Increased activity of EGFR influences multiple aspects of GBM biology such as cell proliferation, motility, invasiveness, drug resistance and angiogenesis; through downstream activation of PI3K—protein kinase B (PKB)—mechanistic target of rapamycin (mTOR), PI3K—Ras homolog (Rho), Ras—rapidly accelerated fibrosarcoma (Raf)—extracellular signal-regulated kinase (Erk) and janus kinase (Jak)—signal transducer and activator of transcription (STAT) signaling pathways (Lund-Johansen et al. 1990, Dunn et al. 2000, Friedman and Bigner 2005, Anand et al. 2011, Ramis et al. 2012).

On the other hand, overexpression and gene amplification of PDGFR-α and β and its ligands have been demonstrated to occur mostly in LGG and secondary glioblastomas. Overexpression of PDGFR ligands, however, varies among the different grades: platelet derived growth factor (PDGF)-A is expressed in all gliomas and PDGF-B is expressed only in high-grade glioblastomas, suggesting that it participates in the conversion of low to high-grade gliomas. In human gliomas, as shown by a proteomic study, the activation of mTOR and Ras—Erk pathways frequently co-occurred in PDGF-B enriched tumors. PDGFR and PDGF-B co-expression has been related to angiogenesis during glioblastoma development (Kesari and Stiles 2006, Jiang et al. 2011, Nazarenko et al. 2012).

6.5 Cell Invasion in CNST

Tumor growth depends on the ability of its cells to expand into the surrounding normal tissues. Tumor cells use migration mechanisms similar to those occurring during normal processes, such as morphogenesis and wound healing. Cell invasion is a complex, multifactorial and dynamic process that has four basic coordinated steps: (1) detachment of invading cells from the primary tumor, (2) cell adhesion to the ECM, (3) ECM degradation and (4) tumor cell motility and cytoskeletal contractility (Fig. 2A) (Tonn and Goldbrunner 2003, Demuth and Berens 2004, Salhia et al. 2006).

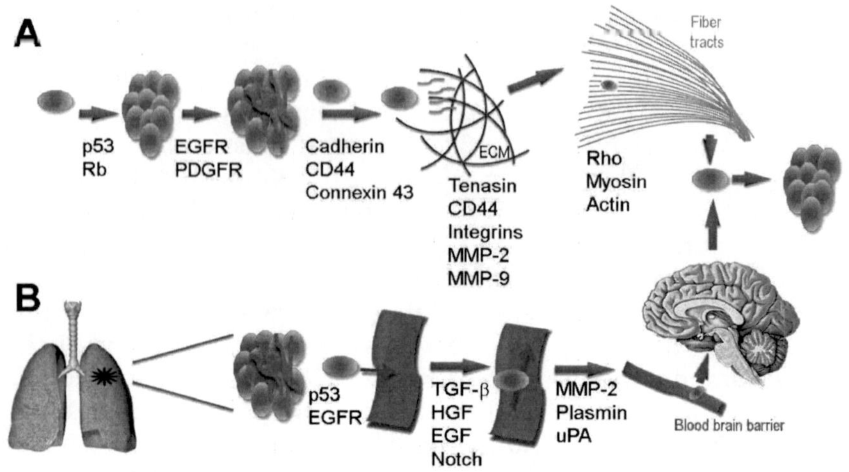

Figure 2. Cell invasion in primary and secondary brain tumors. A) Cell intrinsic genetic alterations of the p53 and retinoblastoma (Rb) pathways lead to cell proliferation in low grade gliomas (LGG). Transformation from LGG to glioblastoma multiforme can occur by mutation or amplifications of epidermal growth factor receptor (EGFR) and platelet derived growth factor receptor (PDGFR) pathways; these changes increase cell proliferation motility and invasiveness of tumor cells by activation of different pathways (see text). Brain parenchyma invasion by glioma cells involves detachment of cells from the primary tumor, adhesion to and degradation of the extracellular matrix (ECM) by tumor cells and cell motility along fiber tracts to settle in another site and continue proliferating. B) Metastatic brain tumors need cell invasion mechanisms to colonize the brain parenchyma. Tumor cells detached from the primary tumor interact and degrade the ECM in order to reach blood vessels. Blood flow transports tumoral cells to the brain parenchyma, where they settle and proliferate. The main molecules that have been implicated in each step of cell proliferation and invasion are shown (for abbreviations of molecules see text).

Color image of this figure appears in the color plate section at the end of the book.

6.5.1 Primary Tumors

In primary tumors, the invasion of normal brain tissue by glioma cells involves close relations between tumor cells, adjacent cells and ECM components, plus important proteolytic activity. Malignant cells move along white matter axonal fiber tracts and the basement membrane of blood vessels faster that in grey matter. The reason of this preference is not understood, but is attributed to a lesser resistance of the tissue, presence of biochemical substrates that mediate cell adhesion and migration or both. Interestingly, GBM cells are unable to penetrate the basement membrane and thus cannot propagate inside or outside the CNS by blood flow (Maher et al. 2001, Ziu et al. 2006).

Detachment of glioma cells implies destabilization and disorganization of cell-cell junctions maintaining the tumor mass together. Tumor cells can disseminate as individual cells (individual cell migration) or expand in solid strands, sheets, filaments or clusters. Glioblastoma cells migrate individually by mesenchymal migration. *In vitro* and *in vivo* studies have shown that adherens junctions are disrupted during cell detachment. Alterations in epithelial, neural and placental cadherins have been reported during glioblastoma progression and are associated with an increase in cell motility and invasiveness (Arnold et al. 1999, Lewis-Tuffin et al. 2010).

Other events associated with the detachment of cells from the primary tumor mass are the loss of connexin 43 expression and cleavage of cluster of designation (CD)44. Connexin 43 is the most abundant protein in gap junctions, which have a major role in CNS communication and differentiation. A decrease in gap junction formation is correlated with increased motility, uncontrolled division and de-differentiation of glioma cells. On the other hand, the CD44 glycoprotein anchors the primary tumor mass to the ECM and its cleavage by ADAM (a disintegrin and metalloproteinase) 10 and 17 promotes cell migration (Goldbrunner et al. 1998, 1999, Nakada et al. 2003, Jimenez et al. 2006, De Vuyst et al. 2009).

The unusual aggressiveness of GBM may be mediated by its unique interactions with the ECM of the brain. *In vivo* and *in vitro* studies have shown that the ECM of gliomas and their migration pathways present higher expression of tenascin, fibronectin, laminin, vitronectin and different types of collagen when compared with normal tissue ECM; a high expression of glycosaminoglycans, mainly in the tumor, has also been reported (Mahesparan et al. 2003, Sarkar et al. 2006, Ziu et al. 2006, Kawataki et al. 2007). The hyaluronic-acid (HA) content in the brain tissue has been shown to increase transiently during tumor-cell migration and is usually found at the interface between the tumor mass and the healthy tissue. In normal tissues, HA is involved in homeostasis, biomechanical integrity and tissue structure. However; in malignant tissues, HA facilitates tumor invasion and migration through two main receptors: CD44 and the receptor for hyaluronan-mediated motility. The mechanisms are still under study, but there is evidence that GBM tumor invasion promoted by HA involves the activation of PI3K—PKB—mTOR, ostepontin and Rho pathways (Nakagawa et al. 1996, Enam et al. 1998, Goldbrunner et al. 1999, Tonn and Goldbrunner 2003).

Glioblastoma progression involves other changes in ECM composition; one change that has been related to the migratory cell phenotype is increased fibronectin around the glio-mesenchymal junction of tumors and in tumor-associated blood vessels. It has been suggested that the fibronectin deposited in the ECM of tumors strongly promotes the migration of glioma cells and that the activation of its receptor plays a critical role in the behavior of tumor cells. In addition, a high abundance of collagen type IV has been detected in the ECM of glioblastomas, as well as in the basement membrane of tumor infiltrating blood vessels (Ohnishi et al. 1997, Treasurywala and Berens 1998, Mahesparan et al. 2003, Rao Malla et al. 2012).

Cell-ECM interactions are mediated mainly by integrin receptors. Integrins are trans-membrane glycoproteins, forming heterodimers between α and β subunits that harmonize interactions between the ECM and the cellular cytoskeleton. When integrins bind to ECM components, actin filaments and signaling proteins move to specific cell membrane domains and stabilize this interaction forming complexes called focal adhesions. Expression of the α5 and β3 integrin subunits has been reported to increase in glioblastoma tumors, compared with normal tissue; this heterodimer binds to fibronectin and promotes cell invasion (Rooprai et al. 1999, Farber et al. 2008, Carbonell et al. 2009, Delamarre et al. 2009).

Glioma tumor advance depends on ECM degradation. The most notable proteases that degrade the ECM during GBM tissue invasion are the matrix metalloproteinases (MMPs), in particular MMP-2 and MMP-9. In addition to MMPs, many studies have identified cathepsin B as a key enzyme responsible of ECM degradation; it is secreted by the GBM during tumor invasion (Wick et al. 2001, Hu et al. 2006, Sarkar et al. 2006).

Directional cell invasion requires cell motility and contractility, driven by cytoplasmic contractile forces, through a series of cytoskeletal reorganizations that create a leading and trailing edge. Glioma cells undergo mesenchymal migration and acquire a fibroblast-like spindle-shaped morphology, which depends on dynamic integrin adhesion and the presence of high-traction forces. Glioma motility requires polymerization and de-polymerization of actin filaments, regulated by Rho signaling pathways, including cell division control protein 42 and Rho. Glioma cells also need contractile forces mediated by the A and B isoforms of myosin II, which is the major source of cytoplasmic contractile force. Myosin II activity is controlled by phosphorylation and de-phosphorylation of its light chain by myosin light chain kinase and myosin light chain phosphatase, respectively (Beadle et al. 2008, Ivkovic et al. 2012, Ramis et al. 2012).

Overexpression of growth factor receptors and their ligands is a common feature of glioma tumors and some are related to invasion and migration mechanisms. Besides epidermal growth factor (EGF)-EGFR and PDGF-PDGFR deregulation; another growth factor implicated in glioblastoma progression is the scatter factor hepatocyte growth factor (SF HGF) and its receptor (HGFR). Expression of these proteins correlates positively with tumor malignancy grade and is associated to neoplasm invasion. Chemokines also play a pivotal role in glioblastoma invasion; chemokine receptor 4 and chemokine ligand 12 promote migration, proliferation and activation of MMPs (Wick et al. 2001, Esencay et al. 2010).

6.5.2 Secondary Tumors

Secondary brain tumor development also depends on cell invasion and angiogenesis (Fig. 2B). In 1889, Paget proposed that tumor cells are like seeds seeking favorable microenvironments and the brain is an attractive site for these cells. This hypothesis considers three principles: (1) a primary tumor with an heterogeneous population of cells with different characteristics; (2) a subset of tumor cells capable of migrating, multiplying and invading new environments and (3) colonization of the new environment that depends on interactions between the tumor cells and the target site (seed-soil) (Santarelli et al. 2007, Fidler et al. 2010, Rahmathulla et al. 2012).

Metastasis can be separated in two main steps, the migration of cells from the primary tumor site and the colonization of a new organ by these cells. The different events between the two main steps are known as the metastasis cascade. Migration of tumor cells depends on their potential to metastasize and their ability to evade the structural organization of normal tissues. Only a subset of cells can survive the pressures of the tumor environment, such as hypoxia, nutrient deprivation, poor bloody supply and immune reactions. These cells acquire the capacity to resist apoptosis and escape from inhibitors of cell proliferation and growth suppressors by a combination of fixed genetic mutations and accumulation of different epigenetic changes. For example, it has been reported that mutations of p53, Kirsten-Ras and EGFR are involved in the metastasis of lung primary carcinoma to the brain (Nicolson et al. 1994, Santarelli et al. 2007, Rahmathulla et al. 2012).

Several studies have focused on the epithelial-mesenchymal transition (EMT); which is considered a temporary and reversible phenomenon wherein cells can dedifferentiate in order to migrate to a distant focus and re-differentiate back to their original state. Activation of the EMT could be intrinsic or induced by extrinsic gene mutations and or growth factor signaling, such as transforming growth factor (TGF)-β, HGF, EGF, insulin-like growth factors, fibroblast growth factor (FGF) and Notch signaling. Recent evidence supports the hypothesis that the EMT enables cancer cells to disseminate from the primary tumor to a distant site by conferring them with the ability to lose adhesion, degrade the ECM and resist apoptosis. Crosstalk between tumor cells and adjacent stromal cells is necessary to activate EMT (Carbonell et al. 2009, Nie et al. 2012).

The spread of tumor cells requires ECM invasion and penetration into the vasculature. Tumor cells degrade the ECM and vascular basement membrane to reach the circulation; this process is regulated by protein-protein interactions between the cell membrane and the ECM, as well as expression of proteolytic enzymes. Several studies have documented interactions between membrane proteins and the ECM in normal and tumor cyto-architecture. E-Cadherin-Catenin complexes and integrins are the main proteins participating in these interactions. Their deregulation allows fragmented, disordered architecture and separation of the tumor mass.

Tumor cell migration depends on their ability to remodel the ECM by proteolysis. The advancing edge of tumor cells can perform proteolytic activity and release signals that promote malignant cell proliferation and angiogenesis cascades. In this respect, it has been reported that proteases such as heparinase are involved in ECM and brain-blood barrier (BBB) degradation by tumor cells. MMP-2 was identified in a series of metastasis without relevance to their site of origin. Another protease involved in metastasis is plasmin, derived from plasminogen activation by urokinase in the urikinase-type plasminogen activator (uPA) system, which has a main role in migration and invasion mechanisms (Gondi et al. 2004, Mendes et al. 2005, Malemud 2006, Mendes et al. 2007, Fidler et al. 2010, Rojiani et al. 2010, Feng et al. 2011, Rao Malla et al. 2012).

Colonization of the brain by tumor cells requires adhesion to the endothelium. During brain metastasis tumor cells reinforce their skeleton, increasing its adherence to the vascular wall and behave like macrophages, creating pseudopodia that allow them to penetrate cell-cell junctions. Invasion of the brain parenchyma by tumor cells starts by opening the BBB; this process has been extensively studied in the metastasis of breast cancer to the brain. The brain metastasis signature consists of 17 genes, indentified by genomic profile and univariate analysis; among them cycloxygenase-2, the EGFR ligand heparin-binding EGF-like growth factor and the α-2, 6-sialyltransferase were identified as proteins that facilitate extravasation and enhance colonization (Nicolson et al. 1994, 1996, Weber and Ashkar 2000, Bos et al. 2009, Fan et al. 2011).

A key step for continued tumor growth is angiogenesis. Failure of vascular growth will compromise the nutrient and oxygen source for the formed tumor. The onset of angiogenesis is influenced by a complex interplay of pro-angiogenic and anti-angiogenic factors (Rahmathulla et al. 2012, Fidler et al. 2010, Carbonell et al. 2009).

6.6 Angiogenesis in CNST

When tumor cells exceed a critical number and a more solid growth pattern occurs, or when tumor cells become metabolically hyperactive, the tumor tissue overcomes pre-existent blood vessels and becomes dependent on angiogenesis (Wesseling et al. 1994). Angiogenesis is the formation of new blood vessels by rerouting or remodeling of existing ones. This essential biological process is highly regulated and involves multiple coordinated steps, such as the production and release of angiogenic factors, directional migration and proliferation of micro-vascular cells, ECM degradation and formation of new blood vessels (Fig. 3) (Wesseling et al. 1997).

In 1971 Folkman proposed that tumors cannot grow beyond 1 to 2 mm^3 without the continuous recruitment of new blood vessels. The same author introduced the concept that tumors probably release molecules that diffuse to stimulate the growth of new blood vessels (Folkman 1971) and proposed the existence of endogenous factors with the ability to inhibit angiogenesis. When the expression of pro-angiogenic and anti-angiogenic factors is balanced, the "angiogenic switch" remains off. However, when pro-angiogenic factors outweigh anti-angiogenic factors in the local tissue environment, the "angiogenic switch" turns on (Fig. 4) (Hanahan and Folkman 1996, Nathoo et al. 2005, Kim and Lee 2009, Tate and Aghi 2009, Onishi et al. 2011).

Numerous inducers of angiogenesis have been identified, including the vascular endothelial growth factor (VEGF) family, angiopoietins (Ang), TGF, PDGF, tumor necrosis factor (TNF)-α, interleukins (IL) and members of the fibroblast growth factor family (Otrock et al. 2007). These molecules can be secreted by cancer cells, as well as endothelial, stromal and blood

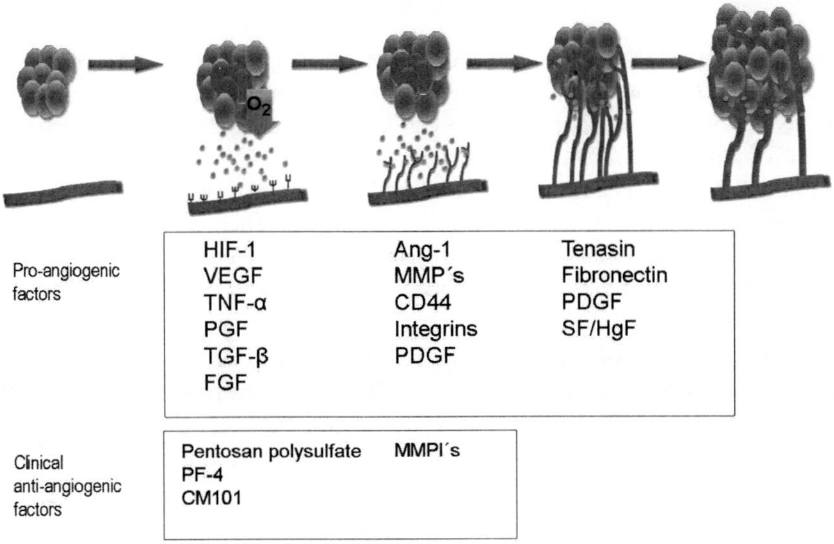

Figure 3. Angiogenesis in primary and secondary brain tumors. Angiogenesis is necessary for tumor growth when its size exceeds the capacity of the original blood vessels to nourish and oxygenate. A decrease in tumor oxygen levels triggers the imbalance between proangiogenic and antiangiogenic factors. In response to hypoxia, tumor cells secrete pro-angiogenic factors that reach the receptors on the endothelial cells of preexisting blood vessels. Formation of new vessels involves remodeling and indeed dissolution of preexisting ones. New vessels surround the tumor providing nutrients and oxygen to cells and maintaining tumor growth. For abbreviations see text.

Color image of this figure appears in the color plate section at the end of the book.

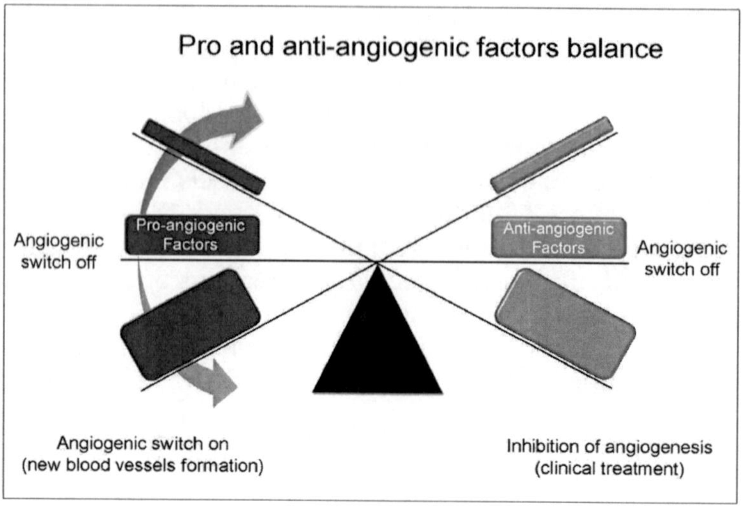

Figure 4. "Angiogenic switch". Angiogenesis is a process triggered by imbalance between pro- and anti-angiogenic factors; when there is equilibrium between them the switch remains off. However, an increase in pro-angiogenic or a decrease in anti-angiogenic factors turns on the process.

cells and the ECM (Kim and Lee 2009). In addition, many factors control and influence angiogenesis including membrane-bound proteins, cell-matrix and cell-cell interactions, as well as environmental factors such as hypoxia (Brahimi-Horn et al. 2001, Otrock et al. 2007).

Tissue hypoxia is the most potent activator of angiogenesis; it has been proposed that angiogenesis in solid tumors, including brain tumors, is triggered by low oxygen concentrations (hypoxia). When the tumor grows fast, tumor cells move away from blood vessels and there are deficits in the blood supply (Kim and Lee 2009). Exposure of brain tumor cells to hypoxia induces expression of hypoxia-inducible factor (HIF)-1, a transcription factor that regulates the expression of VEGF and many other factors that initiate angiogenesis; inducing endothelial cell proliferation, invasion and migration (Kaur et al. 2005, Kim and Lee 2009). In fact, HIF-1α expression is found at the invasive front of glioblastomas and correlates with glioma grade and vessel density (Zagzag et al. 2000).

VEGF is one of the most important pro-angiogenic factors. When tumor cells secrete VEGF and it binds to its receptor (VEGFR)-2 on endothelial cells, it starts a cascade of intracellular molecular events that culminates in endothelial cell proliferation, migration and survival. The biological effects of VEGF upon angiogenesis are mediated through diverse signaling pathways. Activation of Ras—Raf—mitogen activated protein kinase or phospholipase C-γ—protein kinase C signaling pathways promote endothelial cell proliferation (Kowanetz and Ferrara 2006, Brastianos and

Batchelor 2009, Kim and Lee 2009). Initiation of mitogen-activated protein kinase signaling enhances vascular permeability through PKB and the endothelial nitric oxide synthase signaling pathways (Wong et al. 2009). PKB inhibits endothelial cell apoptosis, promotes endothelial cell migration and increases the expression of HIF, leading to enhanced VEGF expression (Hoeben et al. 2004, Brastianos and Batchelor 2009, Kim and Lee 2009). Endothelial cell proliferation is stimulated by VEGF through uPA, tissue-type plasminogen activator and plasminogen activator inhibitor (PAI)-1. In addition, the expression of plasminogen activators can break down and remodel ECM components (Wong et al. 2009).

Because VEGF regulates several steps of angiogenesis, decreased expression of this protein can inhibit angiogenesis induced by brain tumors (Yano et al. 2000). Yano et al. (2000) demonstrated that a 30–50 per cent decrease in the expression levels of VEGF is associated with an inhibition of angiogenesis and metastasis. Decreased VEGF expression may be compensated by increased expression of other pro-angiogenic factors; such as IL-6, IL-8 or basic FGF (bFGF) (Loboda et al. 2012). However, several transcription factors are capable of up-regulating VEGF transcription, including the E26 transformation specific sequence-1 proto-oncogene and STAT-3. In addition, a variety of growth factors can also up-regulate VEGF expression, including TGF-β, EGF, PDGF-B and bFGF (Wong et al. 2009).

The next step of angiogenesis is the formation of new blood vessels from the existing vessels through their dissolution (Tate and Aghi 2009, Onishi et al. 2011). In brain tumors, especially in gliomas, tumor cells accumulate around the existing cerebral blood vessels. The close contact between cancer cells and vascular cells results in the lifting up of the astrocytic foot process by tumor cells, leading to disruption of endothelial cell-ECM interactions (Zagzag et al. 2000). Afterwards, endothelial cells and some tumor cells express Ang-2, an angiogenic factor. Ang-2 binds to the tyrosine kinase with immunoglobulin-like and EGF-like domains (Tie)-2 receptor and induces vessel wall destabilization by reducing the association between pericytes, endothelial cells and the ECM, thus starting the regression of blood vessels seen in early tumor-induced angiogenesis (Maisonpierre et al. 1997, Stratmann et al. 1998, Holash et al. 1999, Zagzag et al. 1999). Eventually, the activation of Tie-2 by Ang-2 increases the expression of MMP-2; which promotes angiogenesis in conjunction with VEGF (Hu et al. 2003). In the absence of VEGF activity, Ang-2 becomes anti-angiogenic by promoting endothelial cell death and blood-vessel regression (Wong et al. 2009).

On the other hand, Tie-2 activation by Ang-1 leads to the stabilization, remodeling and maturation of blood vessels. Ang-1 induces phosphorylation of Tie-2 and the p85 subunit of PI3K as well as increasing PI3K activity in a dose-dependent manner, leading to endothelial cell survival via PKB signaling. In addition, Ang-1 binding to Tie-2 induces vessel sprouting by increased endothelial secretion of plasmin and MMP (Wong et al. 2009). However, Ang-1 also inhibits tumor cell extravasation and maintains the integrity of healthy blood vessels outside the tumors. Thus, Ang-1 and Ang-2 may be pro- or anti-angiogenic factors in response to the microenvironment, making it difficult for them to be considered therapeutic targets against cancer (Carmeliet and Jan 2011).

The next step in the progression of angiogenesis is degradation of the vessel basement membrane and the surrounding ECM to facilitate endothelial cell invasion (Tate and Aghi 2009, Onishi et al. 2011). MMPs are implicated in this process and can be produced by endothelial cells, blood vessels and tumor cells. In brain tumors (especially in gliomas), MMPs are produced and used by tumor cells (Rao et al. 1996, Forsyth et al. 1999, Rojiani et al. 2010). Pro-angiogenic factors, such as the growth factors and cytokines IL-1β, EGF, TGF-β and TNF-α have been shown to regulate the secretion of MMPs (Rao et al. 1996). At the same time, ECM degradation results in the release of pro-angiogenic factors bound to the proteoglycans that abound within the ECM and the basement membrane (Rojiani et al. 2010). MMP-2 (gelatinase A) and MMP-9 (gelatinase B) are highly expressed in brain tumors and their expression levels correlate with the histological tumor grade and its prognosis, especially those of MMP-9 (Rao et al. 1996, Forsyth et al. 1999, Lakka et al. 2005, Wong et al. 2009, Rojiani et al. 2010). In the final phase of angiogenesis, MMP activity must be inhibited for the stabilization of blood vessel; their activity is controlled at the level of gene transcription, zymogen activation by proteolysis and inhibition of the active forms by the tissue inhibitors of metalloproteinases (Forsyth et al. 1999, Otrok et al. 2007).

Endothelial cell proliferation and migration toward tumor cells that express pro-angiogenic factors is the last step in the angiogenic process. The expression of cell surface adhesion or migration proteins, such integrins and CD44, is up-regulated in endothelial cells. The integrin family of cell surface receptors that regulates cell adhesion and migration is involved in intracellular signaling pathways and plays an important role in the interface between the cell surface and its environment (D'Abaco and Kaye 2007). Integrins create a link between the ECM and the cell cytoskeleton, maintaining cell shape and architecture as well as regulating cell migration (Ridley et al. 2003). In brain tumors, increased expression of integrins αvβ3 and α5β1 in endothelial cells initiates a calcium-dependent signaling pathway leading to endothelial cell migration (Leavesley et al. 1992, 1993) and adhesion (Kim et al. 2000).

ECM proteins, expressed by tumor cells, also stimulate angiogenesis and assist tumor cell migration and invasion. Tenascin-C is an ECM protein with a clear role in angiogenesis; it is not expressed in the adult brain, but in gliomas it is localized within sites of angiogenesis at the invading tumor border (Tate and Aghi 2009). The expression of tenascin-C correlates with the tumor grade, high levels of tenascin-C have been observed in grade III and IV tumors, making it a marker of poor prognosis, disease recurrence and shorter survival time (D'Abaco and Kaye 2007). Tenascin-C has also been shown to promote endothelial cell migration during angiogenesis in an autocrine manner. It also promotes VEGF expression and focal adhesion kinase activation, which are important for ongoing angiogenesis. Other examples of ECM proteins involved in angiogenesis are fibronectin and laminin-8 (Tate and Aghi 2009).

At the same time, endothelial cells secrete PDGF that recruits pericytes around the new sprouting vessels, which support construction of a new vessel basement membrane (Ferrara and Krebel 2005). Recently, it has been proposed that platelets are activated and play an important role in regulating angiogenesis by releasing pro-angiogenic factors (Italiano Jr. et al. 2008). The last details in the formation of blood vessels are not fully known; however, the final result of brain tumor angiogenesis is an abnormal vasculature. Tumor vessels are organized in a chaotic manner with avascular hypoxic voids of many sizes and do not follow the hierarchical branching pattern of normal vascular networks. The structure of the vessel wall is also abnormal in tumors; their diameters are uneven and endothelial cells form an imperfect coating, with wide connections in some areas and stacked layers of endothelial cells in others (Jain 2003). The expression of adhesion molecules is also heterogeneous, TNF-α or VEGF up-regulate the expression of intracellular adhesion molecule-1, vascular cell adhesion molecule-1 or E-selectin; while in other conditions TNF-α or Ang-1 down-regulate the expression of adhesion molecule-1 (Jain 2003).

6.6.1 Angiogenesis Promoters

Brain tumor cells produce various growth factors or cytokines, which should balance tumor angiogenesis through paracrine control. FGF, EGF, TGF-α, SF HGF, TGF-β and IL-8 are involved in tumor angiogenesis (Ryuto et al. 1996).

FGF is overexpressed in various types of cancer; for example FGF is expressed by glioma tumor cells and endothelial cells. In gliomas, both acidic FGF (aFGF) and bFGF are up-regulated and stimulate endothelial cell proliferation, migration and resistance of endothelial cells to apoptosis (Zagzag et al. 1990, Schmidt et al. 1999, Otrock et al. 2007). The biological activities of FGFs are initiated by binding to their tyrosine kinase receptors, which are up-regulated. Expression of fibroblast growth factor receptor (FGFR)-1 is increased in endothelial cells, whereas tumor cells overexpress FGFR-1 and 4 (Tate and Aghi 2009, Wong et al. 2009). The action of FGFs involves the activation of parallel signaling pathways as a result of receptor auto-phosphorylation (Otrock et al. 2007). FGF-2 induces a mitogenic response in endothelial cells, which requires the activation of protein kinase C (Simons 2004); while the anti-apoptotic effect of FGF is mediated by a mitogen activated kinase kinase-dependent pathway that increases the expression of B-cell lymphoma (Bcl)-extra large (xL) and Bcl-2 (Pardo et al. 2002). In addition, bFGF induced VEGF (Yoshiji et al. 2002) stimulates expression of VEGFR in endothelial cells (Mandriota and Pepper 1997) and prompts endothelial cell proliferation and migration (Sato and Rifkin 1988).

TGF-β is also involved in angiogenesis, but has both pro- and anti-angiogenic effects. When TGF-β binds to its type II receptor, this stimulates type I receptor phosphorylation and activation of downstream signaling proteins (Otrock et al. 2007, Wong et al. 2009). Expression of its type III receptor, endoglin, is induced by hypoxia in brain tumors; facilitating binding of TGF-β1 to type II receptors (Sugita 2011). At low doses, TGF-β1 up-regulates other pro-angiogenic factors (VEGF, PDGF-A, PDGF-B, αvβ3 integrin) and proteinases (MMP-2); while at high doses, TGF-β1 inhibits endothelial cell growth and promotes basement membrane reformation (Carmeliet 2003, Otrock et al. 2007, Wong et al. 2009).

In brain tumors, expression of SF HGF and its receptor HGFR is increased in cancer cells and endothelial cells, contributing to tumor progression and angiogenesis (Abounader and Laterra 2005). Higher levels of SF HGF correlate significantly with angiogenic activity in malignant gliomas (Schmidt et al. 1999). Inhibition of SF HGF expression or signaling results in suppression of *in vivo* tumor growth and angiogenesis (Abounader et al. 2002).

EGF and TGF-α are two potent angiogenic factors; both are overexpressed in brain tumor endothelial cells (Dunn et al. 2000). These proteins bind to the EGFR, which is rarely present in normal glial cells but is expressed in human gliomas (Dunn et al. 2000). EGF and TGF-α stimulate endothelial cell proliferation and VEGF production in glioma cells, acting in an autocrine and paracrine manner (Dunn et al. 2000, Wong et al. 2009).

TNF-α induces tumor angiogenesis indirectly by activation of other angiogenic factors, the most important of which is VEGF. Expression of VEGF is up-regulated in human glioma cells upon TNF-α treatment; mediated in part by transcription factors such as specificity protein 1 or nuclear factor kappa light chain enhancer in B cells (Ryuto et al. 1996, Yoshida et al. 1997). In addition, TNF-α enhances IL-8 and bFGF production in human micro-vascular endothelial cells and induces tubular morphogenesis *in vitro* (Yoshida et al. 1997). TNF-α is expressed in malignant gliomas and other cells including reactive astrocytes, endothelial cells and infiltrating macrophages; while the TNF-α receptor is overexpressed by glioma and endothelial cells (Wong et al. 2009).

Other pro-angiogenic factors implicated in brain tumor progression are IL-6 and IL-8, both members of the chemokine family (Cohen et al. 1996,

Brat et al. 2005, Wong et al. 2009). Chemokines are specialized cytokines produced and secreted by several cell types, including tumor and pseudopalisading cells (Wong et al. 2009). IL-6 induces transcriptional activation of VEGF and regulates VEGF promoter activity via direct interaction with STAT3 (Cohen et al. 1996, Wong et al. 2009). IL-6 also mediates up-regulation of bFGF by activation of Jak—STAT3 and PI3K—PKB pathways (Jee et al. 2004). The angiogenic activity of IL-8, mediated by activation of IL-8 receptor alpha and beta G-protein coupled proteins in endothelial cells, stimulates endothelial proliferation and capillary tube formation; these effects can be blocked by monoclonal antibodies to IL-8 (Li et al. 2003, Brat et al. 2005). In addition, IL-8 inhibits endothelial cell apoptosis by increasing the levels of the anti-apoptotic factors Bcl-xL and Bcl-2, as well as decreasing the levels of Bcl-2 associated X protein (Brat et al. 2005). Finally, IL-8 stimulates the expression of MMP-2 and MMP-9 mRNAs in endothelial cells, which are required for degradation of the vessel basement membrane and ECM (Rao et al. 1996, Forsyth et al. 1999, Li et al. 2003, Brat et al. 2005).

6.6.2 Angiogenesis Inhibitors

Endogenous angiogenesis inhibitors are molecules, naturally present in body fluids or in the tissues, which possess anti-angiogenic activity (Sund et al. 2005). Interestingly, many malignant tumors can generate inhibitors of angiogenesis, including thrombospondin (Good et al. 1990), angiostatin (O'Reilly et al. 1994) and endostatin (O'Reilly et al. 1997). According to Kerbel and Folkman (2002), there are two types of angiogenesis inhibitors: direct and indirect. Direct anti-angiogenic factors, such as angiostatin, prevent endothelial cells from proliferating, migrating or avoiding cell death in response to several pro-angiogenic proteins including VEGF, bFGF, IL-8 or PDGF. On the other hand, indirect angiogenesis inhibitors generally prevent the expression or block the activity of tumor proteins that activate angiogenesis or block the expression of their receptors on endothelial cells (Kerbel and Folkman 2002).

6.6.2.1 Direct Angiogenesis Inhibitors

Angiostatin, a specific inhibitor of endothelial cell proliferation and migration, is a 38–45 kDa internal fragment of plasminogen containing at least three kringle domains (O'Reilly et al. 1994, Ji et al. 1998, Nyberg et al. 2005). There is evidence that angiostatin can also induce apoptosis of endothelial cells (Lucas et al. 1998). Several MMP family members, including matrilysin (MMP-7), MMP-9, MMP-2, metalloelastase (MMP-12) and stromelysin-1 (MMP-3), hydrolyze plasminogen to generate angiostatin (Nyberg et al. 2005). There are four potential receptors for angiostatin:

integrin αvβ3, adenosine triphosphate (ATP) synthase, angiomotin and chondroitin sulfate proteoglycan (CSPG). When angiostatin binds to integrin αvβ3, it inhibits its interaction with the ECM and reduces cell attachment and adhesion; all of which are necessary for endothelial cell survival and migration (Rege et al. 2005). Angiostatin may interfere with ATP synthase activity, blocking this source of cellular energy and thus inhibiting pro-angiogenic processes (Moser et al. 1999). The angiomotin receptors are expressed by endothelial cells and interact with angiostatin to inhibit endothelial cell migration and capillary tube formation. Finally, the anti-angiogenic activity of angiostatin can be inhibited when this molecule binds to the CSPG receptor, expressed by glioblastoma tumor cells (Chekenya et al. 2002, Rege et al. 2005).

Endostatin is an endogenous angiogenesis inhibitor corresponding to a 20-kDa fragment derived from the carboxy-terminus of type XVIII collagen, a membrane protein found in most vascular basement membranes (O'Reilly et al. 1997, Sund et al. 2005). It is unclear which protease(s) are responsible for the generation of endostatin from collagen type XVIII and it is unknown which receptors mediate their biologic effects. However, endostatin can interact with several cell surface molecules; such as heparan sulfate proteoglycans, integrin α5β1, tropomyosin and VEGFR2 (Rehn et al. 2001, Rege et al. 2005). Recently, it has been reported that endostatin can block endothelial cell motility, induce apoptosis, cause G1 arrest of endothelial cells, block VEGF-mediated signaling by direct interaction with VEGFR2 and down-regulate many genes in endothelial cells; including immediate-early genes, cell cycle-related genes and genes regulating apoptosis inhibitors, among others (O'Reilly et al. 1997, Rehn et al. 2001, Nyberg et al. 2005).

Thrombospondins (TSP) are a family of ECM glycoproteins that regulate cell adhesion, proliferation and survival, as well as protease activation and angiogenesis (Nyberg et al. 2005). TSP-1 was one of the first proteins to be identified as an anti-angiogenic factor (Good et al. 1990). The anti-angiogenic effect of TSP-1 may be mediated by antagonizing the VEGF pathway, inhibiting the release of VEGF from the ECM or interacting directly with VEGF signal transduction (Lawler and Lawler 2012). TSP-1 binds and suppresses the activity of MMP-9, inhibiting VEGF release from the ECM (Rodriguez-Manzaneque et al. 2001, Lawler and Lawler 2012). TSP-1 also directly binds VEGF and FGF-2, inhibiting angiogenesis by two independent pathways (Nyberg et al. 2005, Colombo et al. 2010, Lawler and Lawler 2012). Apoptosis of endothelial cells is induced by binding of TSP-1 and TSP-2 to CD36, release of cytochrome-C and sequential activation of both caspase-9 and caspase-3 (Ren et al. 2009, Lawler and Lawler 2012). The interaction of

TSP-1 and 2 with CD34 inhibits endothelial cell migration and formation of sprouting capillaries (Dawson et al. 1997).

6.6.2.2 Indirect Angiogenesis Inhibitors

Interferons (INF) are cytokines produced by the immune system in response to pathogens, such as viruses and tumor cells. They were the first endogenous anti-angiogenic regulators that were identified (Brouty-Boye and Zetter 1980). When tumor cells are treated with INF-α β there is a reduction in IL-8 secretion, inhibiting endothelial proliferation and capillary tube formation. INF-α β also reduces the action of uPA, PAI -1 and MMP-9. IFN-α β also decreases bFGF activity, resulting in endothelial apoptosis and decreased blood vessel density within the tumors (Sidky and Borden 1987, Dinney et al. 1998, Slaton et al. 1999, Nyberg et al. 2005).

Several members of the IL family indirectly inhibit angiogenesis. IL-4 reduces bFGF-induced neo-vascularization and blocks the migration of endothelial cells *in vitro* (Volpert et al. 1998). The anti-angiogenic effects of IL-12 and IL-18 are mediated by INF-γ induction. IL-12 inhibits VEGF, bFGF and MMP-9 mRNA expression, blocking endothelial cell proliferation (Sgadari et al. 1996, Coughlin et al. 1998, Duda et al. 2000, Wong et al. 2009). IL-18 inhibits endothelial cell proliferation and induces tumor regression (Cao et al. 1999). IL-10 may reduce the synthesis of VEGF, IL-1b, TNF-α, IL-6 and MMP-9 in tumor-associated macrophages (Huang et al. 1999, Wong et al. 2009).

6.7 Clinical Implications of CNST Angiogenesis

There is evidence demonstrating that the degree of angiogenesis in primary tumors is associated with their potential to develop metastatic disease and their impact on overall survival and disease-free survival. The number of vessels per microscopic field (angiogenesis index) is a quantitative determination of the degree of vascularity. It has been shown to be an independent prognostic marker for some tumors, including brain tumors; especially for astrocytomas, in which endothelial proliferation is a grading criterion for anaplasia (Maione et al. 1990). Brain tumors have a complex glomeruloid vessel pattern. Their micro-vasculature has a heterogeneous distribution and caliber. LGG mainly incorporate pre-existing vessels, while glioblastomas develop new vasculature (Jain et al. 2007). The onset of angiogenesis is the main scenario for tumor cell shedding and is quantitatively related to the area of tumor vessels. This is particularly true for glioblastomas, in which the new vessels contribute to the hostile micro-environment, due to structural and functional alterations, and thus to the

Table 4. Pro-angiogenic and Anti-angiogenic Factors (EC = endothelial cell, for other abbrevations see text).

Pro-angiogenic factors		**Anti-angiogenic factors**	
Factor	**Biological function**	**Factor**	**Biological function**
VEGF	↑ EC proliferation, migration and survival ↓ EC apoptosis ↑expression of HIF	Angiostatin	↓ EC proliferation, survival and migration ↓ capillary tube formation ↑ EC apoptosis
Ang-1	↑ EC survival ↑ plasmin and MMP release Stabilizes new blood vessels	Endostatin	↓ EC motility ↑ EC apoptosis Causes G1 arrest of ECs Blocks VEGF
Ang-2	Induces vessel wall destabilization Begins regression of blood vessels in conjunction with VEGF	Thrombospondins (TSP-1, TSP-2)	↑ EC apoptosis ↓ release of VEGF ↓ EC migration ↓ formation of sprouting capillaries
MMP	Degrades ECM components ↑ release of pro-angiogenic factors	Interferons (INF-α β)	↑ EC apoptosis ↓ EC proliferation ↓ capillary tube formation ↓ blood vessel density within tumors
Integrins	↑ EC adhesion and migration	Interleukines (IL-4, IL-10, IL-12, IL-18)	↓ neo-vascularization ↓ EC migration ↓ VEGF, bFGF and MMP-9 mRNA expression ↓ EC proliferation ↓ synthesis of VEGF, IL-1β, TNF-α, IL-6 and MMP-9 ↑ tumor regression
Tenascin-C	↑ EC migration ↑VEGF expression	TGF-β	↓ EC growth ↑ basement membrane reformation
PDGF	Recruits pericytes Stabilizes new blood vessels and the basement membrane		
FGF	↑EC proliferation and migration ↑ EC resistance to apoptosis ↑ VEGF expression		

Table 4. contd....

Table 4. contd.

Pro-angiogenic factors		**Anti-angiogenic factors**	
Factor	**Biological function**	**Factor**	**Biological function**
TGF-β	↑ expression of VEGF, PDGF-A, PDGF-B, αvβ3 integrin and MMP-2		
SF HGF	↑ tumor and EC proliferation		
EGF	↑ EC proliferation ↑ VEGF production		
TGF-α	↑ EC proliferation ↑ VEGF production		
TNF-α	↑ VEGF expression ↑ IL-8 and bFGF production		
Interleukines (IL-6, IL-8)	↑ transcriptional activation of VEGF Regulates VEGF promoter activity ↑ bFGF ↑ EC proliferation ↑ capillary tube formation ↑ MMP-2 and MMP-9 mRNA expression ↓ EC apoptosis		

aggressive phenotype of these tumors. The association between increased angiogenic index, increased metastasis risk and decreased survival may be explained by augmented shedding of highly angiogenic cells into the circulation. Highly angiogenic tumors thereby increase the likelihood of metastatic foci, which are angiogenic and thus capable of rapid growth and further metastasis. These findings support the usefulness of the angiogenesis index as a prognostic factor and its potential usage in brain tumor treatment.

Current scientific evidence supports the use of therapeutic strategies aimed at inhibiting various steps of tumor angiogenesis. Most anti-angiogenic compounds currently in clinical trials interfere with the response of endothelial cells to angiogenic peptides. For the most part they are cytostatic in nature, preventing tumor growth rather that inducing a reduction in tumor size. As a consequence, the usual paradigms for anticancer drug development where tumor response in phase II trials prompt further development, are not appropriate. In addition, anti-angiogenic therapies need to be administered in several different clinical situations: the adjuvant setting for patients with a high risk of relapse or recurrence after the primary disease is under control; as a maintenance therapy for patients with advanced, metastatic or recurrent disease after reduction of the tumor by standard treatment; perhaps even as a chemo-preventive agent for patients at high risk for developing malignancies. Concomitant administration of chemotherapy and radiation therapy is also a desirable property for any anti-angiogenic treatment.

Strategies aimed at inhibiting tumor neovascularization and targeted therapies aimed at directly destroying the tumor neo-vasculature are under development. Therapies affecting an end target or pathway that cannot be circumvented by alternate mechanisms may significantly enhance efficacy and broaden applicability. Furthermore, pre-clinical studies have shown that a combination of an anti-angiogenic factor with a cytotoxic agent may significantly increase the activity of the latter (Teicher et al. 1992, Brem et al. 1993a). These combinations arrested the *in vivo* growth of primary tumors and metastases, reduced the size of established tumors and rendered some mice tumor-free. This combination approach has produced such benefits for several classes of agents; including angiostatic steroids and matrix metalloproteinase inhibitors (MMPIs) or agents, such as fumagillin analogs, that inhibit the response of endothelial cells to angiogenic factors. In addition, anti-angiogenesis agents may also potentiate the effects of radiation therapy (Sotomayor et al. 1992). Thus, anti-angiogenic agents may prove clinically useful when used alone, as adjuvant therapy or in the setting of a small volume disease and when combined with cytotoxic or radiation therapies in patients with advanced or metastatic disease.

The sulfated polysaccharide xylanopolyhydro-gensulfate (pentosan polysulfate) was the first anti-angiogenic agent in the clinical setting. *In vitro*, it inhibited bFGF-induced endothelial cell migration and proliferation (Herbert et al. 1988, Klein-Soyer et al. 1989), diminished the growth of an adrenal cancer cell line with constitutive bFGF production (Wellstein et al. 1990, 1991) and also inhibited the paracrine effects of heparin-binding growth factors secreted by a variety of malignant tumor cell lines (Zugmaier et al. 1992). However, a phase I clinical trial in patients with acquired immunodeficiency syndrome (AIDS)-associated Kaposi's sarcoma showed that it is toxic, since it produced anticoagulation, thrombocytopenia and increased transaminase levels without evident clinical activity (Pluda et al. 1993). Similar results were obtained when pentosan polysulfate was administered subcutaneously to patients with solid tumors (Swain et al. 1995). Thus, there is no solid clinical evidence of its benefits in tumors.

In 1990 Ingber et al. reported the anti-angiogenic properties of fumagillin, a secreted antibiotic product of the fungus *Aspergillus fumigatus fresenius* originally isolated in 1949 (Ingber et al. 1990, Hanson et al. 1949). A more potent and less-toxic fumagillin analog, TNP-470 (AGM-1470), inhibits *in vitro* proliferation and migration of normal, but not transformed, endothelial cells and hinders *in vitro* capillary tube formation at cytostatic but not cytotoxic concentrations (Ingber et al. 1990, Kusaka et al. 1991, Brem et al. 1991, Kusaka et al. 1994, Antoine et al. 1994). Anti-angiogenic activity was shown *in vivo* using a variety of assays. *In vivo* growth inhibition of a variety of human xenografts and murine tumors has also been shown (Brem et al. 1993b, Yanase et al. 1993, Yamaoka et al. 1993, Takamiya et al. 1993). TNP-470 is currently undergoing phase I and early phase II clinical trials in Kaposi's sarcoma, CNST and solid tumors. Preliminary results indicate that the agent is well tolerated but has dose-limiting CNS toxicity, manifested as reversible cerebellar symptoms (Figg et al. 1997, Kudelka et al. 1997).

Platelet factor (PF)-4 is a naturally occurring compound stored in the α-granules of platelets. PF-4 inhibits endothelial cell migration and proliferation at cytostatic but not cytotoxic concentrations (Maione et al. 1990, Sharpe et al. 1990, Sato et al. 1990). PF-4 apparently binds to glycosaminoglycans on the cell surface and blocks binding of bFGF to its receptor (Sato et al. 1990, Watson et al. 1994). Intra-lesion administration of PF-4 also inhibits the *in vivo* growth of human xenografts and murine tumors (Sharpe et al. 1990, Maione et al. 1991). Currently, phase I trials of both systemic and local PF-4 administration are ongoing in patients with solid tumors and AIDS-associated Kaposi's sarcoma. A phase II study of intra-tumor administration of PF-4 began recently in patients with primary brain tumors. The preliminary results indicate the drug is well tolerated, with mild toxicity manifested as a local reaction at sites of intra-lesional injection, mild phlebitis, fatigue and anemia (Lippi and Favaloro 2010). However, there is little evidence of angiogenic inhibition in an orthotopic human glioblastoma model using PF-4 microspheres (Benny et al. 2008).

Tecogalan is a sulfated polysaccharide peptidoglycan complex derived from a bacterial cell wall polysaccharide which inhibits *in vitro* endothelial cell growth. This effect is enhanced when it is administered concomitantly with tamoxifen (Tanaka et al. 1989, 1991), cortisone acetate or tetrahydro S (Inoue et al. 1988). Tecogalan co-administration with cortisone acetate or tetrahydro S also has been reported to possess *in vivo* antitumor activity against human xenografts and murine tumors (Inoue 1988, Tanaka et al. 1989, 1991). This agent has recently undergone phase I clinical testing in patients with solid tumors (Eckhardt et al. 1996). The dose-limiting toxicity was anticoagulation, shown by increased activated partial thromboplastin time. Other toxicities included fever and rigors.

Thalidomide was originally marketed as a sedative hypnotic in the 1950s. If ingested in the first two months of pregnancy, during limb bud formation, it produced severe birth defects; manifested by hypo-plastic and aplastic malformations of the extremities. In addition to its known immuno-modulatory properties, the embryopathy was probably due to its anti-angiogenic activity (D'Amato et al. 1994). *In vitro* thalidomide by itself did not affect bFGF-induced endothelial cell proliferation. *In vivo* it did not have anti-angiogenic activity in the chorioallantoic membrane assay (D'Amato et al. 1994). However, it exhibited potent anti-angiogenic activity in the corneal micro-pocket assay. After evaluating several analogues, D'Amato et al. concluded that a hepatic epoxide metabolite of thalidomide was responsible for its anti-angiogenic activity but not for its immuno-modulatory properties (D'Amato et al. 1994). Based on these findings, clinical trials of thalidomide in patients with AIDS-associated Kaposi's sarcoma, breast cancer, prostate cancer and primary brain tumors are currently underway.

MMPIs have recently entered into clinical trials. Batimastat (BB-94) has broad *in vitro* inhibitory activity against a variety of MMPs, including MMP-2, MMP-3 and MMP-9; but has no effect on the growth of various tumor and fibroblast cell lines *in vitro* (Davies et al. 1993, Chirivi et al. 1994). However, it inhibits the growth and metastatic spread of human xenografts and mouse tumors (Chirivi et al. 1994). Phase I trials administering batimastat intra-peritoneally and intra-pleurally are being performed, while phase II trials are in development. Other MMPIs include COL-3, Neovastat, BMS-255291 and CGS-27023A; which is a broad spectrum MMPI. These agents are administered intra-peritoneally, intra-pleurally or intravenously; but result impractical for patients that require frequent doses and/or long-term administration. Marimastat (BB2516), an oral MMPI, is on clinical evaluation and apparently has good bioavailability. Marimastat or batimastat inhibit MMP activity *in vitro*. Phase I trials in patients with pancreatic, ovarian or prostate carcinoma have shown that marimastat is well tolerated. The main toxicity was musculoskeletal syndrome, which is characterized by pain, stiffness and inflammation in the joints (Mazurek et al. 2011). Marimastat has not improved the survival of patients with glioblastoma or gliosarcoma as a single agent, further studies are needed to study its possible benefit when combined with cytotoxic chemotherapy (Levin et al. 2006).

CM101 is a group B streptococcus polysaccharide toxin which binds to the capillary endothelium of several carcinomas, but not to the normal mature endothelium (Hellerqvist et al. 1993). CM101 might act by binding to tumor-associated neo-vasculature with the subsequent induction of vascular and cellular inflammatory reactions within the tumor; as suggested by the increased levels of the soluble E-selectin (DeVore et al. 1997). When CM101 was tested in a xenograft murine model, it resulted in an intense intra-tumor inflammatory reaction as shown by the large amount of cytokines that were detected; including TNF-α, IL-lα, IL-6 and macrophage inhibitory protein-1 (Hellerqvist et al. 1993, DeVore et al. 1997). It also produced necrosis, hemorrhage and thrombosis. CM101 reduced tumor volume and prolonged overall survival in some mice models

(Thurman et al. 1994, DeVore et al. 1997). Preliminary results of a phase I trial suggest the drug is well tolerated, but the response rate is lower than 20 per cent (DeVore et al. 1997); additionally, no patients with brain tumors were included.

IL-12 is secreted by activated T cells and natural killer (NK) cells. It enhances cytotoxic T-cell and NK-cell activity, as well as inducing IFN-γ production (Trinchieri et al. 1994). IL-12 increases cell-mediated immunity by inducing the differentiation of T-helper (Th)-1 cells from uncommitted T cells, while inhibiting Th-2 cell differentiation (Scott et al. 1993). IL-12 also has shown antitumor and anti-metastatic activities in preclinical models (Brunda et al. 1993). The anti-angiogenic activity of IL-12 is mediated by IFN-γ induction (Voest et al. 1995). Human INF-γ inducible protein 10 is a potent inhibitor of *in vivo* angiogenesis (Strieter et al. 1995, Angiolillo et al. 1996).

6.8 Conclusion

Angiogenesis is a central process for the growth and metastasis of CNS malignancies. The blood vessel network suffers major modifications when a tumor develops by angiogenesis. This is a complex process that involves multiple steps and pathways with positive and negative signaling. The local balance between the molecules that participate in signaling determines if a tumor will grow and spread or remain dormant. Molecular events seems to be interdependent and highly regulated, reflecting a consecutive process that allows the growth, invasion and dissemination of tumor cells into healthy tissues, even very far away from their primary site. Tumor growth and expansion also involves the interaction of malignant and endothelial cells with the surrounding tissue matrix.

Because of its growth and dissemination pattern, GBM has become the main research model for brain tumors. Much of the knowledge about cell invasion and angiogenesis has been obtained by experimental work using animal glioblastoma models; allowing the identification of molecular targets with potential uses in the clinic. Tumor angiogenesis and cell invasion have a great number of pathogenic steps that can be blocked or modified in an effort to inhibit tumor growth and dissemination. However, specific therapeutic tools with a significant impact on the treatment of brain tumors do not exist to this day; despite the efforts, progress in basic research has not been fully reflected in therapeutic options to treat CNS tumors. This could be due to the complexity of the processes that determine the growth and spread of tumors in patients and the lack of tools for the identification of critical times for clinical action. Thus, a greater effort is needed in this research field in order to better understand the interaction between brain tumor cells and the surrounding tissue, especially endothelial cells, to improve therapeutic opportunities.

Editors' Note

Emerging evidence suggests that cancer stem cells (CSC) not only exist, but can trans-differentiate into endothelial cells, smooth muscle-like cells or mural-like cells and pericytes. These new cells contribute to tumor vascularization by sprouting or intussusceptive angiogenesis, vasculogenesis or the recently described vasculogenic mimicry (VM). VM refers to the ability of tumor cells to form functional blood-conducting channel networks lacking endothelial cells. It is unclear whether there is a relationship between VM and mosaic vessels with a lumen lined by both endothelial and tumor cells. VEGFR2 seems to participate in CSC trans-differentiation and VM, but may be independent from VEGF. HIF and vascular-endothelial-cadherin also may play a role in both processes (Mao et al. 2013, Soda et al. 2013, Yao et al. 2013).

References

Abounader, R., B. Lal, C. Luddy, G. Koe, B. Davidson, E.M. Rosen and J. Laterra. 2002. *In vivo* targeting of SF/HGF and c-met expression via U1snRNA/ribozymes inhibits glioma growth and angiogenesis and promotes apoptosis. FASEB J. 16: 108–110.

Abounader, R. and J. Laterra. 2005. Scatter factor/hepatocyte growth factor in brain tumor growth and angiogenesis. Neuro. Oncol. 7: 436–451.

[ACS] American Cancer Society. 2012: http://www.cancer.org/Cancer.

Anand, M.T., E. Van Meter and H.L. Fillmore. 2011. Epidermal growth factor induces matrix metalloproteinase-1 (MMP-1) expression and invasion in glioma cell lines via the MAPK pathway. J. Neurooncol. 104: 679–687.

Angiolillo, A.L., C. Sgadari and G. Tosato. 1996. A role for the interferon-inducible protein 10 in inhibition of angiogenesis by interleukin-12. Ann. NY Acad. Sci. 795: 158–167.

Antoine, N., R. Greimers, C. De Roanne, M. Kusaka, E. Heinen, L.J. Simar and V. Castronovo. 1994. AGM-1470, a potent angiogenesis inhibitor, prevents the entry of normal but not transformed endothelial cells into the G1 phase of the cell cycle. Cancer Res. 54: 2073–2076.

Arnold, S.M., A.B. Young, R.K. Munn, R.A. Patchell, N. Nanayakkara and W.R. Markesbery. 1999. Expression of p53, bcl-2, E-cadherin, matrix metalloproteinase-9, and tissue inhibitor of metalloproteinases-1 in paired primary tumors and brain metastasis. Clin. Cancer Res. 5: 4028–4033.

Barnholtz-Sloan, J.S., A.E. Sloan, F.G. Davis, F.D. Vigneau, P. Lai and R.E. Sawaya. 2004. Incidence proportions of brain metastases in patients diagnosed (1973 to 2001) in the Metropolitan Detroit Cancer Surveillance System. J. Clin. Oncol. 22: 2865–2872.

Beadle, C., M.C. Assanah, P. Monzo, R. Vallee, S.S. Rosenfeld and P. Canoll. 2008. The role of myosin II in glioma invasion of the brain. Mol. Biol. Cell 19: 3357–3368.

Benny, O., S.K. Kim, K. Gvili, I.S. Radzishevsky, A. Mor, L. Verduzco, L.G. Menon, P.M. Black, M. Machluf and R.S. Carroll. 2008. *In vivo* fate and therapeutic efficacy of PF-4/CTF microspheres in an orthotopic human glioblastoma model. FASEB J. 22: 488–499.

Bondy, M.L., M.E. Scheurer, B. Malmer, J.S. Barnholtz-Sloan, F.G. Davis, D. Il'yasova, C. Kruchko, B.J. McCarthy, P. Rajaraman, J.A. Schwartzbaum, S. Sadetzki, B. Schlehofer,

T. Tihan, J.L. Wiemels, M. Wrensch and P.A. Buffler. 2008. Brain tumor epidemiology: consensus from the Brain Tumor Epidemiology Consortium. Cancer 113: 1953–1968.
Bos, P.D., X.H. Zhang, C. Nadal, W. Shu, R.R. Gomis, D.X. Nguyen, A.J. Minn, M.J. van de Vijver, W.L. Gerald, J.A. Foekens and J. Massague. 2009. Genes that mediate breast cancer metastasis to the brain. Nature 459: 1005–1009.
Brahimi-Horn, C., E. Berra and J. Pouyssegur. 2001. Hypoxia: the tumor's gateway to progression along the angiogenic pathway. Trends Cell Biol. 11: S32–S36.
Brastianos, P.K. and T.T. Batchelor. 2009. VEGF inhibitors in brain tumors. Clin. Adv. Hematol. Oncol. 7: 753–760, 768.
Brat, D.J., A.C. Bellail and E.G. Van Meir. 2005. The role of interleukin-8 and its receptors in gliomagenesis and tumoral angiogenesis. Neuro. Oncol. 7: 122–133.
Brem, H., D. Ingber, C.H. Blood, D. Bradley, S. Urioste and J. Folkman. 1991. Suppression of tumor metastasis by angiogenesis inhibition. Surg. Forum 42: 439–441.
Brem, H., I. Gresser, J. Grosfeld and J. Folkman. 1993a. The combination of antiangiogenic agents to inhibit primary tumor growth and metastasis. J. Pediatr. Surg. 28: 1253–1257.
Brem, H. and J. Folkman. 1993b. Analysis of experimental antiangiogenic therapy. J. Pediatr. Surg. 28: 445–451.
Brouty-Boye, D. and B.R. Zetter. 1980. Inhibition of cell motility by interferon. Science 208: 516–518.
Brunda, M., J.L. Luistro, R.R. Warrier, R.B. Wright, B.R. Hubbard, M. Murphy, S.F. Wolf and M.K. Gately. 1993. Antitumor and antimetastatic activity of interleukin 12 against murine tumors. J. Exp. Med. 178: 1223–1230.
Burger, P.C., E.R. Heinz, T. Shibata and P. Kleihues. 1988. Topographic anatomy and CT correlations in the untreated glioblastoma multiforme. J. Neurosurg. 68: 698–704.
Burnet, N.G., S.J. Jefferies, R J. Benson, D.P. Hunt and F.P. Treasure. 2005. Years of life lost (YLL) from cancer is an important measure of population burden—and should be considered when allocating research funds. Br. J. Cancer 92: 241–245.
Cao, R., J. Farnebo, M. Kurimoto and Y. Cao. 1999. Interleukin-18 acts as an angiogenesis and tumor suppressor. FASEB J. 13: 2195–2202.
Carbonell, W., S.O. Ansorge, N. Sibson and R. Muschel. 2009. The vascular basement membrane as "soil" in brain metastasis. PLoS One 4: e5857.
Carmeliet, P. 2003. Angiogenesis in health and disease. Nat. Med. 9: 653–660.
Carmeliet, P. and R.K. Jain. 2011. Molecular mechanisms and clinical applications of angiogenesis. Nature 473: 298–307.
[CBTRUS] Central Brain Tumor Registry of the United States. 2012. http://www.cbtrus.org/reports/reports.html.
Chandana, S., R.S. Movva, M. Arora and T. Singh. 2008. Primary brain tumors in adults. Am. Fam. Physician 77: 1423–1430.
Chekenya, M., M. Hjelstuen, P.O. Enger, F. Thorsen, A.L. Jacob, B. Probst, O. Haraldseth, G. Pilkington, A. Butt, J.M. Levine and R. Bjerkvig. 2002. NG2 proteoglycan promotes angiogenesis-dependent tumor growth in CNS by sequestering angiostatin. FASEB J. 16: 586–588.
Chirivi, R.G., A. Garofalo, M.J. Crimmin, L.J. Bawden, A. Stoppacciaro, P.D. Brown and R. Giavazzi. 1994. Inhibition of the metastatic spread and growth of B16-BL6 murine melanoma by a synthetic matrix metalloproteinase inhibitor. Int. J. Cancer 58: 460–464.
Claes, A., A.J. Idema and P. Wesseling. 2007. Diffuse glioma growth: a guerilla war. Acta Neuropathol. 114: 443–458.
Cohen, T., D. Nahari, L.W. Cerem, G. Neufeld and B.Z. Levi. 1996. Interleukin 6 induces the expression of vascular endothelial growth factor. J. Biol. Chem. 271: 736–741.
Colombo, G., B. Margosio, L. Ragona, M. Neves, S. Bonifacio, D.S. Annis, M. Stravalaci, S. Tomaselli, R. Giavazzi, M. Rusnati, M. Presta, L. Zetta, D.F. Mosher, D. Ribatti, M. Gobbi and G. Taraboletti. 2010. Non-peptidic thrombospondin-1 mimics as fibroblast growth factor-2 inhibitors: an integrated strategy for the development of new antiangiogenic compounds. J. Biol. Chem. 285: 8733–8742.

Coughlin, C.M., K.E. Salhany, M. Wysocka, E. Aruga, H. Kurzawa, A.E. Chang, C.A. Hunter, J.C. Fox, G. Trinchieri and W.M. Lee. 1998. Interleukin-12 and interleukin-18 synergistically induce murine tumor regression which involves inhibition of angiogenesis. J. Clin. Invest. 101: 1441–1452.

D'Abaco, G.M. and A.H. Kaye. 2007. Integrins: molecular determinants of glioma invasion. J. Clin. Neurosci. 14: 1041–1048.

D'Amato, R.J., M.S. Loughnan, E. Flynn and J. Folkman. 1994. Thalidomide is an inhibitor of angiogenesis. Proc. Natl. Acad. Sci. USA 91: 4082–4085.

Davies, B., P.D. Brown, N. East, M.J. Crimmin and F.R. Balkwill. 1993. A synthetic matrix metalloproteinase inhibitor decreases tumor burden and prolongs survival of mice bearing human ovarian carcinoma xenografts. Cancer Res. 53: 2087–2091.

Dawson, D.W., S.F. Pearce, R. Zhong, R.L. Silverstein, W.A. Frazier and N.P. Bouck. 1997. CD36 mediates the *in vitro* inhibitory effects of thrombospondin-1 on endothelial cells. J. Cell Biol. 138: 707–717.

De Vuyst, E., N. Wang, E. Decrock, M. De Bock, M. Vinken, M. Van Moorhem, C. Lai, M. Culot, V. Rogiers, R. Cecchelli, C.C. Naus, W.H. Evans and L. Leybaert. 2009. Ca(2+) regulation of connexin 43 hemichannels in C6 glioma and glial cells. Cell Calcium 46: 176–187.

Delamarre, E., S. Taboubi, S. Mathieu, C. Berenguer, V. Rigot, J.C. Lissitzky, D. Figarella-Branger, L. Ouafik and J. Luis. 2009. Expression of integrin alpha6beta1 enhances tumorigenesis in glioma cells. Am. J. Pathol. 175: 844–855.

Demuth, T. and M.E. Berens. 2004. Molecular mechanisms of glioma cell migration and invasion. J. Neurooncol. 70: 217–228.

DeVore, R.F., C.G. Hellerqvist, G.B. Wakefield, B.D. Wamil, G.B. Thurman, P.A. Minton, H.W. Sundell, H.P. Yan, C.E. Carter, Y.F. Wang, G.E. York, M.H. Zhang and D.H. Johnson. 1997. Phase I study of the antineovascularization drug CM101. Clin. Cancer Res. 3: 365–372.

Dinney, C.P., D.R. Bielenberg, P. Perrotte, R. Reich, B.Y. Eve, C.D. Bucana and I.J. Fidler. 1998. Inhibition of basic fibroblast growth factor expression, angiogenesis, and growth of human bladder carcinoma in mice by systemic interferon-alpha administration. Cancer Res. 58: 808–814.

Duda, D.G., M. Sunamura, L. Lozonschi, T. Kodama, S. Egawa, G. Matsumoto, H. Shimamura, K. Shibuya, K. Takeda and S. Matsuno. 2000. Direct *in vitro* evidence and *in vivo* analysis of the antiangiogenesis effects of interleukin 12. Cancer Res. 60: 1111–1116.

Dunn, I.F., O. Heese and P.M. Black. 2000. Growth factors in glioma angiogenesis: FGFs, PDGF, EGF, and TGFs. J. Neurooncol. 50: 121–137.

Eckhardt, S.G., H.A. Burris, J.R. Eckardt, G. Weiss, G. Rodriguez, M. Rothenberg, D. Rinaldi, R. Barrington, J.G. Kuhn, K. Masuo, K. Sudo, R. Atsumi, T. Oguma, L. Higashi, S. Fields, L. Smetzer and D.D. Von Hoff. 1996. A phase I clinical and pharmacokinetic study of the angiogenesis inhibitor, tecogalan sodium. Ann. Oncol. 7: 491–496.

Enam, S.A., M.L. Rosenblum and K. Edvardsen. 1998. Role of extracellular matrix in tumor invasion: migration of glioma cells along fibronectin-positive mesenchymal cell processes. Neurosurgery 42: 599–608.

Esencay, M., E.W. Newcomb and D. Zagzag. 2010. HGF upregulates CXCR4 expression in gliomas via NF-kappaB: implications for glioma cell migration. J. Neurooncol. 99: 33–40.

Fan, J., B. Cai, M. Zeng, Y. Hao, F.G. Giancotti and B.M. Fu. 2011. Integrin beta-4 signaling promotes mammary tumor cell adhesion to brain microvascular endothelium by inducing ErbB2-mediated secretion of VEGF. Ann. Biomed. Eng. 39: 2223–2241.

Farber, K., M. Synowitz, G. Zahn, D. Vossmeyer, R. Stragies, N. van Rooijen and H. Kettenmann. 2008. An alpha5beta1 integrin inhibitor attenuates glioma growth. Mol. Cell. Neurosci. 39: 579–585.

Feng, S., J. Cen, Y. Huang, H. Shen, L. Yao, Y. Wang and Z. Chen. 2011. Matrix metalloproteinase-2 and -9 secreted by leukemic cells increase the permeability of blood-brain barrier by disrupting tight junction proteins. PLoS One 6: e20599.

Ferrara, N. and R.S. Kerbel. 2005. Angiogenesis as a therapeutic target. Nature 438: 967–974.

Fidler, I., J.K. Balasubramanian, Q. Lin, S.W. Kim and S.J. Kim. 2010. The brain microenvironment and cancer metastasis. Mol. Cells 30: 93–98.

Figg, W.D., J.M. Pluda, R.M. Lush, M.W. Saville, K. Wyvill, E. Reed and R. Yarchoan. 1997. The pharmacokinetics of TNP-470, a new angiogenesis inhibitor. Pharmacotherapy 17: 91–97.

Folkman, J. 1971. Tumor angiogenesis: therapeutic implications. N. Engl. J. Med. 285: 1182–1186.

Forsyth, P.A., H. Wong, T.D. Laing, N.B. Rewcastle, D.G. Morris, H. Muzik, K.J. Leco, R.N. Johnston, P.M. Brasher, G. Sutherland and D.R. Edwards. 1999. Gelatinase-A (MMP-2), gelatinase-B (MMP-9) and membrane type matrix metalloproteinase-1 (MT1-MMP) are involved in different aspects of the pathophysiology of malignant gliomas. Br. J. Cancer 79: 1828–1835.

Friedman, H.S. and D.D. Bigner. 2005. Glioblastoma multiforme and the epidermal growth factor receptor. N. Engl. J. Med. 353: 1997–1999.

[GLOBOCAN] Cancer incidence, martality an prevalence worlwide in 2008: http://globocan.iarc.fr/.

Goldbrunner, R.H., J.J. Bernstein and J.C. Tonn. 1998. ECM-mediated glioma cell invasion. Microsc. Res. Tech. 43: 250–257.

Goldbrunner, R.H., J.J. Bernstein and J.C. Tonn. 1999. Cell-extracellular matrix interaction in glioma invasion. Acta Neurochir. (Wien.) 141: 295–305.

Gondi, C.S., S.S. Lakka, D.H. Dinh, W.C. Olivero, M. Gujrati and J.S. Rao. 2004. Downregulation of uPA, uPAR and MMP-9 using small, interfering, hairpin RNA (siRNA) inhibits glioma cell invasion, angiogenesis and tumor growth. Neuron Glia Biology 1: 165–176.

Good, D.J., P.J. Polverini, F. Rastinejad, M.M. Le Beau, R.S. Lemons, W.A. Frazier and N.P. Bouck. 1990. A tumor suppressor-dependent inhibitor of angiogenesis is immunologically and functionally indistinguishable from a fragment of thrombospondin. Proc. Natl. Acad. Sci. USA 87: 6624–6628.

Hanahan, D. and J. Folkman. 1996. Patterns and emerging mechanisms of the angiogenic switch during tumorigenesis. Cell 86: 353–364.

Hanson, F.R. and T.E. Eble. 1949. An antiphage agent isolated from Aspergillus sp. J. Bacteriol. 58: 527–529.

Hellerqvist, C.G., G.B. Thurman, D.L. Page, Y.F. Wang, B.A. Russell, C.A. Montgomery and H.W. Sundell. 1993. Antitumor effects of GBS toxin: a polysaccharide exotoxin from group B beta-hemolytic streptococcus. J. Cancer Res. Clin. Oncol. 120: 63–70.

Herbert, J.M., M. Cottineau, F. Driot, J.M. Pereillo, J.P. Maffrand. 1988. Activity of pentosan polysulphate and derived compounds on vascular endotheliat cell proliferation and migration induced by acidic and basic FGF *in vitro*. Biochem. Pharmacol. 37: 4281–4288.

Hoeben, A., B. Landuyt, M.S. Highley, H. Wildiers, A.T. Van Oosterom and E.A. De Bruijn. 2004. Vascular endothelial growth factor and angiogenesis. Pharmacol. Rev. 56: 549–580.

Holash, J., P.C. Maisonpierre, D. Compton, P. Boland, C.R. Alexander, D. Zagzag, G.D. Yancopoulos and S.J. Wiegand. 1999. Vessel cooption, regression, and growth in tumors mediated by angiopoietins and VEGF. Science 284: 1994–1998.

Hu, B., P. Guo, Q. Fang, H.Q. Tao, D. Wang, M. Nagane, H.J. Huang, Y. Gunji, R. Nishikawa, K. Alitalo, W.K. Cavenee and S.Y. Cheng. 2003. Angiopoietin-2 induces human glioma invasion through the activation of matrix metalloprotease-2. Proc. Natl. Acad. Sci. USA 100: 8904–8909.

Hu, B., M.J. Jarzynka, P. Guo, Y. Imanishi, D.D. Schlaepfer and S.Y. Cheng. 2006. Angiopoietin 2 induces glioma cell invasion by stimulating matrix metalloprotease 2 expression through the alphavbeta1 integrin and focal adhesion kinase signaling pathway. Cancer Res. 66: 775–783.

Huang, S., S.E. Ullrich and M. Bar-Eli. 1999. Regulation of tumor growth and metastasis by interleukin-10: the melanoma experience. J. Interferon Cytokine Res. 19: 697–703.

Ingber, D., T. Fujita, S. Kishimoto, K. Sudo, T. Kanamaru, H. Brem and J. Folkman. 1990. Synthetic analogues of fumagillin that inhibit angiogenesis and suppress tumor growth. Nature 348: 555–557.

Inoue, K., H. Korenaga, N.G. Tanaka, N. Sakamoto and S. Kadoya. 1988. The sulfated polysaccharide-peptidoglycan complex potently inhibits embryonic angiogenesis and tumor growth in the presence of cortisone acetate. Carbohydr. Res. 181: 135–142.

Italiano, J.E. Jr., J.L. Richardson, S. Patel-Hett, E. Battinelli, A. Zaslavsky, S. Short, S. Ryeom, J. Folkman and G.L. Klement. 2008. Angiogenesis is regulated by a novel mechanism: pro- and antiangiogenic proteins are organized into separate platelet alpha granules and differentially released. Blood 111: 1227–1233.

Ivkovic, S., C. Beadle, S. Noticewala, S.C. Massey, K.R. Swanson, L.N. Toro, A.R. Bresnick, P. Canoll and S.S. Rosenfeld. 2012. Direct inhibition of myosin II effectively blocks glioma invasion in the presence of multiple motogens. Mol. Biol. Cell 23: 533–542.

Jain, R.K. 2003. Molecular regulation of vessel maturation. Nat. Med. 9: 685–693.

Jain, R.K., E. di Tomaso, D.G. Duda, J.S. Loeffler, A.G. Sorensen and T.T. Batchelor. 2007. Angiogenesis in brain tumors. Nat. Rev. Neurosci. 8: 610–622.

Jee, S.H., C.Y. Chu, H.C. Chiu, Y.L. Huang, W.L. Tsai, Y.H. Liao and M.L. Kuo. 2004. Interleukin-6 induced basic fibroblast growth factor-dependent angiogenesis in basal cell carcinoma cell line via JAK/STAT3 and PI3-kinase/Akt pathways. J. Invest. Dermatol. 123: 1169–1175.

Ji, W.R., F.J. Castellino, Y. Chang, M.E. Deford, H. Gray, X. Villarreal, M.E. Kondri, D.N. Marti, M. Llinas, J. Schaller, R.A. Kramer and P.A. Trail. 1998. Characterization of kringle domains of angiostatin as antagonists of endothelial cell migration, an important process in angiogenesis. FASEB J. 12: 1731–1738.

Jiang, Y., M. Boije, B. Westermark and L. Uhrbom. 2011. PDGF-B Can sustain self-renewal and tumorigenicity of experimental glioma-derived cancer-initiating cells by preventing oligodendrocyte differentiation. Neoplasia 13: 492–503.

Jimenez, T., W.P. Fox, C.C. Naus, J. Galipeau and D.J. Belliveau. 2006. Connexin over-expression differentially suppresses glioma growth and contributes to the bystander effect following HSV-thymidine kinase gene therapy. Cell Commun. Adhes. 13: 79–92.

Kaur, B., F.W. Khwaja, E.A. Severson, S.L. Matheny, D.J. Brat and E.G. Van Meir. 2005. Hypoxia and the hypoxia-inducible-factor pathway in glioma growth and angiogenesis. Neuro-oncol. 7: 134–153.

Kawataki, T., T. Yamane, H. Naganuma, P. Rousselle, I. Anduren, K. Tryggvason and M. Patarroyo. 2007. Laminin isoforms and their integrin receptors in glioma cell migration and invasiveness: Evidence for a role of alpha5-laminin(s) and alpha3beta1 integrin. Exp. Cell Res. 313: 3819–3831.

Kerbel, R. and J. Folkman. 2002. Clinical translation of angiogenesis inhibitors. Nat. Rev. Cancer 2: 727–739.

Kesari, S. and C.D. Stiles. 2006. The bad seed: PDGF receptors link adult neural progenitors to glioma stem cells. Neuron 51: 151–153.

Kim, S., K. Bell, S.A. Mousa and J.A. Varner. 2000. Regulation of angiogenesis *in vivo* by ligation of integrin alpha5beta1 with the central cell-binding domain of fibronectin. Am. J. Pathol. 156: 1345–1362.

Kim, W., Y. and H.Y. Lee. 2009. Brain angiogenesis in developmental and pathological processes: mechanism and therapeutic intervention in brain tumors. FEBS J. 276: 4653–4664.

Kleihues, P., F. Soylemezoglu, B. Schauble, B.W. Scheithauer and P.C. Burger. 1995. Histopathology, classification, and grading of gliomas. Glia 15: 211–221.

Klein-Soyer, C., A. Beretz, J.P. Cazenave, E. Wittendorp-Rechenmann, J.L. Vonesch, R.V. Rechenmann, F. Driot and J.P. Maffrand. 1989. Sulfated polysaccharides modulate effects of acidic and basic fibroblast growth factors on repair of injured confluent human vascular endothelium. Arteriosclerosis 9: 147–153.

Kowanetz, M. and N. Ferrara. 2006. Vascular endothelial growth factor signaling pathways: therapeutic perspective. Clin. Cancer Res. 12: 5018–5022.

Kudelka, A., P.T. Levy, C.F. Verschraegen, C.L. Edwards, S. Piamsomboon, W. Termrungruanglert, R.S. Freedman, A.L. Kaplan, D.G. Kieback, C.A. Meyers, K.A. Jaeckle, E. Loyer, M. Steger, R. Mante, G. Mavligit, A. Killian, R.A. Tang, J.U. Gutterman and J.J. Kavanagh. 1997. A phase I study of TNP-470 administered to patients with advanced squamous cell cancer of the cervix. Clin. Cancer Res. 3: 1501–1505.

Kusaka, M., K. Sudo, T. Fujita, S. Marui, F. Itoh, D. Ingber and J. Folkman. 1991. Potent anti-angiogenic action of AGM-1470: comparison to the fumagillin parent. Biochem. Biophys. Res. Commun. 174: 1070–1076.

Kusaka, M., K. Sudo, E. Matsutani, Y. Kozai, S. Marui, T. Fujita, D. Ingber and J. Folkman. 1994. Cytostatic inhibition of endothelial cell growth by the angiogenesis inhibitor TNP-470 (AGM-1470). Br. J. Cancer 69: 212–216.

Lakka, S.S., C.S. Gondi and J.S. Rao. 2005. Proteases and glioma angiogenesis. Brain Pathol. 15: 327–341.

Lawler, P.R. and J. Lawler. 2012. Molecular basis for the regulation of angiogenesis by thrombospondin-1 and -2. Cold Spring Harb. Perspect. Med. 2: a006627.

Leavesley, D.I., G.D. Ferguson, E.A. Wayner and D.A. Cheresh. 1992. Requirement of the integrin beta 3 subunit for carcinoma cell spreading or migration on vitronectin and fibrinogen. J. Cell Biol. 117: 1101–1107.

Leavesley, D.I., M.A. Schwartz, M. Rosenfeld and D.A. Cheresh. 1993. Integrin beta 1- and beta 3-mediated endothelial cell migration is triggered through distinct signaling mechanisms. J. Cell Biol. 121: 163–170.

Levin, V., A.S. Phuphanich, W.K. Yung, P.A. Forsyth, R.D. Maestro, J.R. Perry, G.N. Fuller and M. Baillet. 2006. Randomized, double-blind, placebo-controlled trial of marimastat in glioblastoma multiforme patients following surgery and irradiation. J. Neurooncol. 78: 295–302.

Lewis-Tuffin, L., J.F. Rodriguez, C. Giannini, B. Scheithauer, B.M. Necela, J.N. Sarkaria and P.Z. Anastasiadis. 2010. Misregulated E-cadherin expression associated with an aggressive brain tumor phenotype. PLoS One 5: e13665.

Li, A., S. Dubey, M.L. Varney, B.J. Dave and R.K. Singh. 2003. IL-8 directly enhanced endothelial cell survival, proliferation, and matrix metalloproteinases production and regulated angiogenesis. J. Immunol. 170: 3369–3376.

Lippi, G. and E.J. Favaloro. 2010. Recombinant platelet factor 4: a therapeutic, anti-neoplastic chimera? Semin. Thromb. Hemost. 36: 558–569.

Loboda, A., A. Jozkowicz and J. Dulak. 2012. HIF-1 versus HIF-2—is one more important than the other? Vascul. Pharmacol. 56: 245–251.

Louis, D., N.H. Ohgaki, O.D. Wiestler, W.K. Cavenee, P.C. Burger, A. Jouvet, B.W. Scheithauer and P. Kleihues. 2007. The 2007 WHO classification of tumors of the central nervous system. Acta Neuropathol. 114: 97–109.

Lucas, R., L. Holmgren, I. Garcia, B. Jimenez, S.J. Mandriota, F. Borlat, B.K. Sim, Z. Wu, G.E. Grau, Y. Shing, G.A. Soff, N. Bouck and M.S. Pepper. 1998. Multiple forms of angiostatin induce apoptosis in endothelial cells. Blood 92: 4730–4741.

Lund-Johansen, M., R. Bjerkvig, P.A. Humphrey, S.H. Bigner, D.D. Bigner and O.D. Laerum. 1990. Effect of epidermal growth factor on glioma cell growth, migration, and invasion *in vitro*. Cancer Res. 50: 6039–6044.

Maher, E.A., F.B. Furnari, R.M. Bachoo, D.H. Rowitch, D.N. Louis, W.K. Cavenee and R.A. DePinho. 2001. Malignant glioma: genetics and biology of a grave matter. Genes Dev. 15: 1311–1333.

Mahesparan, R., T.A. Read, M. Lund-Johansen, K.O. Skaftnesmo, R. Bjerkvig and O. Engebraaten. 2003. Expression of extracellular matrix components in a highly infiltrative *in vivo* glioma model. Acta Neuropathol. 105: 49–57.

Maione, T.E., G.S. Gray, J. Petro, A.J. Hunt, A.L. Donner, S.I. Bauer, H.F. Carson and R.J. Sharpe. 1990. Inhibition of angiogenesis by recombinant human platelet factor-4 and related peptides. Science 247: 77–79.

Maione, T.E., G.S. Gray, A.J. Hunt and R.J. Sharpe. 1991. Inhibition of tumor growth in mice by an analogue of platelet factor 4 that lacks affinity for heparin and retains potent angiostatic activity. Cancer Res. 51: 2077–2083.

Maisonpierre, P., C.C. Suri, P.F. Jones, S. Bartunkova, S.J. Wiegand, C. Radziejewski, D. Compton, J. McClain, T.H. Aldrich, N. Papadopoulos, T.J. Daly, S. Davis, T.N. Sato and G.D. Yancopoulos. 1997. Angiopoietin-2, a natural antagonist for Tie2 that disrupts *in vivo* angiogenesis. Science 277: 55–60.

Malemud, C.J. 2006. Matrix metalloproteinases (MMPs) in health and disease: an overview. Front. Biosci. 11: 1696–1701.

Mandriota, S.J. and M.S. Pepper. 1997. Vascular endothelial growth factor-induced *in vitro* angiogenesis and plasminogen activator expression are dependent on endogenous basic fibroblast growth factor. J. Cell Sci. 110: 2293–2302.

Mao, X.G., X.Y. Xue, L. Wang, X. Zhang, M. Yan, Y.Y. Tu, W. Lin, X.F. Jiang, H.G. Ren, W. Zhang and S.J. Song. 2013. CDH5 is specifically activated in glioblastoma stemlike cells and contributes to vasculogenic mimicry induced by hypoxia. Neuro Oncol. 15: 865–879.

Mazurek, S.G., J. Li, G.H. Nabozny, G.A. Reinhart, A.C. Muthukumarana, P.C. Harrison and R.M. Fryer. 2011. Functional biomarkers of musculoskeletal syndrome (MSS) for early *in vivo* screening of selective MMP-13 inhibitors. J. Pharmacol. Toxicol. Methods 64: 89–96.

McComb, R.D. and P.C. Burger. 1985. Pathologic analysis of primary brain tumors. Neurol. Clin. 3: 711–728.

Mendes, O.H., T. Kim and G. Stoica. 2005. Expression of MMP2, MMP9 and MMP3 in breast cancer brain metastasis in a rat model. Clin. Exp. Metastasis 22: 237–246.

Mendes, O.H., T. Kim, G. Lungu and G. Stoica. 2007. MMP2 role in breast cancer brain metastasis development and its regulation by TIMP2 and ERK1/2. Clin. Exp. Metastasis 24: 341–351.

Moser, T.L., M.S. Stack, I. Asplin, J.J. Enghild, P. Hojrup, L. Everitt, S. Hubchak, H.W. Schnaper and S.V. Pizzo. 1999. Angiostatin binds ATP synthase on the surface of human endothelial cells. Proc. Natl. Acad. Sci. USA 96: 2811–2816.

Moulding, H.D., D.P. Friedman, M. Curtis, L. Kenyon, A.E. Flanders, S.H. Paek and D.W. Andrews. 2008. Revisiting anaplastic astrocytomas I: an expansive growth pattern is associated with a better prognosis. J. Magn. Reson. Imaging 28: 1311–1321.

Nakada, M., Y. Okada and J. Yamashita. 2003. The role of matrix metalloproteinases in glioma invasion. Front. Biosci. 8: e261–269.

Nakagawa, T., T. Kubota, M. Kabuto and T. Kodera. 1996. Hyaluronic acid facilitates glioma cell invasion *in vitro*. Anticancer Res. 16: 2917–2922.

Nathoo, N., A. Chahlavi, G.H. Barnett and S.A. Toms. 2005. Pathobiology of brain metastases. J. Clin. Pathol. 58: 237–242.

Nayak, L., E.Q. Lee and P.Y. Wen. 2012. Epidemiology of brain metastases. Curr. Oncol. Rep. 14: 48–54.

Nazarenko, I., S.M. Hede, X. He, A. Hedren, J. Thompson, M.S. Lindstrom and M. Nister. 2012. PDGF and PDGF receptors in glioma. Ups. J. Med. Sci. 117: 99–112.

[NCI] National Cancer Institute. 2012. http://www.cancer.gov/cancertopics.

Nicolson, G.L., D.G. Menter, J. Herrmann, P. Cavanaugh, L. Jia, J. Hamada, Z. Yun, M. Nakajima and D. Marchetti. 1994. Tumor metastasis to brain: role of endothelial cells, neurotrophins, and paracrine growth factors. Crit. Rev. Oncog. 5: 451–471.

Nicolson, G.L., D.G. Menter, J.L. Herrmann, Z. Yun, P. Cavanaugh and D. Marchetti. 1996. Brain metastasis: role of trophic, autocrine, and paracrine factors in tumor invasion and colonization of the central nervous system. Curr. Top. Microbiol. Immunol. 213: 89–115.

Nie, F., J. Yang, S. Wen, Y.L. An, J. Ding, S.H. Ju, Z. Zhao, H.J. Chen, X.G. Peng, S.T. Wong, H. Zhao and G.J. Teng. 2012. Involvement of epidermal growth factor receptor overexpression in the promotion of breast cancer brain metastasis. Cancer 118: 5198–5209.

Nyberg, P., L. Xie and R. Kalluri. 2005. Endogenous inhibitors of angiogenesis. Cancer Res. 65: 3967–3979.

Ohnishi, T., N. Arita, S. Hiraga, T. Taki, S. Izumoto, Y. Fukushima and T. Hayakawa. 1997. Fibronectin-mediated cell migration promotes glioma cell invasion through chemokinetic activity. Clin. Exp. Metastasis 15: 538–546.

Onishi, M., T. Ichikawa, K. Kurozumi and I. Date. 2011. Angiogenesis and invasion in glioma. Brain Tumor Pathol. 28: 13–24.

O'Reilly, M.S., L. Holmgren, Y. Shing, C. Chen, R.A. Rosenthal, M. Moses, W.S. Lane, Y. Cao, E.H. Sage and J. Folkman. 1994. Angiostatin: a novel angiogenesis inhibitor that mediates the suppression of metastases by a Lewis lung carcinoma. Cell 79: 315–328.

O'Reilly, M.S., T. Boehm, Y. Shing, N. Fukai, G. Vasios, W.S. Lane, E. Flynn, J.R. Birkhead, B.R. Olsen and J. Folkman. 1997. Endostatin: an endogenous inhibitor of angiogenesis and tumor growth. Cell 88: 277–285.

Otrock, Z.K., R.A. Mahfouz, J.A. Makarem and A.I. Shamseddine. 2007. Understanding the biology of angiogenesis: review of the most important molecular mechanisms. Blood Cells Mol. Dis. 39: 212–220.

Pardo, O.E., A. Arcaro, G. Salerno, S. Raguz, J. Downward and M.J. Seckl. 2002. Fibroblast growth factor-2 induces translational regulation of Bcl-XL and Bcl-2 via a MEK-dependent pathway: correlation with resistance to etoposide-induced apoptosis. J. Biol. Chem. 277: 12040–12046.

Pluda, J.M., L.E. Shay, A. Foli, S. Tannenbaum, P.J. Cohen, B.R. Goldspiel, D. Adamo, M.R. Cooper, S. Broder and R. Yarchoan. 1993. Administration of pentosan polysulfate to patients with human immunodeficiency virus-associated Kaposi's sarcoma. J. Natl. Cancer Inst. 85: 1585–1592.

Preusser, M., D. Capper, A. Ilhan-Mutlu, A.S. Berghoff, P. Birner, R. Bartsch, C. Marosi, C. Zielinski, M.P. Mehta, F. Winkler, W. Wick and A. von Deimling. 2012. Brain metastases: pathobiology and emerging targeted therapies. Acta Neuropathol. 123: 205–222.

Pytel, P. and R.V. Lukas. 2009. Update on diagnostic practice: tumors of the nervous system. Arch. Pathol. Lab. Med. 133: 1062–1077.

Rahmathulla, G.S., A. Toms and R.J. Weil. 2012. The molecular biology of brain metastasis. J. Oncol. 723541.

Ramis, G., E. Thomas-Moya, S. Fernandez de Mattos, J. Rodriguez and P. Villalonga. 2012. EGFR inhibition in glioma cells modulates Rho signaling to inhibit cell motility and invasion and cooperates with temozolomide to reduce cell growth. PLoS One 7: e38770.

Rao Malla, R., S. Gopinath, K. Alapati, B. Gorantla, C.S. Gondi and J.S. Rao. 2012. Knockdown of cathepsin B and uPAR inhibits CD151 and alpha3beta1 integrin-mediated cell adhesion and invasion in glioma. Mol. Carcinog. doi: 10.1002/mc.21915.

Rao, J.S., M. Yamamoto, S. Mohaman, Z.L. Gokaslan, G.N. Fuller, W.G. Stetler-Stevenson, V.H. Rao, L.A. Liotta, G.L. Nicolson and R.E. Sawaya. 1996. Expression and localization of 92 kDa type IV collagenase/gelatinase B (MMP-9) in human gliomas. Clin. Exp. Metastasis 14: 12–18.

Rege, T.A., C.Y. Fears and C.L. Gladson. 2005. Endogenous inhibitors of angiogenesis in malignant gliomas: nature's antiangiogenic therapy. Neuro-oncol. 7: 106–121.

Rehn, M., T. Veikkola, E. Kukk-Valdre, H. Nakamura, M. Ilmonen, C. Lombardo, T. Pihlajaniemi, K. Alitalo and K. Vuori. 2001. Interaction of endostatin with integrins implicated in angiogenesis. Proc. Natl. Acad. Sci. USA 98: 1024–1029.

Ren, B., K. Song, S. Parangi, T. Jin, M. Ye, R. Humphreys, M. Duquette, X. Zhang, N. Benhaga, J. Lawler and R. Khosravi-Far. 2009. A double hit to kill tumor and endothelial cells by TRAIL and antiangiogenic 3TSR. Cancer Res. 69: 3856–3865.

Ridley, A.J., M.A. Schwartz, K. Burridge, R.A. Firtel, M.H. Ginsberg, G. Borisy, J.T. Parsons and A.R. Horwitz. 2003. Cell migration: integrating signals from front to back. Science 302: 1704–1709.

Rivera Luna, R. 2006. Tumores del sistema nervioso. pp. 235–236. *In*: R. Rivera Luna, R. Cardenas Cardos and R. Paredes Aguilera [eds.]. Hemato-oncología Pediátrica. Principios Generales. Editores de Textos Mexicanos, México D.F., México.

Rodriguez-Manzaneque, J.C., T.F. Lane, M.A. Ortega, R.O. Hynes, J. Lawler and M.L. Iruela-Arispe. 2001. Thrombospondin-1 suppresses spontaneous tumor growth and inhibits activation of matrix metalloproteinase-9 and mobilization of vascular endothelial growth factor. Proc. Natl. Acad. Sci. USA 98: 12485–12490.

Rojiani, M., V.J. Alidina, N. Esposito and A.M. Rojiani. 2010. Expression of MMP-2 correlates with increased angiogenesis in CNS metastasis of lung carcinoma. Int. J. Clin. Exp. Pathol. 3: 775–781.

Rooprai, H., K.T. Vanmeter, C. Panou, S. Schnull, G. Trillo-Pazos, D. Davies and G.J. Pilkington. 1999. The role of integrin receptors in aspects of glioma invasion *in vitro*. Int. J. Dev. Neurosci. 17: 613–623.

Ryuto, M., M. Ono, H. Izumi, S. Yoshida, H.A. Weich, K. Kohno and M. Kuwano. 1996. Induction of vascular endothelial growth factor by tumor necrosis factor alpha in human glioma cells. Possible roles of SP-1. J. Biol. Chem. 271: 28220–28228.

Salhia, B., N.L. Tran, M. Symons, J.A. Winkles, J.T. Rutka and M.E. Berens. 2006. Molecular pathways triggering glioma cell invasion. Expert Rev. Mol. Diagn. 6: 613–626.

Santarelli, J.G., V. Sarkissian, L.C. Hou, A. Veeravagu and V. Tse. 2007. Molecular events of brain metastasis. Neurosurg. Focus 22: E1.

Sarkar, S.R., K. Nuttall, S. Liu, D.R. Edwards and V.W. Yong. 2006. Tenascin-C stimulates glioma cell invasion through matrix metalloproteinase-12. Cancer Res. 66: 11771–11780.

Sato, Y. and D.B. Rifkin. 1988. Autocrine activities of basic fibroblast growth factor: regulation of endothelial cell movement, plasminogen activator synthesis, and DNA synthesis. J. Cell Biol. 107: 1199–1205.

Sato, Y., M. Abe and R. Takaki. 1990. Platelet factor 4 blocks the binding of basic fibroblast growth factor to the receptor and inhibits the spontaneous migration of vascular endothelial cells. Biochem. Biophys. Res. Commun. 172: 595–600.

Scherer, H.J. 1940. The forms of growth in gliomas and their practical significance. Brain. 63: 1–35.

Schmidt, N.O., M. Westphal, C. Hagel, S. Ergun, D. Stavrou, E.M. Rosen and K. Lamszus. 1999. Levels of vascular endothelial growth factor, hepatocyte growth factor/scatter factor and basic fibroblast growth factor in human gliomas and their relation to angiogenesis. Int. J. Cancer 84: 10–18.

Schouten, L.J., J. Rutten, H.A. Huveneers and A. Twijnstra. 2002. Incidence of brain metastases in a cohort of patients with carcinoma of the breast, colon, kidney, and lung and melanoma. Cancer 94: 2698–2705.

Scott, P. 1993. IL-12: initiation cytokine for cell-mediated immunity. Science 260: 496–497.

Sgadari, C.A., L. Angiolillo and G. Tosato. 1996. Inhibition of angiogenesis by interleukin-12 is mediated by the interferon-inducible protein 10. Blood 87: 3877–3882.

Sharpe, R.J., H.R. Byers, C.F. Scott, S.I. Bauer and T.E. Maione. 1990. Growth inhibition of murine melanoma and human colon carcinoma by recombinant human platelet factor 4. J. Natl. Cancer Inst. 82: 848–853.

Sidky, Y.A. and E.C. Borden. 1987. Inhibition of angiogenesis by interferons: effects on tumor- and lymphocyte-induced vascular responses. Cancer Res. 47: 5155–5161.

Siegel, R., D. Naishadham and A. Jemal. 2012. Cancer statistics, 2012. CA Cancer J. Clin. 62: 10–29.

Simons, M. 2004. Integrative signaling in angiogenesis. Mol. Cell. Biochem. 264: 99–102.

Slaton, J., W.P. Perrotte, K. Inoue, C.P. Dinney and I.J. Fidler. 1999. Interferon-alpha-mediated down-regulation of angiogenesis-related genes and therapy of bladder cancer are dependent on optimization of biological dose and schedule. Clin. Cancer Res. 5: 2726–2734.

Soda, Y., C. Myskiw, A. Rommel and I.M. Verma. 2013. Mechanisms of neovascularization and resistance to anti-angiogenic therapies in glioblastoma multiforme. J. Mol. Med. (Berl.) 91: 439–448.

Sotomayor, E.A., B.A. Teicher, G.N. Schwartz, S.A. Holden, K. Menon, T.S. Herman and E. Frei. 3rd. 1992. Minocycline in combination with chemotherapy or radiation therapy *in vitro* and *in vivo*. Cancer Chemother. Pharmacol. 30: 377–384.

Stratmann, A., W. Risau and K.H. Plate. 1998. Cell type-specific expression of angiopoietin-1 and angiopoietin-2 suggests a role in glioblastoma angiogenesis. Am. J. Pathol. 153: 1459–1466.

Strieter, R.M., S.L. Kunkel, D.A. Arenberg, M.D. Burdick and P.J. Polverini. 1995. Interferon gamma-inducible protein 10 (IP-10), a member of the C-X-C chemokine family, is an inhibitor of angiogenesis. Biochem. Biophys. Res. Commun. 210: 51–57.

Sugita Y. 2011. Angiogenesis of primary brain tumors: the role of endoglin (CD105). *In*: M. Garami [ed.]. Molecular Targets of CNS Tumors. InTech. Online. Available from: http://www.intechopen.com/books/molecular-targets-of-cns-tumors/angiogenesis-of-primary-brain-tumors-the-role-of-endoglin-cd105-.

Sund, M., Y. Hamano, H. Sugimoto, A. Sudhakar, M. Soubasakos, U. Yerramalla, L.E. Benjamin, J. Lawler, M. Kieran, A. Shah and R. Kalluri. 2005. Function of endogenous inhibitors of angiogenesis as endothelium-specific tumor suppressors. Proc. Natl. Acad. Sci. USA 102: 2934–2939.

Swain, S.M., B. Parker, A. Wellstein, M.E. Lippman, C. Steakley and R. DeLap. 1995. Phase I trial of pentosan polysulfate. Invest. New Drugs 13: 55–62.

Takamiya, Y.R., M. Friedlander, H. Brem, A. Malick and R.L. Martuza. 1993. Inhibition of angiogenesis and growth of human nerve-sheath tumors by AGM-1470. J. Neurosurg. 78: 470–476.

Tanaka, N.G., N. Sakamoto, K. Inoue, H. Korenaga, S. Kadoya, H. Ogawa and Y. Osada. 1989. Antitumor effects of an antiangiogenic polysaccharide from an Arthrobacter species with or without a steroid. Cancer Res. 49: 6727–6730.

Tanaka, N.G., N. Sakamoto, H. Korenaga, K. Inoue, H. Ogawa and Y. Osada. 1991. The combination of a bacterial polysaccharide and tamoxifen inhibits angiogenesis and tumor growth. Int. J. Radiat. Biol. 60: 79–83.

Tate, M.C. and M.K. Aghi. 2009. Biology of angiogenesis and invasion in glioma. Neurotherapeutics 6: 447–457.

Teicher, B.A., E.A. Sotomayor and Z.D. Huang. 1992. Antiangiogenic agents potentiate cytotoxic cancer therapies against primary and metastatic disease. Cancer Res. 52: 6702–6704.

Thurman, G.B., B.A. Russel, G.E. York, Y.F. Wang, D.L. Page, H.W. Sundell and C.G. Hellerqvist. 1994. Effects of group B Streptococcus toxin on long-term survival of mice bearing transplanted Madison lung tumors. J. Cancer Res. Clin. Oncol. 120: 479–484.

Tonn, J.C. and R. Goldbrunner. 2003. Mechanisms of glioma cell invasion. Acta Neurochir. Suppl. 88: 163–167.

Treasurywala, S. and M.E. Berens. 1998. Migration arrest in glioma cells is dependent on the alphav integrin subunit. Glia 24: 236–243.

Trinchieri, G. 1994. Interleukin-12: a cytokine produced by antigen-presenting cells with immunoregulatory functions in the generation of T-helper cells type 1 and cytotoxic lymphocytes. Blood 84: 4008–4027.

Verhaak, R.G., K.A. Hoadley, E. Purdom, V. Wang, Y. Qi, M.D. Wilkerson, C.R. Miller, L. Ding, T. Golub, J.P. Mesirov, G. Alexe, M. Lawrence, M. O'Kelly, P. Tamayo, B.A. Weir, S. Gabriel, W. Winckler, S. Gupta, L. Jakkula, H.S. Feiler, J.G. Hodgson, C.D. James, J.N. Sarkaria, C. Brennan, A. Kahn, P.T. Spellman, R.K. Wilson, T.P. Speed, J.W. Gray, M. Meyerson, G. Getz, C.M. Perou and D.N. Hayes. 2010. Integrated genomic analysis identifies clinically relevant subtypes of glioblastoma characterized by abnormalities in PDGFRA, IDH1, EGFR, and NF1. Cancer Cell 17: 98–110.

Voest, E.E., B.M. Kenyon, M.S. O'Reilly, G. Truitt, R.J. D'Amato and J. Folkman. 1995. Inhibition of angiogenesis *in vivo* by interleukin 12. J. Natl. Cancer Inst. 87: 581–586.

Volpert, O.V., T. Fong, A.E. Koch, J.D. Peterson, C. Waltenbaugh, R.I. Tepper and N.P. Bouck. 1998. Inhibition of angiogenesis by interleukin 4. J. Exp. Med. 188: 1039–1046.

Walker, C., A. Baborie, D. Crooks, S. Wilkins and M.D. Jenkinson. 2011. Biology, genetics and imaging of glial cell tumors. Br. J. Radiol. 84: S90–S106.

Watson, J.B., S.B. Getzler and D.F. Mosher. 1994. Platelet factor 4 modulates the mitogenic activity of basic fibroblast growth factor. J. Clin. Invest. 94: 261–268.

Weber, G.F. and S. Ashkar. 2000. Molecular mechanisms of tumor dissemination in primary and metastatic brain cancers. Brain Res. Bull. 53: 421–424.

Wellstein, A., R. Lupu, G. Zugmaier, S.L. Flamm, A.L. Cheville, P. Delli Bovi, C. Basilico, M.E. Lippman and F.G. Kern. 1990. Autocrine growth stimulation by secreted Kaposi fibroblast growth factor but not by endogenous basic fibroblast growth factor. Cell Growth Differ. 1: 63–71.

Wellstein, A., G. Zugmaier, J.A. Califano, 3rd, F. Kern, S. Paik and M.E. Lippman. 1991. Tumor growth dependent on Kaposi's sarcoma-derived fibroblast growth factor inhibited by pentosan polysulfate. J. Natl. Cancer Inst. 83: 716–720.

Wesseling, P., D.J. Ruiter and P.C. Burger. 1997. Angiogenesis in brain tumors; pathobiological and clinical aspects. J. Neurooncol. 32: 253–265.

Wesseling, P., J.A. van der Laak, H. de Leeuw, D.J. Ruiter and P.C. Burger. 1994. Quantitative immunohistological analysis of the microvasculature in untreated human glioblastoma multiforme. Computer-assisted image analysis of whole-tumor sections. J. Neurosurg. 81: 902–909.

Wick, W., M. Platten and M. Weller. 2001. Glioma cell invasion: regulation of metalloproteinase activity by TGF-beta. J. Neurooncol. 53: 177–185.

Wong, M., L.A. Prawira, A.H. Kaye and C.M. Hovens. 2009. Tumor angiogenesis: its mechanism and therapeutic implications in malignant gliomas. J. Clin. Neurosci. 16: 1119–1130.

Yamaoka, M., T. Yamamoto, T. Masaki, S. Ikeyama, K. Sudo and T. Fujita. 1993. Inhibition of tumor growth and metastasis of rodent tumors by the angiogenesis inhibitor O-(chloroacetyl-carbamoyl)fumagillol (TNP-470; AGM-1470). Cancer Res. 53: 4262–4267.

Yanase, T., M. Tamura, K. Fujita, S. Kodama and K. Tanaka. 1993. Inhibitory effect of angiogenesis inhibitor TNP-470 on tumor growth and metastasis of human cell lines *in vitro* and *in vivo*. Cancer Res. 53: 2566–2570.

Yano, S., H. Shinohara, R.S. Herbst, H. Kuniyasu, C.D. Bucana, L.M. Ellis, D.W. Davis, D.J. McConkey and I.J. Fidler. 2000. Expression of vascular endothelial growth factor is necessary but not sufficient for production and growth of brain metastasis. Cancer Res. 60: 4959–4967.

Yao, X., Y. Ping, Y. Liu, K. Chen, T. Yoshimura, M. Liu, W. Gong, C. Chen, Q. Niu, D. Guo, X. Zhang, J.M. Wang and X. Bian. 2013. Vascular endothelial growth factor receptor 2 (VEGFR-2) plays a key role in vasculogenic mimicry formation, neovascularization and tumor initiation by Glioma stem-like cells. PLoS One 8: e57188.

Yoshida, S., M. Ono, T. Shono, H. Izumi, T. Ishibashi, H. Suzuki and M. Kuwano. 1997. Involvement of interleukin-8, vascular endothelial growth factor, and basic fibroblast growth factor in tumor necrosis factor alpha-dependent angiogenesis. Mol. Cell. Biol. 17: 4015–4023.

Yoshiji, H., S. Kuriyama, J. Yoshii, Y. Ikenaka, R. Noguchi, D.J. Hicklin, J. Huber, T. Nakatani, H. Tsujinoue, K. Yanase, H. Imazu and H. Fukui. 2002. Synergistic effect of basic fibroblast growth factor and vascular endothelial growth factor in murine hepatocellular carcinoma. Hepatology 35: 834–842.

Zagzag, D., D.C. Miller, Y. Sato, D.B. Rifkin and D.E. Burstein. 1990. Immunohistochemical localization of basic fibroblast growth factor in astrocytomas. Cancer Res. 50: 7393–7398.

Zagzag, D., A. Hooper, D.R. Friedlander, W. Chan, J. Holash, S.J. Wiegand, G.D. Yancopoulos and M. Grumet. 1999. *In situ* expression of angiopoietins in astrocytomas identifies angiopoietin-2 as an early marker of tumor angiogenesis. Exp. Neurol. 159: 391–400.

Zagzag, D., R. Amirnovin, M.A. Greco, H. Yee, J. Holash, S.J. Wiegand, S. Zabski, G.D. Yancopoulos and M. Grumet. 2000. Vascular apoptosis and involution in gliomas precede neovascularization: a novel concept for glioma growth and angiogenesis. Lab. Invest. 80: 837–849.

Ziu, M., N.O. Schmidt, T.G. Cargioli, K.S. Aboody, P.M. Black and R.S. Carroll. 2006. Glioma-produced extracellular matrix influences brain tumor tropism of human neural stem cells. J. Neurooncol. 79: 125–133.

Zugmaier, G., M.E. Lippman and A. Wellstein. 1992. Inhibition by pentosan polysulfate (PPS) of heparin-binding growth factors released from tumor cells and blockage by PPS of tumor growth in animals. J. Natl. Cancer Inst. 84: 1716–1724.

CHAPTER 7

Microvascular Dysfunction after Spinal Cord Injury

Humberto Mestre and *Antonio Ibarra**

7.1 Introduction

Acute traumatic spinal cord injury (SCI) is a growing concern for clinical practice. The vast etiologies of SCI make this pathology difficult to diagnose in an effective and timely manner. Advances in technology and in our understanding of SCI have improved our ability to detect patients who have sustained an injury of this type; however, our knowledge is limited and these patients, in spite of intervention, are left with permanent disabilities. The predominant age group affected by SCI clusters around young productive males, which severely affects the financial, psychological, and social aspects of their lives. Our inability to treat SCI lies in the complex pathophysiology of this disease. This chapter aims at describing the hallmarks of microvasculature dysfunction after SCI, emphasizing the modifications suffered by the endothelial cells of the blood-spinal cord barrier (BSCB). An effort has also been made to address the therapeutic interventions directed towards these vascular mechanisms and the clinical implications of these processes.

7.1.1 Epidemiology

The annual incidence of SCI has been estimated at 40 cases per million people in the U.S. This indicates that there are about 12,000 new cases of

Facultad de Ciencias de la Salud, Universidad Anáhuac México Norte, Av. Universidad Anahuac No. 46, Huixquilucan, Edo. de México, 52786, México.
* Corresponding author: iantonio65@yahoo.com; jose.ibarra@anahuac.mx

SCI every year (Sekhon and Fehlings 2001). The latest statistics suggest that there are approximately 236,000 to 327,000 people living with SCI in the U.S. Similar results are observed in European countries and even in Canada (Ackery et al. 2004). Since 2005, the average age of injury is 41 years with mostly males affected (80.6 per cent). The primary cause of SCI is motor vehicle accidents accounting for 39.2 per cent of cases. The next most common cause is traumatic falls (28.3 per cent), followed by acts of violence (14.6 per cent; mainly gunshot wounds). The most frequent neurological affection observed after SCI is incomplete tetraplegia (40.8 per cent) (NSCISC 2012). The permanent disabilities that ensue after SCI condemn patients to life-long medical attention and treatment. The recent advancements in the medical management of these patients have increased their life expectancy and allow them to lead better quality lives. However, they also contribute to an increase in lifetime and health care expenses; patients who sustain a low level cervical (C5–C8) incomplete injury at 25 years of age will have an estimated lifetime cost of $3.3 million USD (NSCISC 2012). Sustaining a SCI represents a physical, emotional, and financial burden on these individuals and thus a solution must be sought.

7.2 Microvasculature of the Human Spinal Cord

Vascular injury plays an important role in all phases of SCI pathophysiology. In order to better understand these degenerative mechanisms and their effect on the microvasculature a thorough understanding of normal vascular anatomy is required. The arterial blood supply, venous drainage and microvasculature of the human spinal cord are briefly reviewed in the next sections. For a detailed description refer to previous reports (Santillan et al. 2012).

7.2.1 Arterial Blood Supply

The arterial blood supply of the spinal cord can be roughly divided into an anterior and posterior system (Dommisse 1974). The anterior system consists of a longitudinal anastomotic trunk named the anterior spinal artery (ASA). This vessel runs along the median anterior sulcus of the spinal cord (Turnbull et al. 1966). The posterior system is formed by two longitudinal anastomotic trunks known as the posterior spinal arteries (PSA). They travel down the two postero-lateral sulci of the dorsal spinal cord. These longitudinal anastomotic trunks originate, at their cephalic and caudal ends, from tributaries of the subclavian and iliac arteries. Sulcal penetrating or central arteries arise from the ASA to supply the central portion and anterior two thirds of the spinal cord (Gillilan 1958). The PSA give rise to a complex network of pial arteries that irrigate the posterior one-third of the spinal cord (Turnbull 1971). The periphery of the spinal cord is mostly supplied by a pial network of vessels, which feeds off the ASA in anterior portions and the PSA in posterior portions. Segmental arteries originating from the descending thoraco-abdominal aorta feed both the ASA and the PSA. The posterior intercostal arteries of the aorta give off spinal branches that further divide into anterior and posterior radicular arteries; these branches contain anterior and posterior radiculo-medullary arteries that feed into the ASA and PSA, respectively (Santillan et al. 2012). The ASA and the PSA should not be considered as a single large caliber blood vessel, but as a consecutive series of anastomotic vascular loops (Hong et al. 2008).

7.2.2 Venous Drainage

The spinal cord arterial blood supply is highly variable between individuals and the venous system is even more so. The venous drainage of the spinal cord can be divided into extrinsic and intrinsic systems, similar to the arterial blood supply (Santillan et al. 2012). The intrinsic system is responsible for draining the spinal cord parenchyma. This system is subdivided into the sulcal (central) and radial (peripheral) veins. The similarity with central and peripheral arterial systems does not indicate that these drain the same areas (Mautes et al. 2000). The sulcal veins travel into the anterior and posterior median sulci and are responsible for the venous drainage of the medial aspect of both sides of the spinal cord. They travel towards the central canal and create an anastomotic ring around it (Brockstein et al. 1994). The radial veins arise from the outer limits of the gray matter and travel towards the periphery of the spinal cord where they form a pial venous plexus. The extrinsic venous system drains the superficial areas of the spinal cord. This system is constituted by an anterior and posterior median spinal vein, which receives tributaries from the central veins and the pial venous plexus. The median spinal veins drain into the radiculo-medullary veins and continue onto the anterior and posterior radicular veins. These veins communicate with the internal vertebral or epidural venous plexus. Then blood exits through the intervertebral veins and into the external vertebral plexus, where it drains into the segmental veins leading to the ascending lumbar and azygos venous systems (Santillan et al. 2012).

7.2.3 Microvasculature

The microvasculature of the human spinal cord is also known as the intrinsic arterial system. The spinal cord parenchyma is irrigated by two arterial systems: the central or centrifugal system and the peripheral (vasocorona) or centripetal system (Santillan et al. 2012). The central system arises from the central arteries (diameter 0.06–0.40 mm) that travel horizontally into the anterior median fissure from their origin at the ASA (Brockstein et al. 1994). Two to 12 central arteries arise from each centimeter of the ASA, depending on which level of the spinal cord is observed; the least number of branches is in the thoracic spinal cord and the most in the lumbar spinal cord (Turnbull 1971). At their entry into the cord they bifurcate either to the left or to the right in a centrifugal-manner, mostly within the gray matter (Gillilan 1958). Capillaries are much more numerous in the gray matter than in the white matter, possibly due to the greater metabolic requirements of the cell bodies as opposed to neuronal axons (Dohrmann et al. 1971). The peripheral system or vasocorona originates from the pial plexus of the ASA and PSA. The vasocorona has small perforating branches (diameter 0.1–0.2 mm), known as rami perforantes, that travel into the white matter and continue in a centripetal direction (Brockstein et al. 1994, Santillan et al. 2012). The majority of the gray matter is vascularized by the central branches of the ASA (Bingham et al. 1975). The ASA, with its central and pial plexus divisions, accounts for the supply of the anterior two-thirds of the white matter (Tator and Koyanagi 1997). The ASA is responsible for irrigation of the anterior commissure, ventral horns, Clarke's nucleus, anterior sections of the cuneate and gracilis fasciculus, plus the corticospinal and spinothalamic tracts (Santillan et al. 2012). The pial plexus feeders from the PSA account for blood flow of the posterior one-third of the spinal cord (Fairholm and Turnbull 1971). They supply the majority of the dorsal horns, specifically the cuneate and gracilis fasciculus. The lateral corticospinal tracts share arterial blood supply from both the anterior and posterior systems (Tator and Koyanagi 1997). The central and peripheral arterial systems do not anastomose directly; they meet in a watershed zone (Fig. 1). A watershed zone is a region that does not receive a direct blood supply but rather is dependent on overlapping vascular fields (Mautes et al. 2000). Areas that depend on watershed zones are particularly vulnerable to compromise of any of their contributing vascular fields. In humans, this area is predominantly within the outer white matter of the ventral half of the spinal cord (Tator and Koyanagi 1997). The microvascular architecture of the white matter in the spinal cord will be clinically significant further on, when discussing the pathophysiology of SCI.

7.3 Endothelial Cells and the Blood-Brain Barrier

The concept that the central nervous system (CNS) was excluded from the systemic circulation in the form of a blood-brain barrier (BBB) arose at the beginning of the 20th century, espoused by the German microbiologist Ehrlich. The first studies observed that dyes and pharmacologically

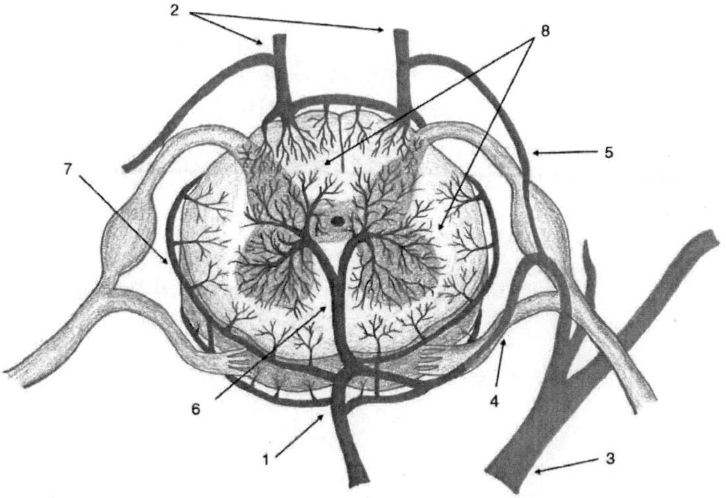

Figure 1. Arterial blood supply and microvasculature of the human spinal cord. The figure separates the anterior arterial system (blue), the posterior arterial system (red) and the segmental arterial system (green). (1) Anterior spinal artery (ASA); (2) Posterior spinal arteries (PSA); (3) Spinal branch from the segmental arteries of the aorta; (4) Anterior radiculomedullary artery; (5) Posterior radiculomedullary artery; (6) Central or sulcal artery; (7) Vasocorona or pial arterial plexus; (8) Watershed zone between central and peripheral arterial systems.

Color image of this figure appears in the color plate section at the end of the book.

active drugs that were injected into the bloodstream would not enter the CNS parenchyma (Bradbury 1985). The BBB hypothesis was discredited after ultrastructural analysis revealed no significant differences between CNS capillaries and capillaries anywhere else in the body. But further studies found that endothelial cells covering the inside of CNS capillaries were structurally and functionally different from other endothelial cells (Goodman et al. 1976). These studies also concluded that the main players for barrier function were endothelial cells and the junctions between them. The BBB is formed by the neurovascular unit that provides a mechanical barrier allowing only passive entrance of small lipophilic molecules or active transport across the endothelium of the brain capillary (Huber et al. 2001). The neurovascular unit is formed by brain-specific endothelial cells that line the lumen of the vessel, forming a complete cylinder. These endothelial cells are joined together by adherens and tight junctions and express important transporters, such as P-glycoprotein (Pgp), on their luminal surface (Rubin and Staddon 1999). The next components of the neurovascular unit are the basal lamina, the pericyte and the astrocyte foot process (Balabanov and Dore-Duffy 1998). For many years, the BBB was thought to isolate the entire neuraxis from the systemic circulation, but recently regional differences have become apparent. The concept that the BBB is functionally different within the spinal cord has given rise to the BSCB. The BSCB is the functional equivalent of the BBB in the spinal cord. It would seem logical if this barrier were a structural extension of the BBB, but subtle differences have been observed between them. Both the BBB and BSCB share the same morphological building blocks (endothelium, pericyte, basal lamina and astrocyte foot process), but they possess seemingly different functions (Bartanusz et al. 2011).

7.3.1 Structural and Functional Components of the Blood-CNS Barrier

7.3.1.1 Endothelial Cell

The endothelial cells (ECs) found lining the brain capillaries are phenotypically different from the rest of the peripheral endothelium. The ECs of the CNS are characterized by tight junctions and lack pinocytic vacuoles (Abbott et al. 2006). Elevated concentrations of cytosolic mitochondria suggest that they are highly energetically active cells, probably due to the requirements of active transport across the cell (Bernacki et al. 2008). Another morphological variation in BBB ECs is that they do not have fenestrations in their plasma membranes and are also selectively permeable to molecules of a certain molecular weight and lipophilicity (Bartanusz et al. 2011). Endothelial cell (EC) structural characteristics restrict transcellular transport through selective permeability and active transport. Paracellular diffusion in spinal cord capillaries is also severely restricted by tight junctions and adherens junctions between individual cells (Abbott et al. 2010). Tight junctions are formed by trans-membrane proteins that encircle the cells' plasma membrane like a continuous belt. The points where adjacent cells are in close proximity are known as zonula occludens. At these points, the membranes of both ECs are fused and they form a penta-laminar structure characteristic of the blood-CNS barrier (Goodman et al. 1976). Several integral proteins form these tight junctions, such as: claudin, occludin and junctional adhesion molecules. In order for these junctions to hold the ECs in place, intracellular proteins known as zonula occludens (ZO)-1, -2, -3 and cingulin bind the trans-membrane proteins to the actin cytoskeleton (Bartanusz et al. 2011). Adherens junctions further tighten the space between ECs and depend on proteins such as cadherins, platelet endothelial cell adhesion molecule-1 and catenins (Abbott et al. 2006). ECs have several specific markers, including γ-glutamyl-transpeptidase, alkaline-phosphatase, von Willebrand factor, glucose transporter-1, endothelial barrier antigen and basigin (Bernacki et al. 2008). The luminal surface of ECs is covered by a glycocalyx which is a negatively charged mesh of proteoglycans, glycosaminoglycans, glycoproteins and glycolipids (Bernacki et al. 2008). Surface transporters and receptors on ECs are primarily N-glycosylated phosphoproteins, like Pgp and multidrug resistance-associated protein (Bernacki et al. 2008).

7.3.1.2 Basal Lamina

The basal lamina or basement membrane surrounds the abluminal surface of the ECs and separates them from the pericytes. The extracellular matrix makes up the basal lamina, which provides thin capillaries with added structural support. There is little research on the specific role that the basement membrane plays in BSCB physiology, as opposed to the cellular component. It seems that all the cells of the neurovascular unit maintain the basal lamina (i.e., ECs, pericytes and astrocytes). They all synthesize proteoglycans, collagen IV, laminins and other extracellular matrix components to keep the integrity of this membrane (Loy et al. 2002, Bartanusz et al. 2011). Destruction of the basal lamina, seen in inflammatory processes, alters the cytoskeletal morphology of ECs and tight junctions compromising the BSCB (Noble et al. 2002).

7.3.1.3 Pericyte

Pericytes are cells that originate developmentally from the mesodermal layer and are relatively different from mesenchymal cells. This subset of cells is found in the vicinity of small caliber vessels, such as capillaries. They are separated from ECs by the basement membrane but are anchored to it by gap junctions (Bernacki et al. 2008). Pericytes are responsible for the regulation of EC activities; such as proliferation, migration and differentiation. They regulate the expression of BBB-specific gene patterns in adjacent ECs and also induce the polarization of astrocyte foot processes (Armulik et al. 2010). These cells also have a role in inflammation, formation of capillary-like structures and capillary diameter (Bernacki et al. 2008). The exact role of pericytes in the BSCB is yet to be determined, but it is clear that they play an important and complex role in the maintenance of the BBB; a similar scenario can be extrapolated to the spinal cord.

7.3.1.4 Astrocyte

Astrocytes derive from the neuroectodermal layer during embryonic development. There are two main types of astrocytic phenotypes: protoplasmic astrocytes found in the gray matter and fibrillary astrocytes found in the white matter. The protoplasmic astrocyte is a cell with a large nucleus and several cytoplasmic appendices, known as feet (Bernacki et al.

2008). These foot-like processes form cap-like structures around neurons on one side and blood vessels on the other, establishing a direct connection between capillaries and neurons. This network assures that every neuronal cell body is in direct contact with at least one capillary (Turnbull 1971). Due to this privileged location, astrocytes express high concentrations of aquaporin 4 and potassium inwardly-rectifying channel 4.1 to aid in ion and volume regulation. Rather than providing structural support to the BCSB, they are critical for the development and maintenance of EC barrier mechanisms by secretory mechanisms (Bartanusz et al. 2011). (For an in-depth review of astrocyte-endothelial interactions in the BBB, please refer to Abbott et al. 2006).

7.3.2 Differences Between BBB and BSCB

The BBB and BSCB share remarkable similarities in ultrastructure and function; however, recent studies have demonstrated that they have morphological and physiological differences. These differences have forced the differentiation of the BSCB as an independent entity from the BBB. Further studies have identified ultra-structural differences between ECs in the brain and the spinal cord. These differences have the potential to make the spinal cord more prone to develop certain pathological conditions (Bartanusz et al. 2011). Some particular features of the BSCB are the presence of glycogen deposits, increased permeability to some molecules as opposed to the BBB, as well as decreased expression of transporter molecules, tight and adherens junction proteins (Bartanusz et al. 2011). The microvasculature of the spinal cord contains glycogen deposits, which are not seen in its cerebral counterpart. Their presence is not fully understood, but they may function as an energy storage reservoir (Sharma 2010). The BSCB also has an increased permeability to inflammatory cytokines, such as interferon (IFN)-α, IFN-γ and tumor necrosis factor-α (Pan et al. 1997). These authors also concluded that there is greater permeability at the distal spinal cord. This suggests associations between regional permeability differences and neuroanatomical functions in the spinal cord (Bartanusz et al. 2011, Campos et al. 2012). Marked differences between ECs in the BBB and in the BSCB also include decreased expression of transporter molecules, like Pgp. Research also indicates that there might be structural differences in BSCB tight junctions, due to the decreased expression of ZO-1 and occludin; changes are also observed in adherens junctions with a reduced synthesis of vascular endothelial cadherin and β-catenin (Ge and Pachter 2006). The minor differences between the BBB and BSCB are enough to predict that their role in disease will be entirely different and should therefore be studied separately.

7.4 Pathophysiology of Spinal Cord Injury

The traumatic origin of SCI causes immediate mechanical damage to the neural tissue. The initial damage causes hemorrhage, ischemia, synaptic disconnection and axonal destruction. However, injury to the CNS possesses unique features compared to the rest of the body. The pathophysiology of SCI is biphasic in nature. The primary injury mechanisms previously mentioned are features inherent to the traumatic force. The secondary phase of injury is a series of auto-destructive degenerative processes that take place after the initial trauma. The secondary injuries further deteriorate the neurological outcome of the patient. The primary and secondary phases of disease are responsible for the dysfunction of the spinal cord microvasculature. In the next sections, special emphasis is made on the pathophysiological processes that directly affect the spinal cord microvasculature, ECs and the BSCB after SCI.

7.4.1 Microvascular Dysfunction After SCI

SCI provokes hemorrhage and alters blood flow to the spinal cord. The kinetic energy applied directly on the spinal cord destroys blood vessels, ECs and their tight junctions, thus breaking down the BSCB. Not to mention that it also results in the functional disconnection of synapses and axons. The site where the injury mechanism concentrates the greatest amount of energy is known as the epicenter. At this location, there is a central hemorrhage characterized by small petechial bleeding normally localized near the central canal (Dohrmann et al. 1971). The extravasation of erythrocytes is secondary to the loss in contiguity of the capillaries. Due to the vascular architecture of the gray matter, the hemorrhage progresses in a centrifugal manner travelling within the highly vascularized gray matter (one hr after SCI) and eventually reaching the white matter (two hr after SCI) (Noble and Wrathall 1989). The hemorrhagic epicenter is directly proportional to the severity of the injury and increases in size from 24 hr to one week post-injury (Noble and Wrathall 1989). Following SCI, gray matter blood flow is obliterated while perfusion in the white matter is paradoxically increased. This suggests that the degeneration of white matter observed after SCI is mainly due to the secondary degenerative cascade (Bingham et al. 1975). The vascular density of white matter in normal subjects is much lower than that seen in the gray matter; this is because of the lower metabolic demands of axons as opposed to cell bodies. This indicates that white matter is probably more resistant to periods of ischemia than gray matter.

The damaged blood vessels allow the influx of blood-borne cells such as: leukocytes, monocytes and macrophages, which activate the innate and specific immune responses. The pro-inflammatory molecules produced by these cell subsets contribute to the added destruction of the neighboring tissue. The extravasation of erythrocytes results in their decomposition into hemoglobin. This process is highly neurotoxic because the released catalytic metal ions aid in the production of free radicals further propagating lipid peroxidation (Hall 2011, Oudega 2012).

Hemorrhage and capillary destruction result in an alteration of spinal cord blood flow. This impedes the arrival of oxygen and glucose to the injured neural tissue. The high metabolic demands of these cells make them strictly aerobic and susceptible to ischemia. During these periods of ischemia adenosine triphosphate (ATP)-dependent ionic regulation fails, increasing intracellular Ca^{2+} concentrations. The change in cytoplasmic pH due to anaerobic respiration activates proteolytic enzymes and caspases that activate programmed cell death or apoptosis. Increased intracellular Ca^{2+} concentrations trigger the uncontrolled release of glutamate, an excitatory neurotransmitter, and permanently depolarize the neuron in a process known as excitotoxicity. All these mechanisms result in the exaggerated production of reactive oxygen species (ROS), eventually overwhelming the cell's antioxidant properties. Ischemia is a degenerative event in the CNS, but recanalization of blood flow may further destroy surviving cells. Reperfusion injury exacerbates production of ROS and increases lipid peroxidation. This phenomenon destroys the cell membranes of neurons, glia, ECs and pericytes, which in unison continue to alter the normal physiology of the spinal cord.

BSCB breakdown secondary to microvascular dysfunction causes the accumulation of fluid in the interstitial space. This vasogenic edema observed after SCI causes tumefaction of the spinal cord. The increase in size within a restricted space (spinal canal) causes the herniation of the spinal cord against the bony structures of the spinal column. This compresses the spinal cord exacerbating and perpetuating the injury mechanism. Severe edema also generates enough pressure to close down small capillaries that may still be patent and delivering blood to the injured tissue, worsening ischemia and increasing the size of the necrotic epicenter.

Injury to the spinal cord can also cause neurogenic shock, spinal shock and autonomic dysreflexia (Tator and Fehlings 1991). These hallmarks of SCI pathology are characterized by bradycardia and hypotension. The drop in cardiac output reduces tissue perfusion, including the spinal cord. The hemodynamic instability of these patients is manifested as a decrease in mean arterial pressure, which can affect the microvascular blood flow of several other organs (Guizar-Sahagun et al. 2004).

7.4.2 Blood-Spinal Cord Barrier Alterations After SCI

The traumatic insult of SCI destroys capillaries and alters homeostatic blood flow. The direct contact of blood-borne cells and molecules with the CNS parenchyma renders the BSCB obsolete. Several mediators of the secondary injury cascade begin to destroy neighboring neurons and glial tissue. BSCB breakdown begins as early as five minutes after SCI and lasts up to 56 days (Popovich et al. 1996, Mautes et al. 2000, Benton et al. 2007, Cohen et al. 2009). Destabilization of barrier mechanisms are initially caused by vascular endothelial growth factor (VEGF)-A, angiopoietin (Ang)-2, as well as matrix metalloproteinase-2 and -9 (Ng et al. 2011). The cellular and molecular mechanisms responsible for BSCB permeability have a compound effect and destabilize the barrier. Traumatic separation of the basal lamina from the endothelium alters tight and adherens junctions, allowing extravasation of leukocytes into the spinal cord (Schnell et al. 1999). Increased vesicular transport across ECs makes them highly permeable (Bartanusz et al. 2011). Oxidative stress after SCI results in uncontrolled synthesis of nitric oxide (NO) (Ibarra et al. 2010). NO suffers a spontaneous reaction with superoxide anion and is converted into the EC-toxic molecule, peroxynitrate (Garcia et al. 2011). This and other free radicals also breakdown the BSCB. Pericyte loss, seven days after SCI, contributes to BSCB instability by loss of direct physical contact with the basement membrane through gap junctions (key and socket) (Balabanov and Dore-Duffy 1998, Benton et al. 2007). During the initial phase of injury there is a decreased expression of Ang-1 and VEGF-B and an antagonistic increased synthesis of VEGF-A and Ang-2 supporting BSCB hyper-permeability (Oudega 2012). Soon thereafter, the expression profile reverts towards a predominance of Ang-1 and VEGF-B reestablishing selective permeability. Despite these physiological repair mechanisms, the barrier still remains incompetent for up to two months (Cohen et al. 2009, Bartanusz et al. 2011). This long-term permeability increases leukocytic transmigration, diffusion of toxic compounds plus a limited arrival of oxygen and nutrients to the injured spinal cord. The destabilization of the BSCB provides the ideal conditions for the secondary injury cascade to develop within the CNS. Several studies have attempted to analyze BSCB re-establishment, but have obtained inconsistent results based on the methods they used (Goodman et al. 1976, Popovich et al. 1996, Jaeger and Blight 1997, Kaptanoglu et al. 2004). The establishment of a re-stabilization time frame varies depending on the tracer molecule used to evaluate extravasation. A study that used horseradish peroxidase concluded that the BSCB was functional 14 days after SCI (Goodman et al. 1976). Another study using the vascular tracer [^{14}C]-α-aminoisobutyric acid observed BSCB permeability over several segments rostral and caudal to the injury epicenter, up to 3 cm in both directions. This study observed a biphasic pattern of BSCB breakdown and concluded that the second dysfunction occurred between 14 to 28 post SCI (Popovich et al. 1996). This suggests that the BSCB repairs itself in a stepwise manner; it progressively recuperates its selective permeability based on molecular size. A recent study used dynamic contrast-enhanced magnetic resonance imaging (DCE-MRI), employing a two-compartment pharmacokinetic model, to measure the extravasation of intravenous gadopentetate-dimeglumine (Gd). This work did a longitudinal series of scans to analyze the spatial and temporal changes of the BSCB after SCI. The study demonstrated, and confirmed their observations with immunohistochemistry, that the BSCB remained compromised even 56 days after SCI (Cohen et al. 2009). Therapies that are aimed at stabilizing the BSCB in an effective and timely manner have shown promise in animal models of SCI.

7.4.3 Endothelial Cell Modifications After SCI

After SCI, EC loss occurs acutely due mainly to the mechanical trauma exerted on the spinal cord microvasculature. These degenerative changes begin after 30 min and continue for three days after injury (Casella et al. 2002, 2006). The ECs that die within the first 24 hr are the result of the primary injury and therefore die in a necrotic process. The ECs that die during the first days after SCI are considered to undergo delayed apoptotic death due to the secondary injury (Casella et al. 2002, 2006). The mechanisms responsible for delayed EC apoptosis are probably a result of ROS and lipid peroxidation (Hall 2011). The hemorrhagic epicenter is a major location for EC apoptosis, due to increased mechanical trauma and the elevated presence of free radicals. Another cause of EC death is a process known as *anoikis*, which is when the EC detaches from the extracellular matrix of the basement membrane and loses integrin stimulation (Oudega 2012). Integrin binding to extracellular matrix (e.g., laminin) is crucial for the structural and functional integrity of ECs. The $\alpha v\beta 3$ integrin is a pro-survival molecule for capillary ECs (Han et al. 2010). Another EC-specific stimulator molecule is Ang-1, which is the ligand for the tyrosine kinase with immunoglobulin-like and epidermal growth factor-like domains (Tie)-2 (Kim et al. 2001). Ang-1 stabilizes blood vessels and EC tight junctions (Thurston et al. 1999). A novel investigation of microvascular EC gene expression, 24 hr after SCI, yielded interesting results. During the period of maximum pathological loss of the spinal vasculature, ECs expressed high mRNA levels of the thrombospondin-1, cysteine-cysteine motif ligand-5 and urokinase plasminogen activator genes (Benton et al. 2008). The precise role of these molecules is yet to be clearly elucidated and may even provide interesting therapeutic targets. After two to three days almost no vasculature is observed at the injury site due to cystic cavitation, this indicates the null presence of ECs (Whetstone et al. 2003).

7.4.4 Vascular Remodeling and Angiogenesis

The angiogenic response seen after SCI is well established. The process of angiogenesis is the formation of new blood vessels from pre-existing functional capillaries. Although it is dependent on ECs, this response occurs in spite of their massive loss after SCI (Casella et al. 2006). The angiogenic response is biphasic and begins with vessels formed at the epicenter and penumbra of a contusive SCI after three to 14 days (Imperato-Kalmar et al. 1997). Vascularity decreases at the epicenter during the second week due to cystic cavitation (Imperato-Kalmar et al. 1997). The formation of new vessels probably continues throughout this time period but accelerates during the second month after injury (Loy et al. 2002). Microscopic analysis shows that these new blood vessels lack association with astrocytes and pericytes (Casella et al. 2006, Oudega 2012). The pro-angiogenic response is probably secondary to travelling angioblasts or a stimulus from spinal cord blood flow (Fassbender et al. 2011). Most of the leaky malfunctioning capillaries are removed during the second and third weeks after SCI (Casella et al. 2006). Molecular identifiers of angiogenic sprouts are isolectin B4, "a disintegrin and metalloprotease" 8 and plasmalemmal vesicle-associated protein (Benton et al. 2007, Mahoney et al. 2009). Traumatic disruption of the spinal cord microvasculature triggers angiogenesis. ECs of the traumatized capillaries begin to express pro-angiogenic growth factors, but many are not detectable in the microenvironment. VEGF, Ang-1 and platelet-derived growth factor (PDGF) are significantly down-regulated after SCI (Bartanusz et al. 2011). The only angiogenic factor with increased expression is hepatocyte growth factor. This demonstrates that ECs are unable to satisfy the concentrations of pro-angiogenic proteins necessary for successful angiogenesis. The result of this unsuccessful endogenous angiogenesis is debatable. Attempts to repair spinal cord perfusion fail due to poor functioning leaky capillaries. This maintains a state of decreased perfusion in the lesion epicenter and surrounding regions resulting in the continued death of neural and glial cell populations. In view of the crucial role of patent functional capillaries, the therapeutic strategies that do not take into account this poorly vascularized injury site will most likely fail. Neuroprotective interventions with the goal of saving neuronal axons or promoting their regeneration will have no chance to succeed if more effective pro-angiogenic therapies are not used in unison. A branch of SCI therapeutics that equally benefits from mature and functional blood vessels within the injury site is cell therapy. Up to 80 per cent of Schwann cells and 50 per cent of bone marrow stromal cells die within the first week after being transplanted into the injured spinal cord (Oudega 2012). This is most likely due to the lack of oxygen and nutrients found in the repaired spinal cord secondary to the faulty repair of the microvasculature.

7.5 Microvasculature-targeted Therapies

Treatment options for acute SCI are extremely limited. Until now, the only compound that is relatively recommended is the glucocorticoid methylprednisolone. Even though this is the mainstay of treatment its use is controversial, as some of the complications of steroid use overshadow its neuroprotective benefits. Alongside methylprednisolone, only a few other drugs have made it to phase III clinical testing such as: naloxone, tirilazad mesylate, ganglioside GM-1 and nimodipine. None showed significant improvements and they never received approval by the Food and Drug Administration (FDA) for their use in SCI. Due to this therapeutic scarcity, the scientific field is desperately looking for new drug classes that will provide the answer in SCI neuroprotection. One area that has received significant attention is microvascular therapeutics. This branch of neurotherapeutics has the aim of minimizing the acute effects of microvascular dysfunction, stabilizing the BSCB and promoting angiogenesis with the goal of diminishing damage to the spinal cord and promoting its regeneration. Studies have shown that preservation of spinal cord tissue and improvement in functional recovery are also associated with increased vascular density and a lesser degree of inflammation and apoptosis (Fassbender et al. 2011). This premise suggested that the ideal therapy would protect the microvasculature during the acute phases of SCI and prevent its continued collapse due to the secondary injury process. Studies that used molecular extravasation as an indirect measure of BSCB dysfunction also concluded that earlier BSCB stabilization yielded a better clinical outcome. Table 1 is adapted from a comprehensive review of microvascular targets in the neuroprotective treatment of SCI and demonstrates some of the compounds being used in animal models (for an in depth review of these therapeutic agents please refer to Fassbender et al. 2011). In the following section we shall provide a brief review of all agents that are directed towards ECs and their role in the pathophysiology of SCI.

One of the principal mechanisms of microvascular and BSCB dysfunction after SCI is EC death. This damage is mediated by ROS and lipid peroxidation of spinal cord microvessels (Ibarra et al. 2010). Compounds that neutralize these ROS and prevent their production, such as steroids (tirilazad mesylate and methylprednisolone) and antioxidants (Tempol and selenium) have all had favorable results maintaining spinal cord blood flow and reducing BSCB leakage (Hall 2011, Bracken 2012). The reduction in ROS synthesis allows for increased survival of ECs and prevents their delayed death during the secondary injury. The survival of ECs is likely favorable for the early stabilization of the BSCB after injury.

Table 1. Microvasculature-Targeted Therapies Tested in Animal Models of SCI.

Target	Treatment	Result
Microvascular stabilization		
Hyperperfusion	Nimodipide + Adrenaline	Increased blood flow and axonal conduction. Limited neurological improvement
	Nimodipide + Dextran	
	Nimodipide + Methylprednisolone	
Oxidative stress	Tirilazad	Increased blood flow and stabilized BSCB
	Selenium	
	Tempol	
Heme oxygenase-1 (HO-1)	HO-1 knockdown	Reduced inflammation and stabilized BSCB
	Hemin	
Nonselective ion channels	Knockouts and knockdowns (RNA interference)	Decreased post-SCI hemorrhage
Tie-2 and αvβ3 integrin	C16 integrin agonist and angiopoietin 1 (Ang-1)	Increased vascular density, reduced inflammation and permeability
Tie-2 and vascular endothelial growth factor (VEGF) receptor (VEGFR)	Adeno-associated virus (AAV)-VEGF and AAV-Ang-1	Decreased MRI lesion volume and vascular stabilization
Protein tyrosine phosphatase	bpV[phen]	Increased vascular density and reduced apoptosis
Angiogenesis		
VEGFR	Zinc finger protein-VEGF isoform A transcription factor	Increased vascular density and reduced apoptosis
	VEGF-producing neural stem cells	
Semaphorin 3A	SM-216289	Increased vascular density
Notch	Agonist	Decreased vascularity yet maintained perfusion
Hepatocyte growth factor receptor	Hepatocyte growth factor	Increased vascular density
αvβ3 integrin	C16 integrin agonist	Increased vascular density
Small-inducible cytokine B10	Antibody	Increased vascular density
Cell therapy	Intravenous CD133+ cells	Increased vascular density

Adapted from: Fassbender et al. 2011.

Membrane channels expressed on the luminal surface of ECs also change after acute SCI and have been investigated for their therapeutic potential. EC expression of nonselective cation channels ($NC_{Ca\text{-}ATP}$), regulated by sulfonylurea receptor 1 (SUR1), is greatly increased after SCI.

These channels mediate progressive hemorrhagic necrosis after injury. During secondary ischemia of the spinal cord ATP levels are very low and ATP-dependent channels open, causing edema and EC death. Use of FDA-approved SUR1-regulated $NC_{Ca\text{-}ATP}$ channel antagonists, employed in the treatment of type II diabetes mellitus (glibenclamide and repaglinide), resulted in white matter sparing, a three-fold reduction in lesion size and neurobehavioral improvement (Simard et al. 2007). Studies using knockout mice and antisense knockdown models confirmed these results (Fassbender et al. 2011).

The most important proponent of the idea that EC death leads to BSCB and microvasculature breakdown is Dr. Theo Hagg at the Kentucky Spinal Cord Injury Research Center. The emphasis on improving EC survival after SCI has yielded promising results. His group worked with Tie-2, a receptor that is almost exclusively expressed by ECs, and its ligand Ang-1. Ang-1 is known to promote EC survival and is considerably lowered after SCI (Han et al. 2010). EC stimulation with Ang-1 decreases inflammation by inhibiting leukocyte diapedesis into the CNS. Ang-1 also down-regulates the expression of leukocyte adhesion molecules, like intercellular adhesion molecule-1, vascular cell adhesion molecule-1 and E-selectin (Kim et al. 2001). The αvβ3 integrin has also been found to contribute to EC survival due to its reciprocal interactions with growth factor receptors (Fassbender et al. 2011). A study using intravenous administration of Ang-1 and an αvβ3 integrin agonist, known as C16, after SCI observed rescued epicenter microvasculature, spared white matter, decreased inflammation and improved locomotor function (Han et al. 2010). This therapeutic strategy has been shown to preserve the integrity of EC tight junctions via an effect of Ang-1 on BSCB stabilization (Thurston et al. 1999). This treatment also showed promise in clinical translation; the therapeutic window was between four hr and 72 hr, which allows enough time for the arrival, diagnosis and treatment of a SCI patient in the clinical setting. Another study used viral vectors of Ang-1 and VEGF for sustained expression after SCI and obtained similar results (Herrera et al. 2010). Ang-2 has opposing effects on blood vessels, being induced mainly during EC barrier breakdown. A recent study, however, showed that higher levels of Ang-2 after SCI correlated with better neurological outcomes. The study also observed that ECs were not producing Ang-2, instead non-vascular cells were producing it. Ang-2 appears to have pro-angiogenic and/or pro-gliogenic properties underlying the improvement in recovery after SCI (Durham-Lee et al. 2012). Other treatment strategies that have ECs as their therapeutic target include: fluoxetine (Lee et al. 2012b), valproic acid (Lee et al. 2012a), VEGF or PDGF (Lutton et al. 2012) and endothelial progenitor cells (Kamei et al. 2012).

The promising results obtained by increasing EC survival indicate that EC death after SCI is a pathophysiological hallmark that accounts for many of the detrimental processes seen. Further research is needed in this area to understand the true role that ECs have on the outcome of SCI. Efforts must continue to find innovative and novel compounds that will provide SCI patients with treatment options. For now, the goal is not necessarily a cure but an intervention that will limit the permanent sequelae of SCI. Groups investigating therapeutic interventions for SCI must make an effort to adhere to clinically translatable models.

7.6 Clinical Considerations for Future Therapeutic Interventions

It is apparent that ECs have a pivotal role in the pathophysiology and outcome of SCI. A number of studies demonstrate that the therapeutic targeting of microvessels and ECs is an incredibly promising field in the treatment of this pathology. The use of these strategies is conceptually and practically coherent with the modern day understanding of SCI. Any agent directed towards the vasculature of the spinal cord has the benefit of having an intravascular administration route. Use of intravenous and intra-arterial routes is customary in any medical institution and is considered routine procedure in any admitted patient. This route is therefore simple, quick and allows the drug to arrive to its target tissue. These treatment protocols are preferred as opposed to neuron-targeted medications that have low BSCB diffusion and obtain low concentrations in the spinal cord. The protection from acute microvascular dysfunction, prevention of delayed EC death and the stimulation of pro-angiogenic mechanisms have all shown to improve the neurological status of the injured animals. The next step in SCI therapeutics is taking these compounds into the clinical setting. With the incredible amount of new drugs being discovered or the novel use of old drugs for SCI, criteria for animal testing models should be created with the goal of shortening the distance between the lab bench and the clinic. Clinical translation is a growing concern in research; with the exponential growth of available agents, stricter design criteria must be implemented. When choosing the therapeutic agent, researchers should try to use drugs that already have FDA approval for other diseases (i.e., glibenclamide for diabetes mellitus II or fluoxetine for clinical depression) (Simard et al. 2007, Lee et al. 2012b). This will hasten the transition from animal models to human clinical testing because the safety of the drug in humans has already been scrutinously verified. If this is not possible, the methodological design should include pharmokinetic, pharmacodynamic and safety studies of the new compound. Human trials of neuroprotective agents have been unable to obtain the benefits observed in animal studies (e.g., tirilazad mesylate and ganglioside GM-1) (Onose et al. 2009). A reason for this is the limited heterogeneity of the sample, which is a poor reflection of the clinical counterpart. SCI in humans is vastly heterogeneous with different etiologies, sex, age, level and completeness of injury; rendering the SCI population highly different in nature. Experimental models should take this into consideration and attempt to study several population types, animal species and injury mechanisms. A scientific transition should take place and new emphasis should be placed on a comprehensive functional behavioral analysis as opposed to the morphological analysis of the injury. Most studies that show a reduction in lesion volume do not necessarily translate into motor or sensory improvement. In clinical trials, more weight is placed on functional improvement than on morphological sparing of microscopic structures, such as white matter and neuronal axons. Another example of cutting the distance between the lab and the clinic is dosing schedules. Therapies should be administered within a realistic time window, providing enough time for the arrival and diagnosis of the SCI patient. Scientists should refrain from pretreatment or agents that require an immediate administration, as these are realistically impossible. New, less invasive methods are required to assess the permeability of the BSCB *in vivo*. This information is relevant towards the diagnosis and the corroboration of therapeutic efficacy in testing new drugs. An example of this is DCE-MRI, which has already demonstrated to be an effective technique (Bilgen et al. 2001, 2002, Cohen et al. 2009). DCE-MRI involves the repeated acquisition of T_1-weighted MR images after the intravenous administration of a paramagnetic contrast agent, such as Gd. Initiatives to improve the accuracy and resolution of this method should be undertaken in order to speed up its inclusion into the clinical setting. The field of microvascular dysfunction after SCI is burgeoning and has demonstrated to hold tremendous potential. The role of ECs in the outcome of SCI has forced science to place increasing importance on their survival. The results obtained from this research have and will continue pushing the limit of SCI therapeutics. All this is done with the hope that one day, we as clinicians and scientists, can provide more for our patients than just a diagnosis.

Acknowledgments

We would like to thank Carolina Musri, a medical student at the University of Mendoza in Argentina, for her incredible artistic skills in the design of the chapter figure.

References

Abbott, N.J., A.A. Patabendige, D.E. Dolman, S.R. Yusof and D.J. Begley. 2010. Structure and function of the blood-brain barrier. Neurobiol. Dis. 37: 13–25.

Abbott, N.J., L. Ronnback and E. Hansson. 2006. Astrocyte-endothelial interactions at the blood-brain barrier. Nat. Rev. Neurosci. 7: 41–53.

Ackery, A., C. Tator and A. Krassioukov. 2004. A global perspective on spinal cord injury epidemiology. J. Neurotrauma 21: 1355–1370.

Armulik, A., G. Genove, M. Mae, M.H. Nisancioglu, E. Wallgard, C. Niaudet, L. He, J. Norlin, P. Lindblom, K. Strittmatter, B.R. Johansson and C. Betholtz. 2010. Pericytes regulate the blood-brain barrier. Nature 468: 557–561.

Balabanov, R. and P. Dore-Duffy. 1998. Role of the CNS microvascular pericyte in the blood-brain barrier. J. Neurosci. Res. 53: 637–644.

Bartanusz, V., D. Jezova, B. Alajajian and M. Digicaylioglu. 2011. The blood-spinal cord barrier: morphology and clinical implications. Ann. Neurol. 70: 194–206.

Benton, R.L., M.A. Maddie, C.A. Worth, E.T. Mahoney, T. Hagg and S.R. Whittemore. 2008. Transcriptomic screening of microvascular endothelial cells implicates novel molecular regulators of vascular dysfunction after spinal cord injury. J. Cereb. Blood Flow Metab. 28: 1771–1785.

Benton, R.L., M.A. Maddie, D.R. Minnillo, T. Hagg and S.R. Whittemore. 2007. Griffonia simplicifolia isolectin B4 identifies a specific subpopulation of angiogenic blood vessels following contusive spinal cord injury in the adult mouse. J. Comp. Neurol. 507: 1031–1052.

Bernacki, J., A. Dobrowolska, K. Nierwinska and A. Malecki. 2008. Physiology and pharmacological role of the blood-brain barrier. Pharmacol. Rep. 60: 600–622.

Bilgen, M., B. Dogan and P.A. Narayana. 2002. *In vivo* assessment of blood-spinal cord barrier permeability: serial dynamic contrast enhanced MRI of spinal cord injury. Magn. Reson. Imaging 20: 337–341.

Bilgen, M., R. Abbe and P.A. Narayana. 2001. Dynamic contrast-enhanced MRI of experimental spinal cord injury: *In vivo* serial studies. Magn. Reson. Med. 45: 614–622.

Bingham, W.G., H. Goldman, S.J. Friedman, S. Murphy, D. Yashon and W.E. Hunt. 1975. Blood flow in normal and injured monkey spinal cord. J. Neurosurg. 43: 162–171.

Bracken, M.B. 2012. Steroids for acute spinal cord injury. Cochrane Database Syst. Rev. 1: 1–52.

Bradbury, M.W. 1985. The blood-brain barrier. Transport across the cerebral endothelium. Circ. Res. 57: 213–222.

Brockstein, B., L. Johns and B.L. Gewertz. 1994. Blood supply to the spinal cord: anatomic and physiologic correlations. Ann. Vasc. Surg. 8: 394–399.

Campos, C.R., C. Schroter, X. Wang and D.S. Miller. 2012. ABC transporter function and regulation at the blood-spinal cord barrier. J. Cereb. Blood Flow Metab. 32: 1559–1566.

Casella, G., M. Bunge and P. Wood. 2006. Endothelial cell loss is not a major cause of neuronal and glial cell death following contusion injury of the spinal cord. Exp. Neurol. 202: 8–20.

Casella, G.T., A. Marcillo, M.B. Bunge and P.M. Wood. 2002. New vascular tissue rapidly replaces neural parenchyma and vessels destroyed by a contusion injury to the rat spinal cord. Exp. Neurol. 173: 63–76.

Cohen, D.M., C.B. Patel, P. Ahobila-Vajjula, L.M. Sundberg, T. Chacko, S.-J. Liu and P.A. Narayana. 2009. Blood-spinal cord barrier permeability in experimental spinal cord injury: dynamic contrast-enhanced MRI. NMR Biomed. 22: 332–341.

Dohrmann, G.J., F.C. Wagner and P.C. Bucy. 1971. The microvasculature in transitory traumatic paraplegia. An electron microscopic study in the monkey. J. Neurosurg. 35: 263–271.

Dommisse, G.F. 1974. The blood supply of the spinal cord. J. Bone Joint Surg. Br. 56B: 225–235.

Durham-Lee, J.C., Y. Wu, V.U.L. Mokkapati, A.A. Paulucci-Holthauzen and O. Nesic. 2012. Induction of angiopoietin-2 after spinal cord injury. Neuroscience 202: 454–464.

Fairholm, D.J. and I.M. Turnbull. 1971. Microangiographic study of experimental spinal cord injuries. J. Neurosurg. 35: 277–286.

Fassbender, J.M., S.R. Whittemore and T. Hagg. 2011. Targeting microvasculature for neuroprotection after SCI. Neurotherapeutics 8: 240–251.

Garcia, E., R. Silva-Garcia, H. Mestre, N. Flores, S. Martinon, E.S. Calderon-Aranda and A. Ibarra. 2011. Immunization with A91 peptide or copolymer-1 reduces the production of nitric oxide and inducible nitric oxide synthase gene expression after spinal cord injury. J. Neurosci. Res. 90: 656–663.

Ge, S. and J.S. Pachter. 2006. Isolation and culture of microvascular endothelial cells from murine spinal cord. J. Neuroimmunol. 177: 209–214.

Gillilan, L.A. 1958. The arterial blood supply of the human spinal cord. J. Comp. Neurol. 110: 75–103.

Goodman, J.H., W.G. Bingham Jr. and W.E. Hunt. 1976. Ultrastructural blood-brain barrier alterations and edema formation in acute spinal cord trauma. J. Neurosurg. 44: 418–424.

Guizar-Sahagun, G., L. Velasco-Hernandez, A. Martinez-Cruz, G. Castañeda-Hernandez, G. Bravo, G. Rojas and E. Hong. 2004. Systemic microcirculation after complete high and low thoracic spinal cord section in rats. J. Neurotrauma 21: 1614–1623.

Hall, E.D. 2011. Antioxidant therapies for acute spinal cord injury. Neurotherapeutics 8: 152–167.

Han, S., S.A. Arnold, S.D. Sithu, E.T. Mahoney, J.T. Geralds, P. Tran, R.L. Benton, M.A. Maddie, S.E. D'Souza, S.R. Whittemore and T. Hagg. 2010. Rescuing vasculature with intravenous angiopoietin-1 and alpha v beta 3 integrin peptide is protective after spinal cord injury. Brain 133: 1026–1042.

Herrera, J.J., L.M. Sundberg, L. Zentilin, M. Giacca and P.A. Narayana. 2010. Sustained expression of vascular endothelial growth factor and angiopoietin-1 improves blood-spinal cord barrier integrity and functional recovery after spinal cord injury. J. Neurotrauma 27: 2067–2076.

Hong, M.K., M.K. Hong, W.R. Pan, D. Wallace, M.W. Ashton and G.I. Taylor. 2008. The angiosome territories of the spinal cord: exploring the issue of preoperative spinal angiography. Laboratory investigation. J. Neurosurg. Spine 8: 352–364.

Huber, J.D., R.D. Egleton and T.P. Davis. 2001. Molecular physiology and pathophysiology of tight junctions in the blood-brain barrier. Trends Neurosci. 24: 719–725.

Ibarra, A., E. Garcia, N. Flores, S. Martinon, R. Reyes, M.G. Campos, M. Maciel and H. Mestre. 2010. Immunization with neural-derived antigens inhibits lipid peroxidation after spinal cord injury. Neurosci. Lett. 476: 62–65.

Imperato-Kalmar, E.L., R.A. McKinney, L. Schnell, B.P. Rubin and M.E. Schwab. 1997. Local changes in vascular architecture following partial spinal cord lesion in the rat. Exp. Neurol. 145: 322–328.

Jaeger, C.B. and A.R. Blight. 1997. Spinal cord compression injury in guinea pigs: structural changes of endothelium and its perivascular cell associations after blood-brain barrier breakdown and repair. Exp. Neurol. 144: 381–399.

Kamei, N., S.M. Kwon, M. Ishikawa, M. Ii, K. Nakanishi, K. Yamada, K. Hozumi, A. Kawamoto, M. Ochi and T. Asahara. 2012. Endothelial progenitor cells promote astrogliosis following spinal cord injury through Jagged1-dependent Notch signaling. J. Neurotrauma 29: 1758–1769.

Kaptanoglu, E., O. Okutan, F. Akbiyik, I. Solaroglu, A. Kilinc and E. Beskonakli. 2004. Correlation of injury severity and tissue Evans blue content, lipid peroxidation and clinical evaluation in acute spinal cord injury in rats. J. Clin. Neurosci. 11: 879–885.

Kim, I., S.O. Moon, S.K. Park, S.W. Chae and G.Y. Koh. 2001. Angiopoietin-1 reduces VEGF-stimulated leukocyte adhesion to endothelial cells by reducing ICAM-1, VCAM-1, and E-selectin expression. Circ. Res. 89: 477–479.

Lee, J.Y., H.S. Kim, H.Y. Choi, T.H. Oh, B.G. Ju and T.Y. Yune. 2012a. Valproic acid attenuates blood-spinal cord barrier disruption by inhibiting matrix metalloprotease-9 activity and improves functional recovery after spinal cord injury. J. Neurochem. 121: 818–829.

Lee, J.Y., H.S. Kim, H.Y. Choi, T.H. Oh and T.Y. Yune. 2012b. Fluoxetine inhibits matrix metalloprotease activation and prevents disruption of blood-spinal cord barrier after spinal cord injury. Brain 135: 2375–2389.

Loy, D.N., C.H. Crawford, J.B. Darnall, D.A. Burke, S.M. Onifer and S.R. Whittemore. 2002. Temporal progression of angiogenesis and basal lamina deposition after contusive spinal cord injury in the adult rat. J. Comp. Neurol. 445: 308–324.

Lutton, C., Y.W. Young, R. Williams, A.C.B. Meedeniya, A. Mackay-Sim and B. Goss. 2012. Combined VEGF and PDGF treatment reduces secondary degeneration after spinal cord injury. J. Neurotrauma 29: 957–970.

Mahoney, E.T., R.L. Benton, M.A. Maddie, S.R. Whittemore and T. Hagg. 2009. ADAM8 is selectively up-regulated in endothelial cells and is associated with angiogenesis after spinal cord injury in adult mice. J. Comp. Neurol. 512: 243–255.

Mautes, A.E., M.R. Weinzierl, F. Donovan and L.J. Noble. 2000. Vascular events after spinal cord injury: contribution to secondary pathogenesis. Phys. Ther. 80: 673–687.

National Spinal Cord Injury Statistics Center. 2012. Spinal cord injury facts and figures at a glance. J. Spinal Cord Med. 35: 197–198.

Ng, M.T., A.T. Stammers and B.K. Kwon. 2011. Vascular Disruption and the Role of Angiogenic Proteins After Spinal Cord Injury. Transl. Stroke Res. 2: 474–491.

Noble, L.J. and J.R. Wrathall. 1989. Distribution and time course of protein extravasation in the rat spinal cord after contusive injury. Brain Res. 482: 57–66.

Noble, L.J., F. Donovan, T. Igarashi, S. Goussev and Z. Werb. 2002. Matrix metalloproteinases limit functional recovery after spinal cord injury by modulation of early vascular events. J. Neurosci. 22: 7526–7535.

Onose, G., A. Anghelescu, D.F. Muresanu, L. Padure, M.A. Haras, C.O. Chendreanu, L.V. Onose, A. Mirea, A.V. Ciurea, W.S. El Masri and K.R. von Wild. 2009. A review of published reports on neuroprotection in spinal cord injury. Spinal Cord 47: 716–726.

Oudega, M. 2012. Molecular and cellular mechanisms underlying the role of blood vessels in spinal cord injury and repair. Cell Tissue Res. 349: 269–288.

Pan, W., W.A. Banks and A.J. Kastin. 1997. Blood-brain barrier permeability to ebiratide and TNF in acute spinal cord injury. Exp. Neurol. 146: 367–373.

Popovich, P.G., P.J. Horner, B.B. Mullin and B.T. Stokes. 1996. A quantitative spatial analysis of the blood-spinal cord barrier. Exp. Neurol. 142: 258–275.

Rubin, L.L. and J.M. Staddon. 1999. The cell biology of the blood-brain barrier. Annu. Rev. Neurosci. 22: 11–28.

Santillan, A., V. Nacarino, E. Greenberg, H.A. Riina, Y.P. Gobin and A. Patsalides. 2012. Vascular anatomy of the spinal cord. J. Neurointerv. Surg. 4: 67–74.

Schnell, L., S. Fearn, H. Klassen, M.E. Schwab and V.H. Perry. 1999. Acute inflammatory responses to mechanical lesions in the CNS: differences between brain and spinal cord. Eur. J. Neurosci. 11: 3648–3658.

Sekhon, L.H. and M.G. Fehlings. 2001. Epidemiology, demographics, and pathophysiology of acute spinal cord injury. Spine 26: S2–12.

Sharma, H.S. 2010. Early microvascular reactions and blood-spinal cord barrier disruption are instrumental in pathophysiology of spinal cord injury and repair: novel therapeutic strategies including nanowired drug delivery to enhance neuroprotection. J. Neural Transm. 118: 155–176.

Simard, J.M., O. Tsymbalyuk, A. Ivanov, S. Ivanova, S. Bhatta, Z. Geng, S.K. Woo and V. Gerzanich. 2007. Endothelial sulfonylurea receptor 1-regulated NCCa-ATP channels mediate progressive hemorrhagic necrosis following spinal cord injury. J. Clin. Invest. 117: 2105–2113.

Tator, C. and I. Koyanagi. 1997. Vascular mechanisms in the pathophysiology of human spinal cord injury. J. Neurosurg. 86: 483–492.

Tator, C. and M.G. Fehlings. 1991. Review of the secondary injury theory of acute spinal cord trauma with emphasis on vascular mechanisms. J. Neurosurg. 75: 15–26.

Thurston, G., C. Suri, K. Smith, J. McClain, T.N. Sato, G.D. Yancopoulos and D.M. McDonald. 1999. Leakage-resistant blood vessels in mice transgenically overexpressing angiopoietin-1. Science 286: 2511–2514.

Turnbull, I.M. 1971. Microvasculature of the human spinal cord. J. Neurosurg. 35: 141–147.

Turnbull, I.M., A. Brieg and O. Hassler. 1966. Blood supply of cervical spinal cord in man. A microangiographic cadaver study. J. Neurosurg. 24: 951–965.

Whetstone, W.D., J.-Y.C. Hsu, M. Eisenberg, Z. Werb and L.J. Noble-Haeusslein. 2003. Blood-spinal cord barrier after spinal cord injury: Relation to revascularization and wound healing. J. Neurosci. Res. 74: 227–239.

CHAPTER 8

Blood Vessel Remodeling After Stroke

Gabriela Moralí[1,*] and *Miguel Cervantes*[2]

8.1 Introduction

Stroke is one of the main worldwide causes of death or disability and poor quality of life in people. It results from a sudden thrombotic occlusion, usually in a branch of the middle cerebral artery, which leads to focal cerebral ischemia. Two regions of brain tissue are differently affected depending on the magnitude of ischemia: the ischemic core is affected by a severe reduction of the cerebral blood flow and immediate generalized neuronal damage, while the peripheral ischemic penumbra is subjected to a mild reduction of the cerebral blood flow and delayed neuronal death of lower magnitude.

Suppression of blood flow due to stroke may be permanent in the ischemic core, unless reperfusion is attempted through appropriate thrombolytic therapy aimed at reestablishing blood flow and reducing the irreversible injury of the infarct zone. In contrast, circulatory adjustments and improvement of blood flow occur in the ischemic penumbra, even if

[1] Unidad de Investigación Médica en Farmacología. UMAE Hospital de Especialidades, Centro Médico Nacional Siglo XXI, Instituto Mexicano del Seguro Social. Av. Cuauhtémoc 330, Col. Doctores, México, D. F. México.
Email: gmorali@hotmail.com
[2] Laboratorio de Neurociencias. División de Estudios de Posgrado, Facultad de Ciencias Médicas y Biológicas "Dr. Ignacio Chávez", Universidad Michoacana de San Nicolás de Hidalgo. Rafael Carrillo y Salvador González Herrejón S/N, Col. Cuauhtémoc, Morelia, Michoacán, México.
Email: miguelcer43@hotmail.com
* Corresponding author

the vessels otherwise supplying it with blood are permanently occluded. This allows the preservation of a viable neural substrate for repair, plasticity and functional recovery.

Ischemia and reperfusion elicit cellular mechanisms leading to brain injury, mostly through neuronal death. Neuronal repair, plasticity and functional recovery of neurons, blood vessels and glia may occur in a sequential or simultaneous manner in the injured brain structures of the penumbra. The latencies and timing of these processes, from minutes to weeks, are important to establish their differential relevance in critical periods of neuronal damage and death, as well as the "window of opportunity" for specific neuroprotective procedures, including those aimed at promoting repair and functional recovery (Okada et al. 1994, Barone and Feuerstein 1999, Lipton 1999, Leker and Shohami 2002, del Zoppo and Mabuchi 2003, Shaller and Graf 2004, Arai et al. 2011).

Among the mechanisms of neuronal repair, plasticity and functional recovery after stroke, blood vessel remodeling in the ischemic penumbra seems to play a relevant role. This occurs mainly through rearrangement of previously existing vessels of the collateral circulation, where arterioles from the resistance vessel class are converted to arteries of the conductance class (arteriogenesis); as well as formation of new capillary vessels sprouting from pre-existing ones as a result of angiogenesis. Evidence supporting this statement has been obtained from clinical studies and from animal models of experimental focal cerebral ischemia, most of them by middle cerebral artery occlusion (MCAO) in rats and mice, which may replicate some aspects of the pathological phenomena leading to brain injury and repair following stroke in humans (Longa et al. 1989, Schaper and Scholz 2003).

This chapter deals with relevant data regarding blood vessel remodeling occurring after stroke, particularly in the ischemic penumbra, which have provided a substantial background for the understanding of the mechanisms of survival and post-ischemic repair of neural structures, as well as the plastic processes taking place in the remaining or newly differentiated cells, usually leading to a better outcome after the ischemic episode.

8.2 Vascular Responses to Ischemia

8.2.1 Arteriogenesis: Collateral Arteriolar Network Remodeling

The vascular responses that occur after an ischemic episode include an increase of collateral circulation in the ischemic penumbra. Mechanical strength on the vascular wall, imposed by a tendency to redirect blood flow from the occluded vessels to the permeable arterial branches, as well as molecular signals from macrophages trigger a complex sequence of cellular and molecular processes. These processes may overlap in time, are involved in the initiation, growth and maturation of the collateral arteriolar network and involve luminal distention, enlargement and remodeling of pre-existing arterioles (Sholz et al. 2001, Wei et al. 2001, Shaper and Sholz 2003).

In experimental models of ischemia by arterial occlusion, remodeling of collateral networks follows a similar pattern. Physical forces exerted on endothelial cells, as a result of the redistribution of pressure and resistance gradients through arteriolar vessels located in the zone affected by arterial occlusion, may provide the stimuli for driving arteriogenesis in collateral vessels. On the other hand, angiogenesis is mainly elicited as a consequence of hypoxia in the ischemic tissue.

A drop in blood pressure can be expected to occur in the vascular network distal to the arterial occlusion; thus, a pressure gradient may establish over interconnecting collaterals between a pre-occlusive high pressure and a post-occlusive low pressure region, tending to redirect blood flow through small preexisting bridging arteries in the vascular territory near the arterial occlusion. This results in a partial reinstatement of blood flow to the ischemic tissue, but the increase in length of the collateral vascular network due to the involvement of additional preexisting collateral vessels and the reduced diameter of these vessels lead to a higher-resistance, low-perfusion vascular network distal to the occlusion and to the persistence of a pressure gradient along these collateral pathways. Under these conditions, increased blood flow through collateral vessels creates pulsatile forces on the arteriolar wall, eliciting shear stress which is sensed by endothelial cells triggering cellular and molecular processes resulting in the growth of collateral arteries (Carmeliet 2000, Ingber 2002, Ingber 2003a,b, Shaper and Sholz 2003, Heil and Schaper 2004, Shaper 2009, Buschmann et al. 2010, Meisner and Price 2010, Bondke-Persson and Buschmann 2011).

Shear stress produces activation of the collateral vessel endothelium characterized by cell swelling, ultra-structural protrusions, increased cellular metabolism, stimulation of transient receptor-potential cation channels and changes in protein expression. The latter initiate and maintain the collateral growth process, including proliferation of endothelial and smooth muscle cells through a pro-inflammatory response mediated by expression of inflammatory cytokines, chemokines such as monocyte chemotactic protein 1 (MCP-1) and adhesion molecules such as selectins and intracellular adhesion molecule (ICAM)-1 that are overexpressed after ischemia (Deindl 2007, Bondke et al. 2008, Troidl et al. 2009, van Royen et al. 2010).

The actions of these molecular mediators may result in activation of intrinsic vascular cells that initiate the remodeling processes. Thus, neutrophils and monocytes invade the vessel wall attracted by the expression of MCP-1 and ICAM-1. They contribute to basement membrane breakdown and extracellular matrix digestion by matrix metalloproteases (MMP)-2 and -9, as well as stimulating proliferation and migration of smooth muscle and endothelial cells. Pro-inflammatory signals induce smooth muscle cells of the collateral artery to differentiate from a contractile phenotype to a synthetic phenotype. They proliferate leading to luminal expansion and subsequent outgrowth of the collateral vessels through remodeling of the vessel wall. These events characterize the growth phase of the arteriolar network. On the other hand, bone marrow-derived endothelial progenitor cells (EPCs) are recruited and appear in the ischemic penumbra. They seem to cooperate in remodeling the collateral vessel network by supplying local factors that favor this process (Cai et al. 2000, 2003, 2004, Kinnaird et al. 2004, Ziegelhoeffer et al. 2004, Grundmann et al. 2007).

Under these conditions, lumen diameter increases leading to a decrease in vascular resistance and stabilization of network conductance, restoration of perfusion in the ischemic tissue distal to the arterial occlusion and normalization of shear stress that ends endothelial pro-inflammatory phenomena, as shown by decreased inflammation and increased angiostatic and anti-inflammatory markers. This decline of inflammatory phenomena coincides with decreased vascular cell proliferation, accretion of macrophages around collateral vessels and return of smooth muscle cells to a contractile phenotype. A mature basement membrane may also be reestablished due to the normal balance between extracellular matrix breakdown and its stabilization (Arras et al. 1998, Shaper and Sholz 2003, Khmelewski et al. 2004, Lee et al. 2004, Meisner and Price 2010, Bondke-Persson and Buschmann 2011). Network hemodynamics and collateral perfusion pressure may lead to bidirectional final remodeling of maturing collateral vessels; the largest, most developed collateral vessels complete maturation and stabilization, but the smaller, less developed collateral vessels may undergo neo-intimal hyperplasia and eventual regression.

Remodeling of the pre-existing vessel network leading to functional collateral vessels that supply blood flow to the hypo-perfused regions in the ischemic penumbra can rescue brain tissue at risk of being damaged due to arterial occlusion. This adaptive spreading of the preexisting collateral vascular network contributes to preventing progress of the ischemic pathophysiology to zones where the remaining neurons can still recover from ischemia (Buschmann et al. 2003, Busch et al. 2003, 2006).

Remodeling of the pre-existing arteriolar network also involves endothelial nitric oxide synthase (NOS) and NOS-derived nitric oxide (NO). They elicit vascular dilation, as well as increasing proliferation and migration of existing endothelial and vascular smooth muscle cells. Thus, they contribute to structural integration of the remodeling vessel wall after stroke (Heil and Schaper 2004, Cui et al. 2009a).

The presence of abundant arteriolar collaterals is associated with better early- and long-term clinical improvement. This is independent of whether patients were treated or not with a thrombolytic after stroke. In contrast, poor development of collateral circulation from pre-existing arterioles is related to a higher mortality after stroke (Wei et al. 2001, Christoforidis et al. 2005).

8.2.2 Angiogenesis: Capillary Vessel Network Remodeling

Another part of vascular remodeling after stroke is a progressive process of angiogenesis. It entails capillary growth by sprouting from or by splitting of preexisting capillary vessels, promoted by hypoxia-inducible factor 1 (HIF-1α) and subsequent expression of vascular endothelial growth factor (VEGF) by neurons and other cellular types. Thus, the development of sprouting angiogenesis involves the proteolytic degradation of the extracellular matrix by endothelial cells in response to pro-angiogenic factors; such as VEGF released by astrocytes, neurons, fibroblasts and smooth muscle cells in the ischemic tissue. This process also includes proliferation and migration of endothelial cells toward pro-angiogenic factors. These factors also elicit splitting of the preexisting capillary lumen by insertion of tissue pillars (intussusception), with the contribution of recruited circulating EPCs. In both cases, angiogenesis through sprouting or through splitting of preexisting capillary vessels leads to formation of new, mature and stabilized vessels that can acquire tissue-specific properties (Risau 1997, Carmeliet 2000, Acker and Acker 2004, Fischer et al. 2006, Kokovay et al. 2006, Ucuzian et al. 2010, Cunningham et al. 2012, Yamashita and Abe 2012).

Angiogenesis is triggered in the ischemic penumbra shortly after the ischemic episode and might persist for several weeks as a result of gene expression, synthesis and release of key molecules from the cells remaining in the ischemic tissue (Krupinski et al. 1994, Hayashi et al. 1997, Plate 1999, Marti 2000, Hayashi et al. 2003). Formation of new vessels after stroke, including angiogenesis (Zhang et al. 2003) and remodeling of the collateral vascular network, is preceded by endothelial proliferation in the ischemic boundary zone, as early as 12–24 hr following stroke, as well as by an increase of circulating EPCs from the bone marrow, which are incorporated to the ischemic penumbra. Angiogenesis can be evidenced three to four days following the ischemic insult (Krupinski 1994, Zhang et al. 2002a, Hayashi et al. 2003, Beck and Plate 2009).

The sequence of cellular events associated with angiogenesis after stroke is highly complex (Hayashi et al. 2006, Arai et al. 2009, Beck and Plate 2009). The development of angiogenesis also includes capillary wall destabilization by transformation of endothelial cells from a quiescent to an active stage, with a high mitotic index and capacity for proliferation, migration and extracellular matrix proteolysis. The transient degradation of the basal membrane surrounding endothelial cells allows endothelial cells to move into the surrounding extracellular matrix. Under these conditions, the junctions between neighboring intimal endothelial cells are disrupted, permitting cell invasion into the basal membrane and the surrounding extracellular matrix. Then they can proliferate and migrate, driven by extracellular matrix signaling to form angiogenic sprouts, multicellular chains and finally cell channels which allow extension of blood flow from preexisting vessels to new capillary networks (Risau 1997, Armulik et al. 2005, Lamalice et al. 2007, Francis et al. 2008).

Endothelial tip cells can sense growth factor gradients, which increase the expression of membrane-type 1 MMP. This protein degrades the basal membrane and extracellular matrix, allowing capillary sprouts from preexisting vessels. Endothelial cells that lose contact with the basal membrane undergo reorganization of their actin cytoskeleton, contributing to sprouting morphogenesis. Thus, mitosis and migration of endothelial cells predominantly at the tip of the growing capillaries lead to formation of the capillary tip and stalk; in fact endothelial cells compete for tip or stalk positions depending on the ratio of their intracellular VEGF receptors (VEGFR1 and VEGFR2).

During migration, endothelial cells interact with a matrix formed by leaked plasma proteins; this interaction is mediated by integrins. Sprouting endothelial cells are guided by angiogenic factors, such as VEGF and its receptor VEGFR2. In fact, VEGF isoforms have several roles during angiogenesis; they increase vascular permeability and aid in extracellular matrix degradation, pericyte recruitment and lumen formation on newly formed capillaries. VEGF isoforms also induce endothelial cell survival, sprouting, proliferation and migration (Ferrara et al. 2003, Chung et al. 2010, Jakobsson et al. 2010). However, these processes have to be precisely regulated, especially following thrombolytic therapy, since their temporal coincidence with overexpression of molecules promoting angiogenesis, such as metalloproteases, may lead to rupture of the blood brain barrier (BBB) and increase the risk of brain hemorrhage after stroke.

Migrating endothelial cells involved in the processes of angiogenesis are genetically programmed to form tubular compartments, either by intracellular vacuolization within endothelial multicellular chains or through formation of endothelial cell cylinders, which lead to an internal extracellular lumen. Data from *in vivo* and *in vitro* studies have provided a model of lumen and tube formation through several cellular phenomena; such as pinocytosis, intracellular vacuolization and cytoplasmic projections that form junctions between neighboring endothelial cells thereby forming complex multicellular capillary tubes (Guldner and Wolff 1973, Davis and Camarillo 1995, 1996, Bayless and Davies 2002, Liu et al. 2002, Davis et al. 2007, Iruela-Arispe and Davis 2009). Other cellular responses that may be involved in angiogenesis are: insertion of tissue columns into forming capillaries, autophagy by lysosomes within endothelial cells, trans-luminal ingrowth of cytoplasmic filopodial projections resulting in a network of luminal structures, exocytosis of vacuoles between closely apposed sprouting endothelial cells resulting in intercellular canalization, apoptosis of centrally located endothelial cells within a cluster; all these processes might lead to tubular neovascular structures. Whether lumen and tubular formations occur simultaneously or one precedes the other remains under discussion, since there is experimental evidence suggesting both possibilities (Davis and Camarillo 1995, Meyer et al. 1997, Djonov et al. 2000, Stromhaug and Klionsky 2001, Peters et al. 2002, Davis and Bayless 2003, Duval et al. 2003, Egginton and Gerritsen 2003, Kamei et al. 2006, Duval et al. 2007).

Once new capillaries are formed, angiogenesis may enter a resolution phase. During this period, the proteolytic phenotype of invading tip cells is deactivated by endothelial cells as a result of their contact with pericytes. Thus, major angiogenic factors are down regulated, tip cell MMP activity is decreased, tissue inhibitor of matrix metalloproteinases (TIMP)-2 and -3 are activated and the levels of platelet-derived growth factor (PDGF), angiopoietin-1(Ang-1) and transforming growth factor (TGF)-β1 are increased (Saunders et al. 2006, Hellstrom et al. 2007, Chung et al. 2010).

During early angiogenesis, new capillaries are formed by endothelial cells, linked by tight and adherens junctions, within a matrix lacking mural cells (pericytes and smooth muscle cells). Thus, the stabilization of these vessels requires recruitment of mural cells and generation of the basal membrane, elastic laminae and surrounding matrices. Afterwards, vessels may acquire tissue-specific morpho-functional specializations, such as the interaction between glial fibrillary acidic protein (GFAP) and angiotensinogen mediating BBB development. In this sense, vascular networks integrated by newly formed capillaries might also require branching, remodeling and removal of superfluous vascular branches. Several molecular mediators have been shown to be involved in these processes, accounting for the better functionality of the new capillaries. Capillary network maturation by angiogenesis involves tissue oxygen tension, mechanical forces, extracellular matrix growth factor gradients, extracellular matrix metalloproteases plus their inhibitors and neuronal growth factors that share roles in angiogenic maturation (Carmeliet 2000, Kluk and Hla 2002, Jain 2003, Lindblom et al. 2003, van Hinsbergh and Koolwijk 2008, Krishnan et al. 2008, Iruela-Arispe and Davis 2009, Chung et al. 2010). Once a functional local vascular network has developed, tissue oxygenation provides negative feedback that inhibits further angiogenesis.

Hemodynamic changes in specific brain structures may have different characteristics and temporal courses related to the angiogenic remodeling processes after stroke. Thus, decreased cerebral blood flow in the ipsilateral and contralateral thalamus two days after focal cerebral ischemia progressively recovers in the ipsilateral thalamus. In fact, it becomes hyperperfused compared to the contralateral side at 30 and 90 days after stroke, associated with increased blood vessel branching as a result of angiogenesis. Angiogenesis was preceded by enhanced expression of vascular adhesion factors (Hayward et al. 2011). Hyper-perfusion may also result from the action of vasoactive agents, such as NO and adenosine released by endothelial and glial cells; accounting for increased cerebral blood flow to serve high-energy demanding cell activity (Iadecola and Nedergaard 2007). Similarly, hemodynamic changes in the peri-infarct cortex are closely related to the temporal course of vascular network remodeling through angiogenesis (Borlongan et al. 2004, Cai et al. 2009, Eve et al. 2009, Hermann and Zechariah 2009).

Bone marrow-derived EPCs, as well as endothelial precursor cells from other ill-defined peripheral sources, are involved in post-stroke angiogenesis. They are recruited from the circulation and incorporated into newly formed vessels, favoring their morphogenesis and stabilization. They also promote development of the BBB. Furthermore, EPCs can release pro-angiogenic factors (Zhang et al. 2002b, Kokovay et al. 2006, Yamashita and Abe 2012). HIF-1α, estrogen, VEGF or erythropoietin released after acute ischemic stroke or trauma may elicit the activation and mobilization of EPCs from the bone marrow, which also depends on the production of NO and the local activity of matrix metalloproteinases, particularly MMP-9 (Heissig et al. 2002, Heeschen et al. 2003, Hoenig et al. 2008, Iwakura et al. 2003, Liman and Endres 2012).

Bone marrow-derived EPCs have stem cell properties and may express the surface markers cluster of differentiation (CD)34, CD133 and VEGFR2, which characterize a specific progenitor cell subtype at a specific maturation stage. The expression of CD34, which defines EPCs, decreases over time as they differentiate towards endothelial cells. They are mobilized to the peripheral blood and reach different organs, where they differentiate into endothelial cells. They participate in vessel regeneration under several physiological or pathophysiological conditions, in response to endogenous chemical signals released by ischemic tissues, exercise or traumatic brain injury (Asahara et al. 1997, Shintani et al. 2001, Zhang et al. 2002b, Adams et al. 2004, Laufus et al. 2004, Taguchi et al. 2004a,b, Greenberg 2007, Gertz et al. 2006, Guo et al. 2009).

The participation of circulating EPCs in neovascularization after stroke may be relevant for angiogenesis and recovery of the ischemic injured neural tissue in the ischemic penumbra. EPCs are known to be mobilized into the peripheral blood in patients after stroke, they are found in a high proportion of newly formed vessels during angiogenesis and their administration results in better functional recovery after experimental focal cerebral ischemia (Hess et al. 2002, Taguchi et al. 2004a,b, Paczkowska et al. 2009, Yamashita and Abe 2012). EPCs may migrate, reaching the ischemic penumbra following stroke via cytokine gradients. They are targeted by several EPCs-activating factors that are also involved in the neovascularization processes. Thus, EPCs either indirectly promote angiogenesis via paracrine mechanisms stimulating endothelial cell proliferation and stabilization or they directly contribute to the formation of new vessels from the pre-existing vasculature through their transformation into endothelial cells (Asahara et al. 1999a,b, Heissig et al. 2002). However, some evidence has suggested that EPCs hardly incorporate into newly formed vessels; rather these cells may support pre-existing edothelium-derived angiogenesis by secreting growth factors, including basic fibroblast growth factor (bFGF) and VEGF (Ziegelhoeffer et al. 2004).

The morphological and functional improvement of the neurovasculature in the ischemic penumbra is partly due to stabilization of new vessels through angiogenesis. This results in capillaries with a monolayer of endothelial cells and appropriate tight junctions supported by surrounding pericytes. Recruited pericytes and astrocytes are involved the development and recovery of an integrated and functional BBB, which had been disrupted after stroke (Zang et al. 2000, Durukan et al. 2009, Kokovay and Cunningham 2006, Slevin et al. 2006, Sandoval and Witt 2008). In fact, early angiogenesis is coincident with a highly leaky BBB; this abnormal vessel permeability decreases over time. This is a very important consequence of vascular network remodeling through angiogenesis.

The BBB is a dynamic structure that separates the peripheral blood from the central nervous system (CNS) microenvironment. The BBB is constituted by endothelial cells, interconnected by specialized tight junctions conforming capillary blood vessels, astrocytic endfeet surrounding and almost completely covering blood vessels and pericytes embedded in the basement membrane, which highly interact with extracellular matrix components. Endothelial cells participate in several BBB functions (Banerjee and Bhat 2007, Fisher 2009, Carvey et al. 2009, Mathiisen et al. 2010, Neuwelt et al. 2011, Nico and Ribatti 2012), such as selective permeability through transport and enzymatic systems or cellular mechanisms restricting the translocation of cells, proteins and water-soluble compounds. This allows the BBB to regulate the composition of the microenvironment by exchanging chemical compounds (nutrients, neurotransmitters) between the blood and the brain (Abbott et al. 2006, Zlokovic 2008, Sofroniew and Vinters, 2010, Zhang and Barres 2010, Allaman et al. 2011, Redzic 2011). Changes in the microenvironment elicit cell adaptive responses that account for survival, proliferation and migration. These responses involve receptor adhesion proteins, such as dystroglycan and integrins, that also importantly participate in cell-cell and cell-extracellular matrix interactions (Hynes 1992, Wang and Milner 2006, Moore and Winder 2010).

BBB permeability increases abnormally after ischemia and reperfusion in stroke patients and in animal models of cerebral ischemia, due to endothelial tight junction loss resulting from the reduction of claudin-5, occludin plus zonula occludens (ZO)-1 and -2 in the plasma membrane of adjacent endothelial cells. In addition, the contacts between astrocytes and vessels are gradually reduced after ischemia in experimental models of stroke by MCAO (from 94 per cent in sham-operated rats to 74 per cent after four hr, 16 per cent after 16 hr, to almost non-existent after 48 hr). Furthermore, swelling of astrocytic endfeet results in vessel lumen obstruction, worsening BBB disruption (Brown and Davis 2002, Mark and Davis 2002, Koto et al. 2007, Kastrup et al. 2008, Baumann et al. 2009, Kwon et al. 2009, Israeli et al. 2010).

The basal membranes of the BBB are disrupted after stroke, mainly by up-regulated and activated MMPs and plasmin proteolysis. Data regarding increased expression of MMP-9 and its temporal course after permanent or transient MCAO have been obtained from experimental models of focal cerebral ischemia and have been related to disruption and increased permeability of the BBB; but with differences due to variations of the experimental conditions in the different studies (Romanic et al. 1998, Rosenberg et al. 1998, Gasche et al. 1999, Planas et al. 2001, Heo et al. 2005, Rosell and Lo 2008, Dong et al. 2009, Jin et al. 2010). Thus, plasma proteins such as fibrinogen that leak into the CNS, as well as new extracellular matrix proteins, such as neurocan and osteopontin secreted by activated glia, lead to further harmful consequences (directly or through inflammation) on BBB integrity that may vary depending on the time that has elapsed since stroke. Some changes in extracellular matrix composition (laminin, fibronectin, collagen IV and their receptors) also correlate with increased BBB permeability, by either inhibiting or initiating remodeling and repair of the ischemic parenchyma (Fukuda et al. 2004, del Zoppo et al. 2006, Baumann et al. 2009, Candelario-Jalil et al. 2009, Kwon et al. 2009, Morancho et al. 2010, Baeten and Akassoglou 2011). Extravasation of fibrinogen through the disrupted BBB may aggravate the damage after stroke due to its interaction with different cell types in the brain. Fibrinogen may activate microglial and resident immune cells, eliciting pro-inflammatory actions. Fibrinogen may also inhibit neurite outgrowth by direct interaction with neurons or by stimulating astrocytic secretion of proteoglycans (Schachtrup et al. 2007, Schachtrup et al. 2010) and promoting formation of a glial scar that may contribute to degeneration of axons (Adams et al. 2004, Adams et al. 2007a,b, Schachtrup et al. 2007, 2010).

Thus, reestablishing the morphologic and functional characteristics of the BBB may be crucial for the formation of neurovascular units and for plastic processes leading to functional recovery in the ischemic penumbra. Indeed, angiogenesis and neurogenesis have close spatial relationships in the ischemic tissue and share common pathways involving molecular mediators that promote their development after stroke (Carmeliet 2005, del Zoppo 2006, Yamashita et al. 2006, Lok et al. 2007, Arai et al. 2009, Beck and Plate 2009, Xiong et al 2010). In fact, neurovascular units (multi-cellular complexes formed by endothelial cells, pericytes, neurons and glial cells functionally linked by growth factors and extracellular matrix proteins) are places where new immature neurons, that migrate from neural progenitor cell (NPC) niches, are closely associated with the remodeling vasculature. Thus, neurovascular units are a relevant component of the structural and neurochemical substrate necessary for the development of neurogenesis and synaptogenesis, both of which may in turn improve functional recovery after stroke (Parent 2003, Zhang et al. 2005a, Greenberg 2007, Lok et al. 2007, Beck and Plate 2009, Guo and Lo 2009, Ohab and Carmichael 2008).

Reactive astrocytes in the peri-infarct cortex may also secrete neurotrophic molecules. These include astrocyte populations that co-express GFAP and the high-mobility group box-1 (HMGB1) protein. The latter may be required for endogenous recovery after stroke; since it can promote endothelial activation and sprouting, as well as increased neurite outgrowth and neuronal survival (Huttunen et al. 2002, Treutiger et al. 2003, Schlueter et al. 2005, Bessis et al. 2007, Hayakawa et al. 2010, 2012). HMGB1 immuno-reactivity has been observed in reactive astrocytes located in the ischemic penumbra; these cells may release HMGB1 into the extracellular medium when activated (Kim et al. 2008), suggesting that HMGB1 may be a link between reactive astrocytes and stroke recovery. In fact, signaling between reactive astrocytes and EPCs through HMGB1 up-regulation has been found to mediate neurovascular remodeling and functional recovery after focal cerebral ischemia in mice (Hayakawa et al. 2012). On the other hand, astrocytes may have a detrimental role in neuronal recovery through the production of extracellular matrix substrates that contribute to glial scar formation, known to inhibit dendritic and axonal growth (Giulian 1993, Fawcett and Asher 1999, Horner and Gage 2000, Silver and Miller 2004).

In addition, experimental data have shown that the density of blood vessels in the peri-infarct cortex increases over time (from 4.36 per cent of the total volume occupied by vessels in controls to 7.25 per cent at six weeks post-stroke in mice). This rise is closely related, spatially and temporally, to plastic changes in neurons; such as an augmented rate of dendritic spine formation after stroke (Brown et al. 2007).

Thus, vascular remodeling and neuronal plasticity are interdependent and occur during recovery after stroke, based on cell-cell interactions between EPCs and NPCs. This allows the generation of new neurons from progenitors and their migration in close association with blood vessels. Blood vessels may participate importantly in providing NPCs with directional guidance toward the damaged peri-infarct regions, where they differentiate into mature neurons. On the other hand, development and stabilization of angiogenesis occurs in response to hypoxia after stroke. The active exchange of molecular signals between vascular and neuronal compartments, which characterizes the neurovascular niche, is partially due to the ability of the capillary endothelium to secrete a rich repertoire of trophic factors, including stromal-derived factor 1 (SDF-1) and Ang-1. These proteins activate neural precursors in the subventricular zone (SVZ) of the forebrain and in the subgranular zone of the dentate gyrus, triggering neurogenesis in the normal brain and during post-stroke recovery (Leventhal et al. 1999, Palmer et al. 2000, Kokaia and Lindvall 2003, Shen et al. 2004, Greenberg and Jin 2005, Ohab et al. 2006, Thored et al. 2006, Yamashita et al. 2006, Chopp et al. 2007, Thored et al. 2007, Zhang et al. 2007, Guo et al. 2008, Zlokovic 2008, Murphy and Corbett 2009, Snapyan 2009, Zhang et al. 2009, Font et al. 2010, Kojima et al. 2010, Liman and Endres 2012).

Together these processes may improve blood supply, macrophage infiltration and scavenging of necrotic tissue, resulting in structural and functional recovery of the neural tissue. In this sense, it has been shown that micro-vessel density in the ischemic penumbra correlates with longer survival of patients after stroke (Krupinski et al. 1994). However, these vascular recovery processes do not occur in brain structures, such as the ischemic core, where massive neuronal death is induced by ischemia, as well as vessel damage and dissociation of neurovascular unit components. Thus, morpho-functional alteration of endothelial tight junctions and the basal membrane, as well as damage of endothelial cells due to the harmful actions of reactive oxygen species, result in permanent injury of the vascular components of the neural tissue in the ischemic core (Okada et al. 1994, Manoonkitiwongsa et al. 2001, del Zoppo and Mabuchi 2003, Hayashi et al. 2006, Slevin et al. 2006, Beck and Plate 2009).

8.2.2.1 Evaluation of Angiogenesis After Stroke

Once the relevance of angiogenesis on the recovery of the neural tissue in the peri-infarct zone was recognized, different procedures to identify, measure and follow up its magnitude and temporal course were developed. Their usefulness as parameters of post-stroke angiogenesis has been tested in animal models of focal cerebral ischemia. Thus, non-invasive image techniques such as magnetic resonance imaging (MRI) or positron emission tomography (PET) allow evaluation of angiogenesis, through the analysis of images based on relaxation times and eco gradients or those produced by radioactively labeled compounds, in experimental models of MCAO (Jiang et al. 2005, Ding et al. 2008a,b, Brea et al. 2009, Cai et al. 2009, Ding et al. 2010, Seevinck et al. 2010, Ding et al. 2011, Dijkhuizen et al. 2012).

Several noninvasive imaging techniques have proven their usefulness for the acute clinical management of stroke, providing valuable data on the time course of irreversible damage and the spared neural tissue of the peri-infarct zone. Ischemic damage may be reversible in this area, since it remains responsive to procedures aimed at promoting vascular and neural remodeling leading to plasticity and functional recovery. Thus, computed tomography and several MRI procedures, such as magnetic resonance angiography, diffusion weighted imaging, diffusion tensor imaging, perfusion weighted imaging and functional MRI may assist in: confirming the clinical diagnosis, locating the thrombus, estimating infarct size, classifying stroke subtype, assessing blood supply and evaluating the risks and benefits of therapy. These imaging techniques can also predict and follow up the functional outcome of stroke patients (Johansen-Berg 2007, Latchaw et al. 2009, Farr and Wegener 2010, Jiang et al. 2010). However, assessment of angiogenesis after stroke through non-invasive imaging techniques remains at the experimental level, aimed at obtaining preclinical knowledge applicable to clinical management of patients after stroke (Dirnagl 2006).

MRI indices of angiogenesis and neurogenesis are highly correlated with neurological function and may be used in the real-time monitoring of recovery after stroke. The combination of MRI, immuno-histochemistry and three-dimensional laser scanning confocal or electronic microscopy allows better experimental study of micro-vascular density in the brain after stroke. Though MRI and PET procedures allow following up the outcome of stroke along weeks, several invasive histological and immuno-histochemical techniques requiring terminal anesthesia of experimental animals have also been useful for the qualitative and quantitative experimental study of angiogenesis (Zhang et al. 2000, Bosomtwi et al. 2008, Chopp et al. 2008, Xiong et al. 2010, Ardelt et al. 2012, Martin et al. 2012, Moisan et al. 2012, Wagner et al. 2012, Wang et al. 2012). Non-invasive techniques can be expected to identify and measure molecular mediators involved in angiogenesis. They may also allow the analysis of specific mechanisms of angiogenesis, the prognosis of stroke outcome and the evaluation of pro-angiogenic neurogenic therapy efficacy.

8.3 Molecular Mediators of Vessel Network Remodeling After Stroke

The regulatory mechanisms involved in angiogenesis after stroke are complex and include up-regulation of several molecular mediators by the HIF pathway, promoting cell-cell and/or cell-extracellular matrix interactions. These molecules have different roles and timing in angiogenesis: NO, growth factors such as VEGF, fibroblast growth factor (FGF), PDGF, TGF-β and insulin-like growth factor 1 (IGF-1), proteolytic enzymes like tissue-type plasminogen activator (tPA), urokinase plasminogen activator (uPA), MMP-2 and -9, their inhibitors plasminogen activator inhibitor-1 and TIMP-1, MCP-1 and SDF-1, pro-inflammatory cytokines such as interleukin (IL)-1β, tumor necrosis factor-alpha (TNF-α), Ang-1 and -2 plus their receptors (Adibhatla et al. 2008, Navaratna et al. 2009, Ucuzian et al. 2010, Cunningham et al. 2012). Thus, genes related to angiogenesis are up-regulated minutes after stroke. HIF and the growth factors may be synthesized by different cell types including endothelial cells, astrocytes, neurons, fibroblasts, smooth muscle cells, platelets and inflammatory cells.

8.3.1 Hypoxia-Inducible Factor

Complex cellular oxygen-sensing systems have evolved to ensure tight regulation of oxygen homeostasis in the brain, inducing an elaborate sequence of adaptive mechanisms in response to oxygen variations that may avoid or minimize brain damage, through short-term (seconds, minutes) and long-term (hours, days) phenomena after stroke. Specific transcription factor systems, such as HIF, have been described to respond to oxygen changes or excess reactive oxygen species by activation of genes encoding metabolic enzymes and pro-angiogenic factors to protect hypoxic tissues (Wang and Semenza 1993, Acker and Acker 2004, Correia and Moreira 2010).

HIF is a heterodimeric transcription factor composed of two subunits: the constitutively expressed HIF-1β subunit and the oxygen-regulated HIF-1α subunit (Wang and Semenza 1993, Correia and Moreira 2010). Under normal oxygen conditions, HIF-1α is hydroxylated by a family of prolyl hydroxylase enzymes, in the presence of ferrous iron and 2-oxoglutarate as cofactors. Then, it is rapidly degraded in an ubiquitin-mediated proteasomal reaction (Wang et al. 1995). Under hypoxic conditions, proteasomal degradation of HIF-1α does not occur and the stabilized protein translocates and accumulates in the nucleus. Once there, it dimerizes with HIF-1β forming the active HIF complex which binds to the hypoxia-responsive element in gene promoters. HIF up-regulates a repertoire of target genes including metabolic enzymes, cytokines, growth factors such as VEGF and inflammatory cytokines such as Ang-2 and their receptors, among others (Wang and Semenza 1993, Sharp and Bernaudin 2004, Correia and Moreira 2010, Cunningham et al. 2012). Consistent with this model, several groups have shown that HIF-1α expression is induced by hypoxic conditions in neurons, astrocytes, ependymal and endothelial cells (Chavez et al. 2000). HIF-1α immunostaining was observed in neurons beginning two hr after hypoxia and persisted for several days following cardiac arrest (Chavez and LaManna 2002). Several reports have shown that HIF-1 activation reduces ischemic injury, while its suppression aggravates brain damage after either focal or global cerebral ischemia (Baranova et al. 2007).

In addition to its actions on the adaptive response to hypoxia, HIF-1α signaling promotes activation and migration of neural stem progenitor cells (NSPCs) toward areas of ischemic injury through SDF-1—fusin signaling and MMP regulation (Ohab et al. 2006, Thored et al. 2006, Cunningham et al. 2012). Thus, besides contributing to maintain the vascular niche by activating constitutive VEGF gene expression and VEGF release, HIF-1α promotes activation and migration of NSPCs through fusin and MMP regulation (Ingraham et al. 2011).

8.3.2 Vascular Endothelial Growth Factor

Among several HIF-induced growth factors, VEGF is one of the most studied because it is involved in several stages of angiogenesis, i.e., endothelial cell migration, mitogenesis, sprouting and tube formation (Margaritescu et al. 2011). VEGF is up-regulated in humans and in experimental animals within hours of stroke and correlates with blood vessel growth in the ischemic penumbra (Krupinski et al. 1994, Hayashi et al. 1997, Plate et al. 1999, Marti et al. 2000, Font et al. 2010, Margaritescu et al. 2011). Increased VEGF and VEGFR mRNA has been detected in the tips of invasive angiogenic sprouts, while antibody blockade of VEGF significantly decreases micro-vessel outgrowth (Gerhardt et al. 2003). On the other hand, transfection of a viral vector for VEGF to ischemic mice increases vascular permeability and outgrowth of blood vessels (Stiver et al. 2004).

VEGF is a homodimeric glycoprotein of 36–46 kDa with four isoforms that result from alternative splicing of a single gene product: VEGF121, VEGF165, VEGF189 and VEGF206. VEGF165 is the most common and active form (Ferrara et al. 2003, Margaritescu et al. 2011). It is known to be up-regulated in endothelial cells and other cell types during active angiogenesis. It may also be released and stored in the vascular basement membrane, by binding to some extracellular matrix proteins. VEGF attachment to extracellular matrix scaffolds, such as fibrin or collagen, has been shown to allow a sustained and controlled release of this growth factor and is associated with the formation of organized vascular networks (Ehrbar et al. 2004). VEGF exists as a soluble form and as a membrane bound form, both bind mostly to two trans-membrane receptor tyrosine kinases VEGFR1 and VEGFR2. VEGFR2 is the main receptor involved in the angiogenic effects of VEGF after stroke, including increased vascular permeability, as well as endothelial cell proliferation, invasion, migration and survival (Ferrara et al. 2003). Neuropilin-1, another trans-membrane receptor that binds to the VEGF-A splice variant of VEGF165, is involved in capillary morphogenesis and may be required for endothelial filopodial tip directionality during angiogenesis (Gerhardt et al. 2004). VEGF also promotes release of von Willebrand factor, expression of integrin, interstitial collagenase, plasminogen activator and plasminogen activator receptor, as well as increasing vascular permeability and fenestration.

Sarcoma (Src) kinases are downstream signaling mediators of VEGF-induced permeability. Pharmacological targeting of Src has been shown to be protective after stroke and to prevent edema formation (Weis and Cheresh 2005). Binding of VEGF to heparin and heparan sulfate proteoglycan contributes to binding between VEGF and VEGFR on the cell surface of endothelial cells, stabilizes the active conformation of VEGF, prevents its degradation and has been shown to greatly increase the angiogenic response of endothelial cells to VEGF in a type-I collagen matrix.

VEGF release and interaction with its receptors, as well as timing of gene expression have been analyzed in order to understand VEGF-dependent effects. Most studies assessing VEGF expression in the brain following acute ischemic stroke have been performed on animal models (Hayashi et al. 1997, Marti et al. 2000, Plate et al. 1999). The VEGF gene is up-regulated minutes after stroke, resulting in increased VEGF mRNA and protein within the first hours and remaining for several days in neurons and up to a few weeks in astrocytes. Augmented VEGF levels have been shown in the serum of patients suffering a stroke at admission, as well as on days one, three, seven and 14 after stroke and have been found to be correlated with the infarct volume and neurological impairment (Slevin et al. 2000). In addition, increased VEGF expression has been found in the ischemic penumbra, compared to the ischemic core and the contralateral region, in patients that died in the acute phase, between the first hours to several days following ischemic stroke (Issa et al. 1999, Margaritescu et al. 2011).

VEGFR1 and VEGFR2 are up-regulated in blood vessels of the ischemic penumbra 48 hr after experimental ischemia and return to basal levels after seven days, while in blood vessels invading the ischemic core they are augmented from day three to days seven or 14 (Plate et al. 1999, Marti et al. 2000). Increased VEGFR2 expression has been found in glial cells, endothelial cells and neurons after ischemia (Hayashi et al. 1997).

As has been mentioned earlier, EPCs are mobilized from the bone marrow, migrate and are recruited to the ischemic areas where they participate in vascular remodeling. This response is mediated by diverse factors, such as the granulocyte and macrophage colony-stimulating factor (GM-CSF), SDF-1 and in particular by VEGF through its interaction with VEGFR2 (Heissig et al. 2002, Liman and Endres 2012). In addition to stimulating the mobilization, proliferation and recruitment of EPCs, VEGF-VEGFR2 promote their survival and reduce their apoptosis through the protein kinase B (PKB) signaling pathway (Barrera-Ocampo et al. 2008). Physical exercise is known to promote angiogenesis and to increase EPCs mobilization either through a NO-dependent reduction of EPCs apoptosis or by VEGF gene up-regulation (Milkiewicz et al. 2006).

In addition to its crucial angiogenic effects, VEGF also exerts direct neurotrophic signaling, neuroprotective and neurogenic effects, as well as promoting hippocampus-dependent memory formation (Jin et al. 2002, Ferrara et al. 2003, Cao et al. 2004). The neurotrophic effects of VEGF are mediated by the VEGFR2 receptor. Exogenous administration of VEGF following focal cerebral ischemia reduces the infarct size and improves neurological performance, through both direct neuroprotective effects and stimulation of angiogenesis (Sun et al. 2003).

8.3.3 Fibroblast Growth Factor

FGF was one of the growth factors initially described for its pro-angiogenic effects (Brea et al. 2009, Ucuzian et al. 2010). FGF is a family of proteins, the most studied are FGF-2 or bFGF and FGF-1 or acidic FGF (aFGF). They bind most commonly to the receptor tyrosine kinases FGFR-1 or FGFR-2 (Gerwins et al. 2000). FGF is stored in the vascular basement membrane and is also up-regulated during active angiogenesis. Unlike VEGF165, FGF binds with a high degree of affinity to extracellular matrix proteins, such as fibrin, which can potentiate its angiogenic capacity (Sahni et al. 2006). FGF-2 has a role in extracellular matrix remodeling during angiogenesis by stimulating the proteolysis of matrix components (via up-regulation of urokinase receptors) and inducing the synthesis of collagen, fibronectin and proteoglycans by endothelial cells (Pepper et al. 1993).

FGF-1 is not detected in the normal brain, but is up-regulated 24 hr after experimental ischemia in neurons and macrophages and reaches its maximum at 14 days. On the other hand, FGF-2 is constitutively expressed in neurons of non-ischemic animals, it increases in peri-infarct areas 24 and 48 hr after MCAO and is evidenced during the following days in endothelial cells of new capillaries extending into the infarct, as well as in reactive astrocytes and macrophages accompanying them (Hara et al. 1994, Lin et al. 1997). In the human brain, increased FGF-2 expression has been shown in the ischemic penumbra (Issa et al. 2005). FGF binding to FGFR-1 increases endothelial cell migration *in vitro* and promotes capillary morphogenesis, activating signaling pathways mediated by protein kinase-C and phospholipase A2, among others. Synergism and cross talk have been demonstrated between FGF and other growth factor pathways, since FGF-2 enhances endogenous VEGF production (Kanda et al. 1996, Ucuzian et al. 2010). FGF-2 administration 2 hr after brain ischemia reduces the infarct volume (Jiang et al. 1996). Its administration at later periods does not affect infarct volume but increases the expression of growth associated protein-43, a protein involved in neuronal plasticity.

8.3.4 Angiopoietin

The angiopoietins are a family of extracellular ligands that bind to the primarily endothelial-specific tyrosine kinase with immunoglobulin-like and epidermal growth factor-like domains (Tie) receptors which, similarly to other growth factor receptors, are tyrosine kinases that function primarily during vascular remodeling and angiogenesis (Ucuzian et al. 2010). Although Ang-1 and Ang-2 bind the same receptor (Tie-2), they elicit very different responses. Ang-1 binding to Tie-2 causes its rapid trans-phosphorylation and subsequent downstream activation of the PKB-forkhead box O1 pathway, promoting endothelial cell survival and inhibiting Ang-2 expression (Daly et al. 2004). Since Ang-2 is a key signaling molecule involved in endothelial cell activation and initiation of angiogenesis, its inhibition is relevant. The early role of Ang-2 in angiogenesis is evidenced by its high up-regulation in endothelial cells after exposure to hypoxia and VEGF (Pichiule and LaManna 2002). Ang-2 expression is associated with the formation of neo-vessels during cerebral ischemia. Ang-2 released from endothelial cells induces an autocrine response in the presence of pro-angiogenic signals such as VEGF, preventing Tie-2 receptor phosphorylation and signaling by antagonizing Ang-1. Ang-2 is involved in the disruption of the endothelial monolayer and promotes endothelial cell survival, vessel sprouting and angiogenesis. On the other hand, experimental evidence suggests that Ang-1 and Tie-2 are up-regulated after stroke by a NO donor and subsequently promote neuroblast cell migration towards the ischemic area (Ucuzian et al. 2010). The role of angiopoietins on brain damage and repair is difficult to define because their functions vary according to the relative levels of Ang-1, Ang-2 and VEGF under specific physiologic and pathologic conditions.

8.3.5 Tumor Necrosis Factor-α

TNF-α is an inflammatory cytokine secreted primarily by activated macrophages during inflammation and the immune response, but it has also been associated with endothelial cells, fibroblasts, myocytes and the neuronal tissue. It has many diverse functions, such as stimulation of GM-CSF and IL-1. It has been suggested to act *in vivo* on endothelial cells directly, by inducing cell differentiation, or indirectly, by stimulating other cells to produce angiogenic factors (Ucuzian et al. 2010).

8.3.6 Matrix Metalloproteinases

MMPs are a family of zinc- and calcium-dependent proteolytic enzymes responsible for the turnover and degradation of the extracellular matrix and vascular basal membranes. MMP activity leads to BBB breakdown and vasogenic edema after stroke, but they also have beneficial effects on stroke recovery by promoting vascular remodeling, cellular migration and sprout formation (Yong 2005, Adhibatla and Hatcher 2008, Dong et al. 2009, Ucuzian et al. 2010). MMPs contain several distinct domains responsible for secretion, latency, catalysis and substrate recognition. All MMPs share one pre-domain and one catalytic domain. The N-terminal pre-domain has a conserved cysteine which chelates the catalytic Zn^{2+} and keeps the pro-MMP in an inactive, latent state. Disruption of this interaction leads to proteolytic cleavage of the pro-MMP and production of the mature active enzyme.

Regulation of MMP expression is a complex process involving interactions among several cell types. In humans, these zinc-containing endopeptidases may be either secreted or membrane bound (van Hinsberg and Koolwijk 2008). Most MMPs are secreted into the extracellular space as inactive zymogens (pro-MMPs) and may be activated by membrane bound MMPs and other proteases, such as plasmin, uPA and tPA. Nitration or oxidation of the inhibitory cysteine can also unmask the catalytic site, thereby activating MMPs in the absence of pro-domain cleavage. The activities of MMPs are further regulated by interaction with TIMPs.

MMP expression is generally very low to undetectable in the adult brain, but many MMPs are up-regulated in response to injury. Neurons, astrocytes, oligodendrocytes, microglia and endothelial cells have all been shown to express MMPs after injury, but the brain regions and cellular sources of expression differ according to the specific MMP and the type of injury (Adibhatla et al. 2008). Stroke is associated with a biphasic disruption of the BBB, leading to vasogenic edema and hemorrhage. Experimental studies have indicated that BBB breakdown and hemorrhage result from expression and activation of MMPs. However, cell-specific expression of MMPs in the neurovascular unit results in interactions that determine the final outcome. Different MMPs are expressed in different conditions and at different times following stroke, with MMP-2 being amongst the first activated, followed by MMP-9 in later stages of inflammation and repair.

Recent research indicates that bone-marrow derived cells are the major source of MMP-9 in cerebral ischemia (Wang et al. 2009). Remarkably, MMP-9 promotes NPC migration towards the lesioned brain area after photo-thrombotic ischemia in a transgenic mice model. On the other hand, clinical studies conducted in patients with various types of stroke have revealed that the MMP-2 gene is associated with development of lacunar strokes. However, it rises in parallel with MMP-9 in a rat model of transient cerebral ischemia (Adhibatla and Hatcher 2008, Dong et al. 2009, Ucuzian et al. 2010).

MMP-9 is known to promote BBB injury, formation of vasogenic edema, increased infarct size and hemorrhagic transformation during the acute phase after cerebral ischemia in animal models and humans. But MMP-9 may have a different role during the delayed phase after stroke. MMP-9 is up-regulated in the peri-infarct cortex at seven–14 days after stroke and is co-localized with markers of neurovascular remodeling, mediating brain regeneration during the later repair phase. Thus, treatment with TIMPs 14 days after stroke suppresses neurovascular remodeling, increases ischemic brain injury and impairs functional recovery (Zhao et al. 2006, Dong et al. 2009).

MMP-9 also mediates proliferation and migration of NPCs from the SVZ toward the damaged tissue during the two week recovery period after transient focal cerebral ischemia in mice, whereas MMP inhibitors suppress neurogenic migration from the SVZ into the striatum (Lee et al. 2006, Zhao et al. 2006, Dong et al. 2009). These data suggest that MMP-9 is involved in endogenous mechanisms of neurovascular remodeling and neuronal repair in the peri-infarct cortex after injury.

8.3.7 Platelet-Derived Growth Factor-β

PDGF-β and its receptor PDGFR-β were initially identified as neuroprotective agents in brain ischemia (Iihara et al. 1997, Brea et al. 2009). However, they play an essential role in recruiting pericytes to brain capillaries during embryonic development. In addition, PDGF-β expression is increased in neurons and macrophages after brain ischemia in both rats and humans; the mRNA is detected around cystic infarction sites weeks after stroke, associated with the formation of new capillaries (Krupinski et al. 1997). Although the precise mechanism for this effect has not been elucidated, PDGF-β binding to PDGFR-β may promote the interaction between endothelial cells and pericytes, allowing vascular remodeling after brain ischemia.

8.3.8 Transforming Growth Factor-β

The TGF-β superfamily includes many structurally similar growth factors, among them the TGF-β isotypes 1 to 3. They are secreted in an inactive latent form and are activated after proteolytic cleavage of the latency-associated peptide domain (Ucuzian et al. 2010). TGF-β controls cellular proliferation and differentiation of a variety of cellular types and particularly stimulates proliferation of endothelial cells. The role of endogenous TGF-β in the brain after ischemia has not been clearly elucidated; however, when locally administered it induces angiogenesis and exerts neuroprotective effects, reducing the infarct volume when injected before but not after ischemia (Prehn et al. 1994). TGF-β is not expressed in the brain of adult non-ischemic rats; but is increased after ischemia in activated astrocytes and microglia, as well as in micro-vessels (Wiessner et al. 1993, Yamashita et al. 1999, Brea et al. 2009). TGF-β has been also detected in vessels growing into the infarct area. In the human brain, augmented TGF-β has been observed in the ischemic penumbra (Krupinski et al. 1996). TGF-β promotes extracellular matrix deposition and integrin receptor up-regulation *in vivo* and participates in endothelial cell-pericyte interactions and vessel stabilization (Ucuzian et al. 2010).

8.3.9 Stromal-Derived Factor-1α

Complex bi-directional signaling occurs between NSPCs and the vasculature under non-pathological and stroke conditions. In the adult brain, NSPCs reside in specialized micro-vascular niche environments that are important for supporting many aspects of neural stem cell function (Goldberg and Hirschi 2009, Madri 2009, Xiong et al. 2010, Cunningham et al. 2012). Within the adult SVZ, primitive stem cell astrocytes (type B cells), transit amplifying cells (type C) and neuroblasts (type A cells), display unique anatomical associations with endothelial cells. Type B astrocyte stem cells maintain contact with SVZ endothelial cells through endfoot-like processes. Transition of astrocyte stem cells to proliferative transit amplifying cells is thought to involve activation of fusin signaling in type B astrocytes by SDF-1 released from endothelial cells (Kokovay et al. 2010).

NSPCs are also found in close association with the vasculature during migration into ischemic brain regions following stroke (Ohab et al. 2006). NSPC re-routing toward areas of ischemic injury is dependent on SDF-1—fusin signaling (Ohab et al. 2006, Robin et al. 2006, Thored et al. 2006). SDF-1 induces migration of adult NSPCs mediated by MMPs (Barkho et al. 2008). Components of the wingless-related integration site, β-catenin pathway are also expressed within the adult SVZ and subgranular zone and are up-regulated in the SVZ following stroke (Morris et al. 2007, Cunningham et al. 2012).

8.3.10 Nitric Oxide (NO)

Activation of endothelial NOS increases release of NO, which has a crucial role in the regulation of vascular tone and in arteriogenesis. NO increases arterial density and diameter, which produces a better functional outcome after stroke (Prior et al. 2003, Cui et al. 2009a). NO also promotes vascular remodeling and angiogenesis after ischemia, increasing endothelial cell proliferation and migration, as well as smooth muscle and arterial-venous differentiation in several ischemic tissues including the brain

(Papapetropoulos et al. 1997, Murohara et al. 1998, Rudic et al. 1998, Amano et al. 2003, Cai et al. 2004, Gertz et al. 2006, Luque Contreras et al. 2006).

8.4 Pro-angiogenic Therapies for Neuro-protection and/or Neuro-recovery

Neuro-protection aimed at counteracting several pathophysiological mechanisms triggered by ischemia, which result in neuronal death and brain damage shortly after stroke, has not been successful in the clinical setting. This in spite of the strong support each of the potential neuro-protective procedures had from data obtained in preclinical studies. Neuro-protective treatments (usually mono-therapy) may not coincide with the narrow window of opportunity during which they are expected to target the harmful actions of excitotoxicity, free radicals, inflammation and pro-apoptotic molecules on neuronal structure and function. The challenge for single drugs to act against the multiple pathophysiological mechanisms of ischemic brain injury, the deficient access of drugs to brain regions where blood flow is absent or insufficient and the presence of concomitant pathologies (diabetes, hypertensive disease) have been considered important factors that account for the failure of neuro-protection. Thus, neurons are irreversibly damaged and the outcome does not improve even though neuro-protection has been attempted.

In addition, though recombinant tPA for the treatment of clinical ischemic stroke shows therapeutic effects, a low percentage of patients are candidates for this treatment because of its narrow therapeutic time window and related risks of hemorrhage. In the clinic, stroke patients are often not treated soon after the event due to delays in seeking medical care and diagnosis, as well as comorbidities that may interfere with treatment effects (Shaller and Graf 2004, Weintraub 2006, Shuaib et al. 2007, Endres et al. 2008, Herman and Chopp 2012).

In addition to the pathophysiological mechanisms of brain damage that occur shortly after stroke, ischemia may trigger long term cellular processes of neuro-repair by remodeling several components of the remaining neural tissue. Usually, this repair process is not sufficient per se to achieve complete recovery of neurological functions dependent on the integrity of the damaged brain structures, as shown in vulnerable hippocampal and cortical regions after global cerebral ischemia (Gonzalez-Burgos et al. 2007, Letechipia-Vallejo et al. 2007, Garcia-Chavez et al. 2008, Cervantes et al. 2008). However, even in the absence of an effective neuro-protective or neuro-repair treatment, many patients show some degree of neurological improvement. Under these conditions, increasing knowledge of endogenous recovery processes is being obtained. Some therapeutic strategies attempting to amplify these restorative mechanisms have been designed and experimentally assessed in order to support translational proposals for clinical trials (Zhang and Chopp 2009, Schabitz et al. 2010, Zorner and Schwab 2010, Dimyan and Cohen 2011, Savitz et al. 2011, Dirk and Chopp 2012).

Presently a substantial amount of experimental work has shifted from neuro-protection in the acute period after stroke to treatments promoting neurological recovery administered during the stroke recovery phase, within an extended, though still not well defined, opportunity window. In this period, remodeling of brain tissue mainly located in the ischemic penumbra could be influenced by pharmacological or cellular therapies, even if initiated days or weeks after stroke. These neurological recovery experimental treatments could potentially benefit most stroke patients (Zhang et al. 2002a, Carmichael 2006, Shen et al. 2007a, Kilic et al. 2008, Bacigaluppi et al. 2009, Murphy 2009, Zhang and Chopp 2009, Andres et al. 2011, Reitmeir et al. 2011, Tsai et al. 2011, Hermann and Chopp 2012).

However, a major challenge for experimental neuro-protection and neuro-repair in stroke is the existence of biphasic responses and dual effects of several molecular mediators involved in immediate ischemic brain damage and during the later recovery phase. Thus, more knowledge is needed about the magnitude, timing and dynamic interaction among the different cell types and molecular mediators in the ischemic brain. The possibility of collateral harmful effects of applying potential treatments outside the appropriate time window has been emphasized by the results of several experiments (NogoA deactivation, VEGF and metalloproteinase modulators) attempting to modulate molecular mediators involved in neuro-restorative processes (Zhang et al. 2000, Rosell and Lo 2008, Kilic et al. 2010, Arai et al. 2011). In fact, the detrimental role of chemical mediators involved in specific pathophysiological phenomena of brain damage during the immediate hours after stroke may change to an opposite beneficial role during later processes of neuro-recovery; thus, the effect of neuro-protective or neuro-restorative treatment has to change concomitantly. This is the case for MMP-9 overexpression which accelerates matrix degradation, disrupts the BBB and may lead to hemorrhagic phenomena in the acute phase after stroke in animal models or in human patients. Thus, immediate neuro-protective treatments have been experimentally assessed through administration of natural MMP inhibitors, MMP monoclonal antibodies and other drugs effective at inhibiting MMP activity: broad-spectrum MMP inhibitors such as batimastat, minocycline and doxycycline, AT1 receptor blockers such as olmesartan and trandolapril (before MCAO in rats), non-steroidal anti-inflammatory drugs, quercetin and melatonin. All of them protect the BBB, reduce the infarct size and decrease brain damage in models of cerebral ischemia (Ashai et al. 2000, Pfefferkorn and Rosenberg

2003, Machado et al. 2006, Burggraf et al. 2007, Hosomi et al. 2005, Cho et al. 2006, Tanaka et al. 2007, Wang et al. 2007, Nagel et al. 2008, Dong et al. 2009). In contrast, MMP activation should favor neuroblast migration, neurogenesis and neuro-recovery in the ischemic penumbra during later periods after stroke (Rosell and Lo 2008). Biphasic effects may also result from VEGF-directed therapies that under certain conditions can promote neuro-restoration either directly as a neuro-protective agent or indirectly by inducing angiogenesis; but also may lead to BBB leakage, brain edema and further brain damage (Schoch et al. 2002, Sun et al. 2003, Manoonkitiwongsa et al. 2006, Ding et al. 2010).

Neurons, including their dendritic cyto-architecture and axonal pathways, have been considered the most important cells in the brain; however, they are anatomically and functionally linked to a complex network of multiple cell types (astrocytes, oligodendrocytes, microglia, endothelial cells and perycites forming micro-vascular structures) immersed in a dynamic intercellular matrix. In this sense, a conceptual model of the "neurovascular unit" has emphasized the important interaction between neuronal and vascular components, their functional integration and cell-cell or cell-matrix signaling between these cells; all of which are importantly disrupted by ischemia, making neurovascular units important therapeutic targets for repair and regeneration of the brain after stroke (Abbott et al. 2006, Font et al. 2010).

Experimental studies of angiogenesis have shown that new vessel formation is an important process for remodeling the ischemic brain after stroke. Thus, angiogenesis importantly contributes to stroke recovery since micro-vascular sprouting and neurogenesis reciprocally promote each other by strong, mainly synergistic, interactions between endothelial cells, neuroblasts and the remaining viable neurons. This occurs at least during the first month after stroke, when brain capillary sprouting and activation of glial cells may create an appropriate environment for neuronal growth and plasticity which may underlie ensuing successful neurological functional recovery. Thus, among the interrelated endogenous restorative responses including neurogenesis and neuronal plasticity, angiogenesis can be targeted by pharmacological or cell-based treatments to stimulate neurological recovery after stroke. In this sense, data concerning the cellular and molecular mechanisms of angiogenesis and vessel network remodeling after stroke and their relationship with neurogenesis and plasticity have allowed several conceptual and experimental approaches for pro-angiogenic treatments. These therapies may promote structural and functional recovery after stroke by modulating endogenous mediators involved in angiogenesis, neurogenesis and neuroplasticity (Chen et al. 2005a, Carmichael 2006, Wang et al. 2006, Murphy and Corbett 2009, Navaratna et al. 2009, Hermann and Zechariah 2009, Zhang and Chopp 2009, Liman and Edres 2012).

Pharmacological interventions have been evaluated in experimental models of stroke to promote angiogenesis, neurogenesis and plasticity for neuro-recovery, including treatment with drugs acting through modulation of molecules participating in angiogenesis or neurogenesis. This is the case for deferoxamine, which promotes recovery by activating HIF-1, which, in turn, increases VEGF expression and angiogenesis (Freret et al. 2006, Shi 2009). Direct administration of VEGF or its increase by means of erythropoietin or atorvastatin promotes neuro-restoration (Sun et al. 2003, Chen et al. 2005b, Ding et al. 2010). Simvastatin alone or simvastatin plus bone marrow stromal cells (BMSCs), lovastatin or rosuvastatin promote and modulate angiogenesis by increasing SDF-1 and fusin in BMSCs and their migration into the ischemic brain (Werner et al. 2002, Chen et al. 2003, Elkind et al. 2008, Cui et al. 2009b, Elewa et al. 2010). Direct administration of granulocyte colony-stimulating factor (G-CSF) increases angiogenesis after stoke (Shyu et al. 2004, Lee et al. 2005, Yanquing et al. 2006, Sehara et al. 2007, Strecker et al. 2010). Activated protein C (APC) reduces ischemic brain injury through several mechanisms: increasing the number of replicating epithelial cells, promoting new vessel formation and better blood flow, as well as preventing BBB breakdown and endothelial barrier leakage. APC also significantly increases post-ischemic NPC proliferation in the SVZ and migration of neuroblasts from this structure to the ischemic penumbra (Shibata et al. 2001, Cheng et al. 2003, Zlokovic et al. 2005, Cheng et al. 2006, Thiyagarajan et al. 2008). Drugs that inhibit the type-5 phosphodiesterase (reducing breakdown of cyclic guanosine monophosphate), such as sildenafil or tadalafil, also increase angiogenesis in the peri-infarct zone after a stroke (Zhang et al. 2002a, Zhang et al. 2005b, 2006, Li et al. 2007, Ding et al. 2008b, Silver 2009, Ding et al. 2011).

Experiments have assessed therapies oriented at increasing the mobilization and recruitment of bone marrow derived cells into the remodeling ischemic penumbra and their effects on angiogenesis and neuro-recovery after stroke. G-CSF, angiotensin-converting enzyme inhibitors and angiotensin II blockers may increase angiogenesis and functional recovery through these mechanisms (Shen et al. 2007b, Antonio et al. 2010, Zacharek et al. 2010, Savitz et al. 2011, Onteniente and Polentes 2011, Liman and Edres 2012). Direct administration or implantation of progenitor cells (Chen et al. 2001, Shyu et al. 2006, Ukai et al. 2007, Fan et al. 2010, Shen et al. 2010, Thwaites et al. 2012) also favors angiogenesis in the ischemic penumbra and promotes functional recovery.

Results from experimental models of focal cerebral ischemia in laboratory animals, mainly rodents, show that therapeutic interventions may induce synergic responses in neurons, micro-vascular network cells and glia. This may restore endogenous environmental conditions that promote neurological recovery after stroke, even in old animals or animals with vascular risk factors such as diabetes, hypertension or metabolic alterations. Substantial advances have been made in the field of neo-vascularization and neuro-recovery after ischemia. However, new vessel formation after stroke depends on a complex interaction of many molecular and cellular factors; thus, important aspects of these processes remain unknown (Darsalia et al. 2005, Marcus et al. 2005, Zhang et al. 2005a, Shen et al. 2007b, Chang et al. 2009, Gao et al. 2009, Hennigan et al. 2009, Gilliani et al. 2010, Hwang et al. 2010, Chen et al. 2011, Hermann and Chopp 2012, Liman and Endres 2012).

8.5 Conclusions

Substantial evidence shows that the cellular and molecular mechanisms involved in remodeling of vascular networks after stroke, including collateral vessel arrangements and angiogenesis, are important for the functional recovery of the neural tissue mainly located in the ischemic penumbra. However, experimental findings obtained from animal models of focal cerebral ischemia also show the complexity of cell-cell and cell-extracellular matrix interactions that achieve remodeling of vascular networks and the relationship between the cells that integrate functional neurovascular units.

Nevertheless, knowledge of the cellular and molecular phenomena taking place during angiogenesis has supported proposals for restorative therapies (instead of neuro-protective procedures against immediate patho-physiological mechanisms of ischemic brain damage). These treatments include pharmacological modulation of the molecular mediators of angiogenesis and/or bone marrow precursor cell activation or administration. Most of these therapeutic proposals and their effectiveness in promoting neuro-recovery have been assessed in experimental models of stroke, even if treatments are initiated days or weeks after stroke.

Conceptual and operational aspects of effective therapies for neuro-recovery after stroke must be properly translated, including their timing, from experimental findings to the clinic. They should act upon the spared biological tissue potentially underlying functional improvements, which seems to be preserved in many, even elderly, stroke patients. Otherwise, promising approaches for neuro-recovery through pharmacological or cell therapies may fail in clinical trials. The latter must be designed taking into account the presence and possible adverse effect of genetic characteristics, age and co-morbidities (diabetes, hypertensive disease), as well as the impact of infarct size, localization and acute thrombolytic treatment on the final results of the angiogenic-mediated neuro-recovery treatment. These strategies could allow the design of effective pharmacological or cell-based restorative therapies for stroke patients.

References

Abbott, N.J., L. Ronnback and E. Hansson. 2006. Astrocyte-endothelial interactions at the blood-brain barrier. Nat. Rev. Neurosci. 7: 41–53.

Acker, T. and H. Acker. 2004. Cellular oxygen sensing need in the CNS function: Physiological and pathological implications. J. Exp. Biol. 207: 3171–3188.

Adams, R.A., C. Schachtrup, D. Davalos, I. Tsigelny and K. Akassoglou. 2007b. Fibrinogen signal transduction as a mediator and therapeutic target in inflammation: Lessons from Multiple Sclerosis. Curr. Med. Chem. 14: 2925–2936.

Adams, R.A., J. Bauer, M.J. Flick, S.L. Sikorski, T. Nuriel, H. Lassmann, J.L. Degen and K. Akassoglou. 2007a. The fibrin-derived gamma377–395 peptide inhibits microglia activation and suppresses relapsing paralysis in central nervous system autoimmune disease. J. Exp. Med. 204: 571–582.

Adams, R.A., M. Passino, B.D. Sachs, T. Nuriel and K. Akassoglou. 2004. Fibrin mechanisms and functions in nervous system pathology. Mol. Interv. 4: 163–176.

Adams, V., K. Lenk, A. Linke, D. Lenz, S. Erbs, M. Sandri, A. Tarnok, S. Gielen, G. Schuler and R. Hambrecht. 2004. Increase of circulating endothelial progenitor cells in patients with coronary artery disease after exercise-induced ischemia. Arterioscler. Thromb. Vasc. Biol. 24: 684–690.

Adibhatla, R.M. and J.F. Hatcher. 2008. Tissue plasminogen activator (tPA) and matrix metalloproteinases in the pathogenesis of stroke: Therapeutic strategies. CNS Neurol. Disord. Drug Targets 7: 243–253.

Al Qteishat, A., J.J. Gaffney, J. Krupinski and M. Slevin. 2006. Hyaluronan expression following middle cerebral artery occlusion in the rat. Neuroreport 17: 1111–1114.

Allaman, I., M. Belanger and P.J. Magistretti. 2011. Astrocyte-neuron metabolic relationships: for better and for worse. Trends Neurosci. 34: 76–87.

Amano, K., H. Matsubara, O. Iba, M. Okigaki, S. Fujiyama, T. Imada, H. Kojima, Y. Nozawa, S. Kawashima, M. Yokoyama and T. Iwasaka. 2003. Enhancement of ischemia-induced angiogenesis by eNOS overexpression. Hypertension 41: 156–162.

Andres, R.H., N. Horie, W. Slikker, H. Keren-Gill, K. Zhan, G. Sun, N.C. Manley, M.P. Pereira, L.A. Sheikh, E.L. McMillan, B.T. Schaar, C.N. Svendsen, T.M. Bliss and G.K. Steinberg. 2011. Human neural stem cells enhance structural plasticity and axonal transport in the ischaemic brain. Brain 134: 1777–1789.

Antonio, N., R. Fernandes, N. Rodriguez-Losada, M.F. Jimenez-Navarro, A. Paiva, G. de Teresa Galvan, L. Goncalves, C.F. Ribeiro and L.A. Providencia. 2010. Stimulation of endothelial progenitor cells: a new putative effect of several cardiovascular drugs. Eur. J. Clin. Pharmacol. 66: 219–230.

Arai, J., G. Jin, D. Navaratna and E.H. Lo. 2009. Brain angiogenesis in developmental and pathological processes: neurovascular injury and angiogenic recovery after stroke. FEBS J. 276: 4644–4652.

Arai, K., J. Lok, S. Guo, K. Hayakawa, C. Xing and E.C. Lo. 2011. Cellular mechanisms of neurovascular damage and repair after stroke. J. Child Neurol. 26: 1193–1198.

Ardelt, A.A., R.S. Carpenter, M.R. Lobo, H. Zeng, R.B. Solanki, A. Zhang, P. Kulesza and M.M. Pike. 2012. Estradiol modulates post-ischemic cerebral vascular remodeling and improves long-term functional outcome in a rat model of stroke. Brain Res. 1461: 76–86.

Armulik, A., A. Abramsson and C. Betsholtz. 2005. Endothelial/Pericyte Interactions. Circ. Res. 97: 512–523.

Arras, M., W.D. Ito, D. Scholz, B. Winkler, J. Schaper and W. Schaper. 1998. Monocyte activation in angiogenesis and collateral growth in the rabbit hind limb. J. Clin. Invest. 101: 40–50.

Asahara, T., H. Masuda, T. Takahashi, C. Kalk, C. Pastore, M. Silver, M. Kearne, M. Magner and J.M. Isner. 1999a. Bone marrow origin of endothelial progenitor cells responsible for postnatal vasculogenesis in physiological and pathological neovascularization. Circ. Res. 85: 221–228.

Asahara, T., T. Murohara, A. Sullivan, M. Silver, R. van der Zee, T. Li, B. Witzenbichler, G. Schatteman and J.M. Isner. 1997. Isolation of putative progenitor endothelial cells for angiogenesis. Science 275: 964–967.

Asahara, T., T. Takahashi, H. Masuda, C. Kalka, D. Chen, H. Iwaguro, Y. Ina, M. Silver and J.M. Isner. 1999b. VEGF contributes to postnatal neovascularization by mobilizing bone marrow-derived endothelial progenitor cells. EMBO J. 18: 3964–3972.

Asahi, M., K. Asahi, J.C. Jung, G.J. del Zoppo, M.E. Fini and E.H. Lo. 2000. Role for matrix metalloproteinase 9 after focal cerebral ischemia: effects of gene knockout and enzyme inhibition with BB-94. J. Cereb. Blood Flow Metab. 20: 1681–1689.

Bacigaluppi, M., S. Pluchino, L. Peruzzotti-Jametti, E. Kilic, Ü. Kilic, G. Salani, E. Brambilla, M.J. West, G. Comi, G. Martino and D.M. Hermann. 2009. Delayed post-ischaemic neuroprotection following systemic neural stem cell transplantation involves multiple mechanisms. Brain 132: 2239–2251.

Baeten, A.M. and K. Akassoglou. 2011. Extracellular matrix receptors and matrix receptors in blood barrier formation and stroke. Dev. Neurobiol. 71: 1018–1039.

Banerjee, S. and M.A. Bhat. 2007. Neuron-glial interactions in blood-brain barrier formation. Annu. Rev. Neurosci. 30: 235–258.

Baranova, O., L.F. Miranda, P. Pichiule, I. Dragatsis, R.S. Johnson and J.C. Chavez. 2007. Neuron-specific inactivation of the hypoxia inducible factor 1alpha increases brain injury in a mouse model of transient focal cerebral ischemia. J. Neurosci. 27: 6320–6332.

Barkho, B.Z., A.E. Munoz, X. Li, L. Li, L.A. Cunningham and X. Zhao. 2008. Endogenous matrix metalloproteinase (MMP)-3 and MMP-9 promote the differentiation and migration of adult neural progenitor cells in response to chemokines. Stem Cells 26: 3139–3149.

Barone, F.C. and G.Z. Feurestein. 1999. Inflammatory mediators and stroke: new opportunities for novel therapeutics. J. Cereb. Blood Flow Metab.19: 819–834.

Barrera-Ocampo, A.A., A.E. Cespedes-Rubio and G.P. Cardona-Gomez. 2008. Mecanismo potencial de neuroprotección y plasticidad sináptica inducidas por el estradiol a través de PI3K/GSK3beta en la isquemia cerebral. Rev. Neurol. 46: 32–39.

Baumann, E., E. Preston, J. Slinn and D. Stanimirovic. 2009. Post-ischemic hypothermia attenuates loss of the vascular basement membrane proteins, agrin and SPARC, and the blood-brain barrier disruption after global cerebral ischemia. Brain Res. 1269: 185–197.

Bayless, K.J. and G.E. Davis. 2002. The Cdc42 and Rac1 GTPases are required for capillary lumen formation in three-dimensional extracellular matrices. J. Cell. Sci. 115: 1123–1136.

Beck, H. and K.H. Plate. 2009. Angiogenesis after cerebral ischemia. Acta Neuropathol. 117: 481–496.

Bessis, A., C. Béchade, D. Bernard and A. Roumier. 2007. Microglial control of neuronal death and synaptic properties. Glia 55: 233–238.

Bondke, A., P. Hillmeister, A. Duelsner, C. Gruber, I.R. Buschmann and K. Lehmann. 2008. Early-phase cerebral arteriogenesis: evidence of endothelial activation, vascular proliferation and local inflammation. FASEB J. 22: 1147.13.

Borlongan, C.V., J.G. Lind, O. Dillon-Carter, G. Yu, M. Hadman, C. Cheng, J. Carroll and D.C. Hess. 2004. Bone marrow grafts restore cerebral blood flow and blood brain barrier in stroke rats. Brain Res. 1010: 108–116.

Bosomtwi, A., Q. Jiang, G.L. Ding, L. Zhang, Z.G. Zhang, M. Lu, J.R. Ewing and M. Chopp. 2008. Quantitative evaluation of microvascular density after stroke in rats using MRI. J. Cereb. Blood Flow Metab. 28: 1978–1987.

Brea, D.,T. Sobrino, P. Ramos-Cabrer and J. Castillo. 2009. Reorganización de la vascularización cerebral tras la isquemia. Rev. Neurol. 49: 645–654.

Brown, C.E., P. Li, D. Jamie, J.D. Boyd, K.R. Delaney and T.H. Murphy. 2007. Extensive turnover of dendritic spines and vascular remodeling in cortical tissues recovering from stroke. J. Neurosci. 27: 4101–4109.

Brown, R.C. and T.P. Davis. 2002. Calcium modulation of adherens and tight junction function: a potential mechanism for blood-brain barrier disruption after stroke. Stroke 33: 1706–1711.

Burggraf, D., A. Trinkl, M. Dichgans and G.F. Hamann. 2007. Doxycycline inhibits MMPs via modulation of plasminogen activators in focal cerebral ischemia. Neurobiol. Dis. 25: 506–513.

Busch, H.J., I. Buschmann, E. Schneeloch, C. Bode, G. Mies and K.A. Hossmann. 2006. Therapeutically induced arteriogenesis in the brain.A new approach for the prevention of cerebral ischemia with vascular stenosis. Der. Nervenarzt. 77: 215–220.

Busch, H.J., I.R. Buschmann, G. Mies, C. Bode and K.A. Hossmann. 2003. Arteriogenesis in hypoperfused rat brain. J. Cereb. Blood Flow Metab. 23: 621–628.

Buschmann, I., A. Pries, B. Styp-Rekowska, P. Hillmeister, L. Loufrani, D. Henrion, Y. Shi, A. Duelsner, I. Hoefer, N. Gatzke, H. Wang, K. Lehmann, L. Ulm, Z. Ritter, P. Hauff, R. Hlushchuk, V. Djonov, T. vanVeen and F. le Noble. 2010. Pulsatile shear and Gja5 modulate arterial identity and remodeling events during flow-driven arteriogenesis. Development 137: 2187–2196.

Buschmann, I.R., H.J. Busch, G. Mies and K.A. Hossmann. 2003. Therapeutic induction of arteriogenesis in hypoperfused rat brain via granulocyte-macrophage colony-stimulating factor. Circulation 108: 610–615.

Cai, W., R. Guzman, A.R. Hsu, H. Wang, K. Chen, G. Sun, A. Gera, R. Choi, T. Bliss, L. He, Z.B. Li, A.L. Maag, N. Hori, H. Zhao, M. Moseley, G.K. Steinberg and X. Chen. 2009. Positron emission tomography imaging of poststroke angiogenesis. Stroke 40: 270–277.

Cai, W., R. Vosschulte, A. Afsah-Hedjri, S. Koltai, E. Kocsis, D. Scholz, S. Kostin, W. Schaper and J. Schaper. 2000. Altered balance between extracellular proteolysis and antiproteolysis is associated with adaptive coronary arteriogenesis. J. Mol. Cell. Cardiol. 32: 997–1011.

Cai, W.J., E. Kocsis, X. Luo, W. Schaper and J. Schaper. 2004. Expression of endothelial nitric oxide synthase in the vascular wall during arteriogenesis. Mol. Cell. Biochem. 264: 193–200.

Cai, W.J., E. Kocsis, X.Wu, M. Rodriguez, X. Luo, W. Schaper and J. Schaper. 2004. Remodeling of the vascular tunica media is essential for development of collateral vessels in the canine heart. Mol. Cell. Biochem. 264: 201–210.

Cai, W.J., S. Koltai, E. Kocsis, D. Scholz, S. Kostin, X. Luo, W. Schaper and J. Schaper. 2003. Remodeling of the adventitia during coronary arteriogenesis. Am. J. Physiol. Heart Circ. Physiol. 284: H31–H40.

Candelario-Jalil, E., Y. Yang and G.A. Rosenberg. 2009. Diverse roles of matrix metalloproteinases and tissue inhibitors of metalloproteinases in neuroinflammation and cerebral ischemia. Neuroscience 158: 983–994.

Cao, L., X. Jiao, D.S. Zuzga, Y. Liu, D.M Fong, D. Young and M.J. During. 2004. VEGF links hippocampal activity with neurogenesis, learning and memory. Nat. Genet. 36: 827–835.

Carmeliet, P. 2000. Mechanisms of angiogenesis and arteriogenesis. Nat. Med. 6: 389–395.

Carmeliet, P. 2005. Angiogenesis in life, disease and medicine. Nature 438: 932–936.

Carmichael, S.T. 2006. Cellular and molecular mechanisms of neural repair after stroke: making waves. Ann. Neurol. 59: 735–742.

Carvey, P.M., B. Hendey and A.J. Monahan. 2009. The blood-brain barrier in neurodegenerative disease: a rhetorical perspective. J. Neurochem. 111: 291–314.

Cervantes, M., G. Morali and G. Letechipia-Vallejo. 2008. Melatonin and ischemia-reperfusion injury of the brain. J. Pineal Res. 45: 1–7.

Chang, E.H., A. Rigotti and P.T. Huerta. 2009. Age-related influence of the HDL receptor SR-BI on synaptic plasticity and cognition. Neurobiol. 30: 407–419.

Chavez, J.C. and J.C. LaManna. 2002. Activation of hypoxia-inducible factor-1 in the rat cerebral cortex after transient global ischemia: potential role of insulin-like growth factor-1. J. Neurosci. 22: 8922–8931.

Chavez, J.C., F. Agani, P. Pichiule and J.C. LaManna. 2000. Expression of hypoxia-inducible factor-1alpha in the brain of rats during chronic hypoxia. J. Appl. Physiol. 89: 1937–1942.

Chen, J., A. Zacharek, C. Zhang, H. Jiang, Y. Li, C. Roberts, M. Lu, A. Kapke and M. Chopp. 2005a. Endothelial nitric oxide synthase regulates brain-derived neurotrophic factor expression and neurogenesis after stroke in mice. J. Neurosci. 25: 2366–2375.

Chen, J., P.R. Sanberg, Y. Li, L.Wang, M. Lu, A.E. Willing, J. Sanchez-Ramos and M. Chopp. 2001. Intravenous administration of human umbilical cord blood reduces behavioral deficits after stroke in rats. Stroke 32: 2682–2688.

Chen, J., X.Ye, T. Yan, C. Zhang, X.P. Yang, X. Cui, Y. Cui, A. Zacharek, C. Roberts, X. Liu, X. Dai, M. Lu and M. Chopp. 2011. Adverse effects of bone marrow stromal cell treatment of stroke in diabetic rats. Stroke 42: 3551–3558.

Chen, J., Z.G. Zhang, Y. Li, Y. Wang, L. Wang, H. Jiang, C. Zhang, M. Lu, M. Katakowski, C.S. Feldkamp and M. Chopp. 2003. Statins induce angiogenesis, neurogenesis, and synaptogenesis after stroke. Ann. Neurol. 53: 743–751.

Chen, J., C. Zhang, H. Jiang, Y. Li, L. Zhang, A. Robin, M. Katakowski, M. Lu and M. Chopp. 2005b. Atorvastatin induction of VEGF and BDNF promotes brain plasticity after stroke in mice. J. Cereb. Blood Flow Metab. 25: 281–290.

Cheng, T., A.L. Petraglia, Z. Li, M. Thiyagarajan, Z. Zhong, Z. Wu, D. Liu, S.B. Maggirwar, R. Deane,J.A. Fernandez, B. Larue, J.H. Griffin, M. Chopp and B.V. Zlokovic. 2006. Activated protein C inhibits tissue plasminogen activator-induced brain hemorrhage. Nat. Med. 12: 1278–1285.

Cheng, T., D. Liu, J.H. Griffin, J.A. Fernandez, F. Castellino, E.D. Rosen, K. Fukudome and B.V. Zlokovic. 2003. Activated protein C blocks p53-mediated apoptosis in ischemic human brain endothelium and is neuroprotective. Nat. Med. 9: 338–342.

Cho, J.Y., I.S. Kim, Y.H. Jang, A.R. Kim and S. R. Lee. 2006. Protective effect of quercetin, a natural flavonoid against neuronal damage after transient global cerebral ischemia. Neurosci.Lett. 404: 330–335.

Chopp, M., Y. Li and J. Zhang. 2008. Plasticity and remodeling of brain. J. Neurol. Sci. 265: 97–101.

Chopp, M., Z.G. Zhang and Q. Jiang. 2007. Neurogenesis, angiogenesis, and MRI indices of functional recovery from stroke. Stroke 38: 827–831.

Christoforidis, G.A., Y. Mohammad, D. Kehagias, B. Avutu and A.P. Slivka. 2005. Angiographic assessment of pial collaterals as a prognostic indicator following intra-arterial thrombolysis for acute ischemic stroke. AJNR 26: 1789–1797.

Chung, A.S., J. Lee and N. Ferrara. 2010. Targeting the tumour vasculature: insights from physiological angiogenesis. Nat. Rev. Cancer 10: 505–514.

Correia, S.C. and P.I. Moreira. 2010. Hypoxia-inducible factor 1: a new hope to counteract neurodegeneration? J. Neurochem. 112: 1–12.

Cui, X., M. Chopp, A. Zacharek, C. Zhang, C. Roberts and J. Chem. 2009a. Role of endothelial nitric oxide synthetase in arteriogenesis after stroke in mice. Neuroscience 159: 744–750.

Cui, X., M. Chopp, A. Zacharek, X. Cui, M. Chopp, A. Zacharek, C. Roberts, M. Lu, S. Savant-Bhonsale and J. Chen. 2009b. Chemokine, vascular and therapeutic effects of combination Simvastatin and BMSC treatment of stroke. Neurobiol. Dis. 36: 35–41.

Cunningham, L.A., K. Candelario and L. Li. 2012. Roles for HIF-1α in neuronal stem function and regenerative response to stroke. Behav. Brain Res. 227: 410–417.

Daly, C., V. Wong, E. Burova, Y. Wei, S. Zabski, J. Griffiths, K.M. Lai, H.C. Lin, E. Ioffe, G.D. Yancopoulos and J.S. Rudge. 2004. Angiopoietin-1 modulates endothelial cell function and gene expression via the transcription factor FKHR (FOXO1). Genes Dev. 18: 1060–1071.

Darsalia, V., U. Heldmann, O. Lindvall and Z. Kokaia. 2005. Stroke-induced neurogenesis in aged brain. 36: 1790–1795.

Davis, G.E. and C.W. Camarillo. 1995. Regulation of endothelial cell morphogenesis by integrins, mechanical forces, and matrix guidance pathways. Exp. Cell. Res. 216: 113–123.

Davis, G.E. and C.W. Camarillo. 1996. An alpha 2 beta 1 integrin-dependent pinocytic mechanism involving intracellular vacuole formation and coalescence regulates capillary lumen and tube formation in three-dimensional collagen matrix. Exp. Cell. Res. 224: 39–51.

Davis, G.E. and K.J. Bayless. 2003. An integrin and Rho GTPase-dependent pinocytic vacuole mechanism controls capillary lumen formation in collagen and fibrin matrices. Microcirculation 10: 27–44.

Davis, G.E., W. Koh and A. Stratman. 2007. Mechanisms controlling human endothelial lumen formation and tube assembly in three-dimensional extracellular matrices. Birth Defects Res. 81: 270–285.

Deindl, E. 2007. Arteriogenesis: a focus on signal transduction cascades and transcription factors. Thromb. Haemost. 98: 940–943.

Del Zoppo, G.J. 2006. Stroke and neurovascular protection. New Eng. J. Med. 354: 553–555.

Del Zoppo, G.J. and T. Mabuchi. 2003. Cerebral microvessel responses to focal ischemia. J. Cereb. Blood Flow Metab. 23: 879–894.

Del Zoppo, G.J., R. Milner, T. Mabuchi, S. Hung, X. Wang and J.A. Koziol. 2006. Vascular matrix adhesion and the blood-brain barrier. Biochem. Soc. Trans. 34: 1261–1266.

Dijkhuizen, R.M., K. van der Marel, M. Willem, W.M. Otte, E.I. Hoff, J.P. van der Zijden, A. van der Toorn and M.P. van Meer. 2012. Functional MRI and diffusion tensor imaging of brain reorganization after experimental stroke. Transl. Stroke Res. 3: 36–43.

Dimyan, M.A. and L.G. Cohen. 2011. Neuroplasticity in the context of motor rehabilitation after stroke. Nat. Rev. Neurol. 7: 76–85.

Ding, G., Q. Jian, L. Li, L. Zhang, Z. Zhang, M. Lu, Q. Li, S. Gu, J.R. Ewing and M. Chopp. 2011. Longitudinal magnetic resonance imaging of sildenafil treatment of embolic stroke in aged rats. Stroke 42: 3537–3541.

Ding, G., Q. Jiang, L. Li, L. Zhang, Y. Wang, Z.G. Zhang, M. Lu, S. Panda, Q. Li, J.R. Ewing and M. Chopp. 2010. Cerebral tissue repair and atrophy after embolic stroke in rat: an MRI study of Erythropoietin therapy. J. Neurosci. Res. 88: 3206–3214.

Ding, G., Q. Jiang, L. Li, L. Zhang, Z.G. Zhang, K.A. Ledbetter, L. Gallopalli, S. Panda, Q. Li, J.R. Ewing and R. Chopp. 2008a. Angiogenesis detected after embolic stroke in rat brain using magnetic resonance T2*WI. Stroke 39: 1563–1568.

Ding, G., Q. Jiang, L. Li, L. Zhang, Z.G. Zhang, K.A. Ledbetter, S. Panda, S.P.N. Davarani, H. Athiraman, Q. Li, J.R. Ewing and M. Chopp. 2008b. Magnetic resonance imaging investigation of axonal remodeling and angiogenesis after embolic stroke in sildenafil-treated rats. J. Cereb. Blood Flow Metab. 28: 1440–1448.

Dirk, M.H. and M. Chopp. 2012. Promoting brain remodelling and plasticity for stroke recovery: therapeutic promise and potential pitfalls of clinical translation. Lancet Neurol. 11: 369–380.

Dirnagl, U. 2006. Bench to bedside: the quest for quality in experimental stroke research. J. Cereb. Blood Flow Metab. 26: 1465–1478.

Djonov, V.G., A.B. Galli and P.H. Burri. 2000. Intussusceptive arborization contributes to vascular tree formation in the chick chorio-allantoic membrane. Anat. Embryol. (Berl.) 202: 347–357.

Dong, X., Y.N. Song, W.G. Liu and X.L. Guo. 2009. MMP-9, a potential target for cerebral ischemic treatment. Curr. Neuropharmacol.7: 269–275.

Durukan, A., I. Marinkovic, D. Strbian, M. Pitkonen, E. Pedrono, L. Soinne, U. Abo-Ramadan and T. Tatlisumak. 2009. Post-ischemic blood-brain barrier leakage in rats: one-week follow-up by MRI. Brain Res. 1280: 158–165.

Duval, H., M. Harri, J. Li, N. Johnson, and C. Print. 2003. New insights into the function and regulation of endothelial cell apoptosis. Angiogenesis 6: 171–183.

Duval, H., N. Johnson N, J. Li, A. Evans, S. Chen, D. Licence, J. Skepper, D.S. Charnock-Jones, S. Smith and C. Print. 2007. Vascular development is disrupted by endothelial cell-specific expression of the anti-apoptotic protein Bcl-2. Angiogenesis 10: 55–68.

Egginton, S. and M. Gerritsen. 2003. Lumen formation: *in vivo* versus *in vitro* observations. Microcirculation 10: 45–61.

Ehrbar, M., V.G. Djonov, C. Schnell, S.A. Tschanz, G. Martiny-Baron, U. Schenk, J. Wood, P.H. Burri, J.A. Hubbell and A.H. Zisch. 2004. Cell-demanded liberation of VEGF121 from fibrin implants induces local and controlled blood vessel growth. Circ. Res. 94: 1124–1132.

Elewa, H.F., A.B. El-Remessy, P.R. Somanath and S.C. Fagan. 2010. Diverse effects of statins on angiogenesis: new therapeutic avenues. Pharmacotherapy 30: 169–176.

Elkind, M.S., R.L. Sacco, R.B. MacArthur, D.J. Fink, E. Peerschke, H. Andrews, G. Neils, J. Stillman, T. Corporan, D. Leifer and K. Cheung. 2008. The Neuroprotection with Statin Therapy for Acute Recovery Trial (NeuSTART): an adaptive design phase I dose-escalation study of high-dose lovastatin in acute ischemic stroke. Int. J. Stroke 3: 210–218.

Endres, M., B. Engelhardt, J. Koistinaho, O. Lindvall, S. Meairs, J.P. Mohr, A. Planas, N. Rothwell, M. Schwaninger, M.E. Schwab, D. Vivien, T. Wieloch and U. Dirnagl. 2008. Improving outcome after stroke: overcoming the translational roadblock. Cerebrovasc. Dis. 25: 268–278.

Eve, D.J., J. Musso III, D.H. Park, C. Oliveira, K. Pollock, A. Hope, M.O. Baradez, J.D. Sinden and P.R. Sanberg. 2009. Methodological study investigating long term laser Doppler measured cerebral blood flow changes in a permanently occluded rat stroke model. J. Neurosci. Methods 180: 52–56.

Fan, Y., F. Shen, T. Frenzel, W. Zhu, J. Ye, J. Liu, Y. Chen, H. Su, W.L. Young and G.Y. Yang. 2010. Endothelial progenitor cell transplantation improves longterm stroke outcome in mice. Ann. Neuro. 67: 488–497.

Farr, T.D. and S. Wegener. 2010. Use of magnetic resonance imaging to predict outcome after stroke: a review of experimental and clinical evidence. J. Cereb. Blood Flow Metab. 30: 703–717.

Fawcett, J.W. and R.A. Asher. 1999. The glial scar and central nervous system repair. Brain Res. Bull. 49: 377–391.

Ferrara, N., H.P. Gerber and J. Le Couter. 2003. The biology of VEGF and its receptors. Nat. Med. 9: 669–676.

Fischer, C., M. Schneider and P. Carmeliet. 2006. Principles and therapeutic implications of angiogenesis, vasculogenesis and arteriogenesis. Handb. Exp. Pharmacol. 176: 157–212.

Fisher, M. 2009. Pericyte signaling in the neurovascular unit. Stroke 40: S13–S15.

Font, M.A., A. Arborix and J. Krupinski. 2010. Angiogenesis, neurogenesis and neuroplasticity in ischemic stroke.Curr. Cardiol. Rev. 6: 238–244.

Francis, M.E., S. Uriel and E.M. Brey. 2008. Endothelial cell-matrix interactions in neovascularization. Tissue Eng. Part B Rev. 14: 19–32.

Freret, T., S. Valable, L. Chazalviel, R. Saulnier, E.T. Mackenzie, E. Petit, M. Bernaudin, M. Boulouard and P. Schumann-Bard. 2006. Delayed administration of deferoxamine reduces brain damage and promotes functional recovery after transient focal cerebral ischemia in the rat. Eur. J. Neurosci. 7: 1757–1765.

Fukuda, S., C.A. Fini, T. Mabuchi, J.A. Koziol, L.L. Eggleston Jr. and G.J. del Zoppo. 2004. Focal cerebral ischemia induces active proteases that degrade microvascular matrix. Stroke 35: 998–1004.

Gao, P., F. Shen, R.A. Gabriel, D. Law, E. Yang, G.Y. Yang, W.L. Young and H. Su. 2009. Attenuation of brain response to vascular endothelial growth factor-mediated angiogenesis and neurogenesis in aged mice. Stroke 40: 3596–3600.

Garcia-Chavez, D., I. Gonzalez-Burgos, G. Letechipia-Vallejo, E. Lopez-Loeza, G. Morali and M. Cervantes. 2008. Long-term evaluation of cytoarchitectonic characteristics of prefrontal cortex yramidal neurons, following global cerebral ischemia and neuroprotective melatonin treatment, in rats. Neurosci. Lett. 448: 148–152.

Gasche, Y., M. Fujimura, Y. Morita-Fujimura, J.C. Copin, M. Kawase, J. Massengale and P.H. Chan. 1999. Early appearance of activated matrix metalloproteinase-9 after focal cerebral ischemia in mice: A possible role in blood-brain barrier dysfunction. J. Cereb. Blood Flow Metab. 19: 1020–1028.

Gerhardt, H., C. Ruhrberg, A. Abramsson, H. Fujisawa, D. Shima and C. Betsholtz. 2004. Neuropilin-1 is required for endothelial tip cell guidance in the developing central nervous system. Dev. Dyn. 231: 503–509.

Gerhardt, H., M. Golding, M. Fruttiger, C. Ruhrberg, A. Lundkvist, A. Abramsson, M. Jeltsch, C. Mitchell, K. Alitalo, D. Shima and C. Betsholtz. 2003. VEGF guides angiogenic sprouting utilizing endothelial tip cell filopodia. J.Cell Biol. 161: 1163–1177.

Gertz, K., J. Priller, G. Kronenberg, K.B. Fink, B. Winter, H. Schrock, S. Ji, M. Milosevic, C. Harms, M. Bohm, U. Dirnagl, U. Laufs and M. Endres. 2006. Physical activity improves long-term stroke outcome via endothelial nitric oxide synthase-dependent augmentation of neovascularization and cerebral blood flow. Circ. Res. 99: 1132–1140.

Gerwins, P., E. Skoldenberg and L. Claesson-Welsh. 2000. Function of fibroblast growth factors and vascular endothelial growth factors and their receptors in angiogenesis. Crit. Rev. Oncol. Hematol. 34: 185–194.

Gillani, R.L., S.Y. Tsai and D.G. Wallace. 2010. Cognitive recovery in the aged rat after stroke and anti-Nogo-A immunotherapy. Behav. Brain Res. 208: 415–424.

Giulian, D. 1993. Reactive glia as rivals in regulating neuronal survival. Glia 7: 102–110.

Goldberg, J.S. and K.K. Hirschi. 2009. Diverse roles of the vasculature within the neural stem cell niche. Regen. Med. 4: 879–897.

Gonzalez-Burgos, I., G. Letechipia-Vallejo, E. Lopez Loeza, G. Morali and M. Cervantes. 2007. Long-term study of dendritic spines from hippocampal CA1 pyramidal cells, after neuroprotective melatonin treatment following global cerebral ischemia in rats. Neurosci. Lett. 423: 162–166.

Greenberg, D.A. 2007. Neurogenesis and stroke. CNS Neurol. Disord. Drug Targets 6: 321–325.

Greenberg, D.A. and K. Jin. 2005. From angiogenesis to neuropathology. Nature 438: 954–959.

Grundmann, S., J.J. Piek, G. Pasterkamp and I.E. Hoefer. 2007. Arteriogenesis: basic mechanisms and therapeutic stimulation. Eur. J. Clin. Invest. 37: 755–766.

Guldner, F.H. and J.R. Wolff. 1973. Seamless endothelia as indicators of capillaries developed from sprouts. Bibl.Anat. 12: 120–123.

Guo, S. and E.H. Lo. 2009. Dysfunctional cell-cell signaling in the neurovascular unit as a paradigm for central nervous system disease. Stroke 40: S4–S7.

Guo, S., W.J. Kim, J. Lok, S.R. Lee, E. Besancon, B.H. Luo, M.F. Stins, X. Wang, S. Dedhar and E.H. Lo. 2008. Neuroprotection via matrix-trophic coupling between cerebral endothelial cells and neurons. Proc. Natl. Acad. Sci. USA 105: 7582–7587.

Guo, X., L. Liu, M. Zhang, B. Angela and J. Zhang. 2009. Correlation of CD34+ cells with tissue angiogenesis after traumatic brain injury in a rat model. J. Neurotrauma 26: 1337–1344.

Hara, Y., I. Tooyama, O. Yasuhara, H. Akiyama, P.L. McGeer, J. Handa and H. Kimura. 1994. Acidic fibroblast growth factor-like immunoreactivity in rat brain following cerebral infarction. Brain Res. 664: 101–107.

Hayakawa, K., T. Nakano, K. Irie, S. Higuchi, M. Fujioka, K. Orito, K. Iwasaki, G. Jin, E.H. Lo, K. Mishima and M.M. Fujiwar. 2010. Inhibition of reactive astrocytes with fluorocitrate retards neurovascular remodeling and recovery after focal cerebral ischemia in mice. J. Cereb. Blood Flow Metab. 30: 871–882.

Hayashi, T., K. Abe, H. Suzuki and Y. Itoyama. 1997. Rapid induction of vascular endothelial growth factor gene expression after transient middle cerebral artery occlusion in rats. Stroke 28: 2039–2044.

Hayashi, T., K. Deguchi, S. Nagotani, H. Zhang, Y. Sehara, A. Tsuchiya and K. Abe. 2006. Cerebral ischemia and angiogenesis. Curr. Neurovasc. Res. 3: 119–129.

Hayashi, T., N. Noshita, T. Sugawara and P.H. Chan. 2003. Temporal profile of angiogenesis and expression of related genes in the brain after ischemia. J. Cereb. Blood Flow Metab. 23: 166–180.

Hayward, N.M., P. Yanev, A. Haapasalo, R. Miettinen, M. Hiltunen, O. Grohn and J. Jolkkonen. 2011. Chronic hyperperfusion and angiogenesis follow subacute hipoperfusion in the thalamus of rats with focal cerebral ischemia. J. Cereb. Blood Flow Metab. 31: 1119–1132.

Heeschen, C., A. Aicher, R. Lehmann, S. Fichtlscherer, M. Vasa, C. Urbich, C. Mildner-Rihm, H. Martin, A.M. Zeiher and S. Dimmeler. 2003. Erythropoietin is a potent physiologic stimulus for endothelial progenitor cell mobilization. Blood 102: 1340–1346.

Heil, M. and W. Schaper. 2004. Influence of mechanical, cellular, and molecular factors on collateral artery growth (arteriogenesis). Circ. Res. 95: 449–458.

Heissig, B., K. Hattori, S. Dia, M. Friedrich, B. Ferris, N.R. Hackett, R.G. Crystal, P. Besmer, D. Lyden, M.A. Moore, Z. Werb and S. Rafii. 2002. Recruitment of stem and progenitor

cells from the bone marrow niche requires MMP-9 mediated release of kit-ligand. Cell 109: 625–637.
Hellstrom, M., L.K. Phng, J.J. Hofmann, E. Wallgard, L. Coultas, P. Lindblom, J. Alva, A.K. Nilsson, L. Karlsson, N. Gaiano, K. Yoon, J. Rossant, M.L. Iruela-Arispe, M. Kalén, H. Gerhardt and C. Betsholtz. 2007. Dll4 signaling through Notch1regulates formation of tip cells during angiogenesis. Nature 445: 776–780.
Hennigan, A., C.K. Callaghan, J. Kealy, J. Rouine and A.M. Kelly. 2009. Deficits in LTP and recognition memory in the genetically hypertensive rat are associated with decreased expression of neurotrophic factors and their receptors in the dentate gyrus. Behav. Brain Res. 197: 371–377.
Heo, J.H., S.W. Han and S.K. Lee. 2005. Free radicals as triggers of brain edema formation after stroke. Free Rad. Biol. Med. 39: 51–70.
Hermann, D.M. and A. Zechariah. 2009. Implications of vascular endothelial growth factor for postischemic neurovascular remodeling. J. Cereb. Blood Flow Metab.29: 1620–1643.
Hermann, D.M. and M. Chopp. 2012. Promoting brain remodelling and plasticity for stroke recovery: therapeutic promise and potential pitfalls of clinical translation. Lancet Neurol. 11: 369–680.
Hess, D.C., W.D. Hill, A. Martin-Studdard, J. Carroll, J. Brailer and J. Carothers. 2002. Bone marrow as a source of endothelial cells and NeuN-expressing cells after stroke. Stroke 33: 1362–1368.
Hoenig, M.R., C. Bianchi and F.E. Sellke. 2008. Hypoxia inducible factor-1 alpha, endothelial progenitor cells, monocytes, cardiovascular risk, wound healing, cobalt and hydralazine: a unifying hypothesis. Curr. Drug Targets 9: 422–435.
Horner, P.J. and F.H. Gage. 2000. Regenerating the damaged central nervous system. Nature 407: 963–970.
Hosomi, N., A. Nishiyama, C.R. Ban, T. Naya, T. Takahashi, M. Kohno and J.A. Koziol. 2005. Angiotensin type 1 receptor blockage improves ischemic injury following transient focal cerebral ischemia. Neuroscience 134: 225–231.
Hung, Y.C., T.J. Chen, E.J. Lee, W.L. Chen, S.Y. Huang , W.T. Lee, M.Y. Lee, H.Y. Chen and T.S. Wu. 2008. Melatonin decreases matrix metalloproteinase-9 activation and expression and attenuates reperfusion-induced hemorrhage following transient focal cerebral ischemia in rats. J. Pineal Res. 45: 459–467.
Huttunen, H.J., J. Kuja-Panula and H. Rauvala. 2002. Receptor for advanced glycation end products (RAGE) signaling induces CREB-dependent chromogranin expression during neuronal differentiation. J. Biol. Chem. 41: 38635–38646.
Hwang, L.L., C.H. Wang, T.L. Li, S.D. Chang, L.C. Lin, C.P. Chen, C.T. Chen, K.C. Liang, I.K. Ho, W.S. Yang and L.C. Chiou. 2010. Sex differences in high-fat diet-induced obesity, metabolic alterations and learning, and synaptic plasticity deficits in mice. Obesity18: 463–469.
Hynes, R.O. 1992. Integrins: versatility, modulation, and signaling in cell adhesion. Cell 69: 11–25.
Iadecola, C. and M. Nedergaard. 2007. Glial regulation of the cerebral microvasculature. Nat. Neurosci. 10: 1369–1376.
Iihara, K., N. Hashimoto, T. Tsukahara, M. Sakata, H. Yanamoto and T. Taniguchi. 1997. Platelet-derived growth factor-BB, but not -AA, prevents delayed neuronal death after forebrain ischemia in rats. J. Cereb. Blood Flow Metab. 17: 1097–1106.
Ingber, D.E. 2002. Mechanical signaling and the cellular response to extracellular matrix in angiogenesis and cardiovascular physiology. Circ. Res. 91: 877–887.
Ingber, D.E. 2003a. Tensegrity I. Cell structure and hierarchical systems biology. J. Cell. Sci. 116: 1157–1173.
Ingber, D.E. 2003b. Tensegrity II. How structural networks influence cellular information processing networks. J. Cell. Sci. 116: 1397–1408.
Ingraham, C.A., G.C. Park, H.P. Makarenkova and K.L. Crossin. 2011. Matrix metalloproteinase (MMP)-9 induced by Wnt signaling increases the proliferation and migration of embryonic neural stem cells at low O_2 levels. J. Biol. Chem. 286: 17649–17657.

Iruela-Arispe, M.L. and G.E. Davis. 2009. Cellular and molecular mechanisms of vascular lumen formation. Dev. Cell 16: 222–231.

Israeli, D., D. Tanne, D. Daniels, D. Last, R. Shneor, D. Guez, E. Landau, Y. Roth, A. Ocherashvilli, M. Bakon, C. Hoffman, A. Weinberg, T. Volk and Y. Mardor. 2010. The application of MRI for depiction of subtle blood brain barrier disruption in stroke. Int. J. Biol. Sci. 7: 1–8.

Issa, R., A. AlQteishat, N. Mitsios, M. Saka, J. Krupinski, E. Tarkowski, J. Gaffney, M. Slevin, S. Kumar and P. Kumar. 2005. Expression of basic fibroblast growth factor mRNA and protein in the human brain following ischaemic stroke. Angiogenesis 8: 53–62.

Issa, R., J. Krupinski, T. Bujny, S. Kumar, J. Kaluza and P. Kumar. 1999. Vascular endothelial growth factor and its receptor, KDR, in human brain tissue after ischemic stroke. Lab. Invest. 79: 417–425.

Iwakura, A., C. Luedemann, S. Shastry, A. Hanley, M. Kearney, R. Aikawa, J.M. Isner, T. Asahara and D.W. Losordo. 2003. Estrogen-mediated, endothelial nitric oxide synthase-dependent mobilization of bone marrow-derived endothelial progenitor cells contributes to reendothelialization after arterial injury. Circulation 108: 3115–3121.

Jain, R.K. 2003.Molecular regulation of vessel maturation. Nat. Med. 9: 685–693.

Jakobsson, L., C.A. Franco, K. Bentley, R.T. Collins, B. Ponsioen, I.M. Aspalter, I. Rosewell, M. Busse, G. Thurston, A. Medvinsky, S. Schulte-Merker and H. Gerhardt. 2010. Endothelial cells dynamically compete for the tip cell position during angiogenic sprouting. Nat. Cell. Biol. 12: 943–953.

Jiang, N., S.P. Finklestein, T. Do, C.G. Caday, M. Charette and M. Chopp. 1996. Delayed intravenous administration of basic fibroblast growth factor (bFGF) reduces infarct volume in a model of focal cerebral ischemia/reperfusion in the rat. J. Neurol. Sci. 139: 173–179.

Jiang, Q., Z.G. Zhang and M. Chopp. 2010. MRI of Stroke Recovery. Stroke 41: 410–414.

Jiang, Q., Z.G. Zhang, G.L. Ding, L. Zhang, J.R. Ewing, L. Wang, R. Zhang, L. Li, M. Lu, H. Meng, A.S. Arbab, J. Hu, Q.J. Li, D.S. Pourabdollah Nejad, H. Athiraman and M. Chopp. 2005. Investigation of neural progenitor cell induced angiogenesis after embolic stroke in rat using MRI. Neuroimage 28: 698–707.

Jin, K., Y. Zhu, Y. Sun, X.O. Mao, L. Xie and D.A. Greenberg. 2002. Vascular endothelial growth factor (VEGF) stimulates neurogenesis *in vitro* and *in vivo*. Proc. Natl. Acad. Sci. USA 99: 11946–11950.

Jin, R., G. Yang and G. Li. 2010. Molecular insights and therapeutic targets for blood-brain barrier disruption in ischemic stroke: critical role of matrix metalloproteinases and tissue type plasminogen activator. Neurobiol. Dis. 38: 376–385.

Johansen-Berg, H. 2007. Functional imaging of stroke recovery: what have we learnt and where do we go from here? Int. J. Stroke 2: 7–16.

Kamei, M., W.B. Saunders, K.J. Bayless, L. Dye, G.E. Davis and B.M. Weinstein. 2006. Endothelial tubes assemble from intracellular vacuoles *in vivo*. Nature 442: 453–456.

Kanda, S., E. Landgren, M. Ljungstrom and L. Claesson-Welsh. 1996. Fibroblast growth factor receptor 1-induced differentiation of endothelial cell line established from tsA58 large T transgenic mice. Cell Growth Differ. 7: 383–395.

Kastrup, A., K. Groschel, T.M. Ringer, C. Redecker, R. Cordesmeyer, O.W. Witte and C. Terborg. 2008. Early disruption of the blood-brain barrier after thrombolytic therapy predicts hemorrhage in patients with acute stroke. Stroke 39: 2385–2387.

Khmelewski, E., A. Becker, T. Meinertz and W.D. Ito. 2004. Tissue resident cells play a dominant role in arteriogenesis and concomitant macrophage accumulation. Circ. Res. 95: E56–E64.

Kilic, E., A. ElAli, U. Kilic, Z. Guo, M. Ugur, U. Uslu, C.L. Bassetti, M.E. Schwab and D.M. Hermann. 2010. Role of Nogo-A in neuronal survival in the reperfused ischemic brain. J. Cereb. Blood Flow Metab. 30: 969–984.

Kilic, E., U. Kilic, M. Bacigaluppi, Z. Guo, N.B. Abdallah, D.P. Wolfer, R.J. Reiter, D.M. Hermann and C.L. Bassetti . 2008. Delayed melatonin administration promotes neuronal survival, neurogenesis and motor recovery, and attenuates hyperactivity and anxiety after mild focal cerebral ischemia in mice. J. Pineal. Res. 45: 142–148.

Kim, J.B. C.M. Lim, Y.M. Yu and J.K. Lee. 2008. Induction and subcellular localization of high-mobility group box-1 (HMGB1) in the postischemic rat brain. J. Neurosci. Res. 86: 1125–1131.

Kinnaird, T., E. Stabile, M.S. Burnett, M. Shou, C.W. Lee, S. Barr, S. Fuchs and S.E. Epstein. 2004. Local deliveryof marrow-derived stromal cells augments collateral perfusion through paracrine mechanisms. Circulation 109: 1543–1549.

Kluk, M.J. and T. Hla. 2002. Signaing of sphingosine-1-phosphate via the S1P/EDG-family of G-protein-coupled receptors. Biochim. Biophys. Acta 1582: 72–80.

Kojima, T., Y. Hirota, M. Ema, S. Takahashi, I. Miyoshi, H. Okano and K. Sawamoto. 2010. Subventricular zone-derived neural progenitor cells migrate along a blood vessel scaffold toward the post-stroke striatum. Stem Cells 28: 545–554.

Kokaia, Z. and O. Lindvall. 2003. Neurogenesis after ischaemic brain insults. Curr. Opin. Neurobiol. 13: 127–132.

Kokovay, E., L. Li and L.A. Cunningham. 2006. Angiogenic recruitment of pericytes from bone marrow after stroke. J. Cereb. Blood Flow Metab. 26: 545–555.

Kokovay, E., S. Goderie, Y. Wang, S. Lotz, G. Lin, Y. Sun, B. Roysam, Q. Shen and S. Temple. 2010. Adult SVZ lineage cells home to and leave the vascular niche via differential responses to SDF1/CXCR4 signaling. Cell Stem Cell 7: 163–173.

Koto, T., K. Takubo, S. Ishida, H. Shinoda, M. Inoue, K. Tsubota, Y. Okada and E. Ikeda. 2007. Hypoxia disrupts the barrier function of neural blood vessels through changes in the expression of claudin-5 in endothelial cells. Am. J. Pathol. 170: 1389–1397.

Krishna, K.and C. Redies. 2009. Expression of cadherin superfamily genes in brain vascular development. J. Cereb. Blood Flow Metab. 29: 224–229.

Krishnan, L., C.J. Underwood, S. Maas, B.J. Ellis, T.C. Kode, J.B. Hoying and J.A.Weiss. 2008. Effect of mechanical boundary conditions on orientation of angiogenic microvessels. Cardiovasc. Res. 78: 324–332.

Krupinski, J., J. Kaluza, P. Kumar, S. Kumar and J.M. Wang. 1994. Role of angiogenesis in patients with cerebral ischemic stroke. Stroke 25: 1794–1798.

Krupinski, J., J. Kaluza, P. Kumar and S. Kumar. 1996. Immunocytochemical studies of cellular reaction in human ischemic brain stroke MAB anti-CD68 stains macrophages, astrocytes and microglial cells in infarcted area. Folia Neuropathol. 34: 17–24.

Krupinski, J., P. Kumar, S. Kumar and J. Kaluza. 1996. Increased expression of TGFbeta 1 in brain tissue after ischemic stroke in humans. Stroke 27: 852–857.

Krupinski, J., P. Stroemer, M. Slevin, E. Marti, P. Kumar and F. Rubio. 2003. Three-dimensional structure and survival of newly formed blood vessels after focal cerebral ischemia. Neuroreport 14: 1171–1176.

Krupinski, J., R. Issa, T. Bujny, M. Slevin, P. Kumar, S. Kumar and J. Kaluza. 1997. A putative role for platelet-derived growth factor in angiogenesis and neuroprotection after ischemic stroke in humans. Stroke 28: 564–573.

Kwon, I., E.H. Kim, G.J. del Zoppo and J.H. Heo. 2009. Ultrastructural and temporal changes of the microvascular basement membrane and astrocyte interface following focal cerebral ischemia. J. Neurosci. Res. 87: 668–676.

Lamalice, L., F. Le Boeuf and J. Huot. 2007. Endothelial cell migration during angiogenesis. Circ. Res. 100: 782–794.

Latchaw, R.E., M.J. Alberts, M.H. Lev, J.J. Connors, R.E. Harbaugh, R.T. Higashida, R. Hobson, C.S. Kidwell, W.J. Koroshetz, V. Mathews, P. Villablanca, S. Warach and B. Walters. 2009. Recommendations for imaging of acute ischemic stroke. A scientific statement from the American Heart Association. Stroke 40: 3646–3678.

Laufs, U., N. Werner, A. Link, M. Endres, S. Wassmann, K. Jurgens, E. Miche, M. Böhm and G. Nickenig. 2004. Physical training increases endothelial progenitor cells, inhibits neointima formation, and enhances angiogenesis. Circulation 109: 220–226.

Lee, C.W., E. Stabile, T. Kinnaird, M. Shou, J.M. Devaney, S.E. Epstein and M.S. Burnett. 2004. Temporal patterns of gene expression after acute hindlimb ischemia in mice: insights into the genomic program for collateral vessel development. J. Am. Coll. Cardiol. 43: 474–482.

Lee, S.R., H.Y. Kim, J. Rogowska, B.Q. Zhao, P. Bhide, J.M. Parent and E.H. Lo. 2006. Involvement of matrix metalloproteinase in neuroblast cell migration from the subventricular zone after stroke. J. Neurosci. 26: 3491–3495.
Lee, S.T., K. Chu, K.H. Jung, S.Y. Ko, E.H. Kim, D.I. Sinn, Y.S. Lee, E.H. Lo, M. Kim and J.K. Roh. 2005. Granulocyte colony-stimulating factor enhances angiogenesis after focal cerebral ischemia. Brain Res. 1058: 120–128.
Leker, R.R. and E. Shohami. 2002. Cerebral ischemia and trauma—different etiologies yet similar mechanisms: neuroprotective opportunities. Brain Res. Rev. 39: 55–73.
Letechipia-Vallejo, G., V. Espinoza-Gonzalez, E. Lopez Loeza, I. Gonzalez.Burgos, M.E. Olvera-Corte, G. Morali and M. Cervantes. 2007. Long-term morphological and functional evaluation of the neuroprotective effects of post-ischemic treatment with melatonin in rats. J. Pineal. Res. 47: 139–146.
Leventhal, C., S. Rafii, D. Rafii, A. Shahar and S.A. Goldman. 1999. Endothelial trophic support of neuronal production and recruitment from the adult mammalian subependyma. Mol. Cell. Neurosci. 13: 450–464.
Li, L., Q. Jiang, L. Zhang, G. Ding, Z. Gang Zhang, Q. Li, J.R. Ewing, M. Lu, S. Panda, K.A. Ledbetter, P.A. Whitton and M. Chopp. 2007. Angiogenesis and improved cerebral blood flow in the ischemic boundary area detected by MRI after administration of sildenafil to rats with embolic stroke. Brain Res. 1132: 185–192.
Liman, T.J. and M. Endres. 2012. New Vessels after stroke: Postischemic Neovascularization and regeneration. Cerebrovasc. Dis. 33: 492–499.
Lin, T.N., J. Te, M. Lee, G.Y. Sun and C.Y. Hsu. 1997. Induction of basic fibroblast growth factor (bFGF) expression following focal cerebral ischemia. Brain Res. Mol. Brain Res. 49: 255–65.
Lindblom, P., H. Gerhardt, S. Liebner, A. Abramsson, M. Enge, M. Hellstrom, G. Backstrom, S. Fredriksson, U. Landegren, H.C. Nystrom, G. Bergstrom, E. Dejana, A. Ostman, P. Lindahl and C. Betsholtz. 2003. Endothelial PDGF-B retentions required for proper investment of pericytes in the microvessel wall. Genes Dev. 17: 1835–1840.
Ling, L., J. Zeng, Z. Pei, R.T. Cheung, Q. Hou, S. Xing and S. Zhang. 2009. Neurogenesis and angiogenesis within the ipsilateral thalamus with secondary damage after focal cortical infarction in hypertensive rats. J. Cereb. Blood Flow Metab. 29: 1538–1546.
Lipton, P. 1999. Ischemic cell death in brain neurons. Physiol. Rev. 79: 1431–1568.
Liu, J., X.B. Wang, D.S. Park and M.P. Lisanti. 2002. Caveolin-1 expression enhances endothelial capillary tubule formation. J. Biol. Chem. 277: 10661–10668.
Lok, J., P. Gupta, S. Guo, W.J. Kim, M.J. Whalen, K. van Leyen and E.H. Lo. 2007. Cell-cell signaling in the neurovascular unit. Neurochem. Res. 32: 2032–2045.
Longa, E.Z., P.R. Weinstein, S. Carlson and R. Cummins. 1989. Reversible middle cerebral artery occlusion without craniectomy in rats. Stroke 20: 84–91.
Luque Contreras, D., H. Vargas Robles, E. Romo, A. Rios and B. Escalante. 2006. The role of nitric oxide in the postischemic revascularization process. Pharmacol Ther. 112: 553–563.
Machado, L.S., A. Kozak, A. Ergul, D.C. Hess, C.V. Borlongan and S.C. Fagan. 2006. Delayed minocycline inhibits ischemia activated matrix metalloproteinases 2 and 9 after experimental stroke. BMC Neurosci. 7: 56.
Madri, J.A. 2009. Modeling the neurovascular niche: implications for recovery from CNS injury. J. Physiol. Pharmacol. 60: 95–104.
Manoonkitiwongsa, P.S., C. Jackson-Friedman, P.J. McMillan, R.L. Schultz and P.D. Lyden. 2001. Angiogenesis after stroke is correlated with increased numbers of macrophages: the clean-up hypothesis. J. Cereb. Blood Flow Metab. 21: 1223–1231.
Manoonkitiwongsa, P.S., R.L. Schultz, E.F. Whitter and P.D. Lyden. 2006. Contraindications of VEGF based therapeutic angiogenesis: Effects on macrophage density and histology of normal and ischemic brains. Vascul. Pharmacol. 44: 316–325.
Margaritescu, O., D. Pirici and C. Margaritescu. 2011. VEGF expression in human brain tissue after acute ischemic stroke. Rom. J. Morphol. Embryol. 52: 1283–1292.

Mark, K.S. and T.P. Davis. 2002. Cerebral microvascular changes in permeability and tight junctions induced by hypoxia-reoxygenation. Am. J. Physiol. Heart Circ. Physiol. 282: H1485–H1494.

Markus, T.M., S.Y. Tsai, M.R. Bollnow, R.G. Farrer, T.E. O'Brien, D.R. Kindler-Baumann, M. Rausch, M. Rudin, C. Wiessner, A.K. Mir, M.E. Schwab and G.L. Kartje. 2005. Recovery and brain reorganization after stroke in adult and aged rats. Ann. Neurol. 58: 950–953.

Marti, H.J., M. Bernaudin, A. Bellail, H. Schoch, M. Euler, E. Petit and W. Risau. 2000. Hypoxia-induced vascular endothelial growth factor expression precedes neovascularization after cerebral ischemia. Am. J. Pathol. 156: 965–976.

Martin, A., E. Mace, R. Boisgard, G. Montaldo, B. Thézé, M. Tanter and B. Tavitian. 2012. Imaging of perfusion, angiogenesis, and tissue elasticity after stroke. J. Cereb. Blood Flow. Metab. 32: 1496–1507.

Mathiisen, T.M., K.P. Lehre, N.C. Danbolt and O.P. Ottersen. 2010. The perivascular astroglial sheath provides a complete covering of the brain microvessels: an electron microscopic 3D reconstruction. Glia 58: 1094–1103.

Meisner, J.K. and R.J. Price. 2010. Spatial and Temporal Coordination of Bone Marrow-Derived Cell Activity During Arteriogenesis: Regulation of the Endogenous Response and Therapeutic Implications. Microcirculation 17: 583–599.

Meyer, G.T., L.J. Matthias, L. Noack, M.A. Vadas and J.R. Gamble. 1997. Lumen formation during angiogenesis *in vitro* involves phagocytic activity, formation and secretion of vacuoles, cell death, and capillary tube remodelling by different populations of endothelial cells. Anat. Rec. 249: 327–340.

Milkiewicz, M., E. Ispanovic, J.L. Doyle and T.L. Haas. 2006. Regulators of angiogenesis and strategies for their therapeutic manipulation. Int. J. Biochem. Cell Biol. 38: 333–357.

Moisan, A., N. Pannetier, E. Grillon, M.J. Richard, F. de Fraipont, C. Remy, E.L. Barbier and O. Detante. 2012. Intracerebral injection of human mesenchymal stem cells impacts cerebral microvasculature after experimental stroke: MRI study. NMR Biomed. 25: 1340–1348.

Moore, C.J. and S.J. Winder. 2010. Dystroglycan versatility in cell adhesion: a tale of multiple motifs. Cell Commun. Signal. 8: 3.

Morancho, A., A. Rosell, L. Garcia-Bonilla and J. Montaner. 2010. Metalloproteinase and stroke infarct size: role for anti-inflammatory treatment? Ann. N. Y. Acad. Sci. 1207: 123–133.

Murohara, T., T. Asahara, M. Silver, C. Bauters, H. Masuda, C. Kalka, M. Kearney, D. Chen, J.F. Symes, M.C. Fishman, P.L. Huang and J.M. Isner. 1998. Nitric oxide synthase modulates angiogenesis in response to tissue ischemia. J. Clin. Invest. 101: 2567–2578.

Murphy, T.H. and D. Corbett. 2009. Plasticity during stroke recovery: from synapse to behaviour. Nat. Rev. Neurosci. 10: 861–872.

Nagel, S., Y. Su, S. Horstmann, S. Heiland, H. Gardner, J. Koziol, F.J. Martinez-Torres and S. Wagner. 2008. Minocycline and hypothermia for reperfusion injury after focal cerebral ischemia in the rat—Effects on BBB breakdown and MMP expression in the acute and subacute phase. Brain Res. 1188: 198–206.

Navaratna, D., S. Guo, K. Arai and E.H. Lo. 2009. Mechanisms and targets for angiogenic therapy after stroke. Cell Adh. Migr. 3: 216–223.

Neuwelt, E.A., B. Bauer, C. Fahlke, G. Fricker, C. Iadecola, D. Janigro, L. Leybaert, Z. Molnar, M.E. O'Donnell, J.T. Povlishock, N.R. Saunders, F. Sharp, D. Stanimirovic, R.J. Watts and L.R. Drewes. 2011. Engaging neuroscience to advance translational research in brain barrier biology. Nat. Rev. Neurosci. 12: 169–182.

Nico, B. and D. Ribatti. 2012. Morphofunctional aspects of the blood-brain barrier. Curr. Drug Metab. 13: 50–60.

Ohab, J.J. and S.T. Carmichael. 2008. Poststroke neurogenesis: emerging principles of migration and localization of immature neurons. Neuroscientist 14: 369–380.

Ohab, J.J., S. Fleming, A. Blesch and S.T. Carmichael. 2006. A neurovascular niche for neurogenesis after stroke. J. Neurosci. 26: 13007–13016.

Okada, Y., B.R. Copeland, R. Fitridge, J.A. Kosiol and G.J. del Zoppo. 1994. Fibrin contributes to microvascular obstructions and parenchymal changes during early focal cerebral ischemia and reperfusion. Stroke 25: 1847–1854.

Onteniente, B. and J. Polentes. 2011. Regenerative medicine for stroke—are we there yet? Cerebrovasc. Dis. 31: 544–551.

Paczkowska, E., M. Kucia, D. Koziarska, M. Halasa, K. Safranow, M. Masiuk, A. Karbicka, M. Nowik, P. Nowacki, M.Z. Ratajczak and B. Machalinski. 2009. Clinical evidence that very small embryonic-like stem cells are mobilized into peripheral blood in patients after stroke. Stroke 40: 1237–1244.

Palmer, .T.D., A.R. Willhoite and F.H. Gage. 2000. Vascular niche for adult hippocampal neurogenesis. J. Comp. Neurol. 425: 479–494.

Papapetropoulos, A., G. Garcia-Cardena, J.A. Madri and W.C. Sessa. 1997. Nitric oxide production contributes to the angiogenic properties of vascular endothelial growth factor in human endothelial cells. J. Clin. Invest. 100: 3131–3139.

Parent, J.M. 2003. Injury-induced neurogenesis in the adult mammalian brain. Neuroscientist 9: 261–272.

Pepper, M.S., A.P. Sappino, R. Stocklin, R. Montesano, L. Orci and J.D. Vassalli. 1993. Upregulation of urokinase receptor expression on migrating endothelial cells. J. Cell. Biol. 122: 673–680.

Persson, A.B. and I.R. Buschmann. 2011. Vascular growth in health and disease. Front. Mol. Neurosci. 4: 14.

Peters, K., D. Troyer, S. Kummer, C.J. Kirkpatrick and J. Rauterberg. 2002. Apoptosis causes lumen formation during angiogenesis *in vitro*. Microvasc. Res. 64: 334–338.

Pfefferkorn, T. and G.A. Rosenberg. 2003. Closure of the blood-brain barrier by matrix metalloproteinase inhibition reduces rtPAmediated mortality in cerebral ischemia with delayed reperfusion. Stroke 34: 2025–2030.

Pichiule, P. and J.C. LaManna. 2002. Angiopoietin-2 and rat brain capillary remodeling during adaptation and deadaptation to prolonged mild hypoxia. J. Appl. Physiol. 93: 1131–1139.

Planas, A.M., S. Sole and C. Justicia. 2001. Expression and activation of matrix metalloproteinase-2 and-9 in rat brain after transient focal cerebral ischemia. Neurobiol. Dis. 8: 834–846.

Plate, K.H., H. Beck, S. Danner, P.R. Allegrini and C. Wiessner. 1999. Cell type specific upregulation of vascular endothelial growth factor in an MCA occlusion model of cerebral infarct. J. Neuropathol. Exp. Neurol. 58: 654–666.

Prehn, J.H., C. Backhauss and J. Krieglstein. 1993. Transforming growth factor-beta 1 prevents glutamate neurotoxicity in rat neocortical cultures and protects mouse neocortex from ischemic injury *in vivo*. J. Cereb. Blood Flow Metab. 13: 521–525.

Prior, B.M., P.G. Lloyd, J. Ren, Z. Li, H.T. Yang, M.H. Laughlin and R.L. Terjung. 2003. Arteriogenesis: role of nitric oxide. Endothelium 10: 207–216.

Redzic, Z. 2011. Molecular biology of the blood-brain and the blood-cerebrospinal fluid barriers: similarities and differences. Fluids Barriers CNS 8: 3.

Reitmeir, R., E. Kilic, U. Kilic, M. Bacigaluppi, A. ElAli, G. Salani, S. Pluchino, M. Gassmann and D.M. Hermann. 2011. Post-acute delivery of erythropoietin induces stroke recovery by promoting perilesional tissue remodelling and contralesional pyramidal tract plasticity. Brain 134: 84–99.

Risau, W. 1997. Mechanisms of angiogenesis. Nature 386: 671–674.

Robin, A.M., Z.G. Zhang, L. Wang, R.L. Zhang, M. Katakowski, L. Zhang, Y. Wang, C. Zhang and M. Chopp. 2006. Stromal cell-derived factor 1alpha mediates neural progenitor cell motility after focal cerebral ischemia. J. Cereb. Blood Flow Metab. 26: 125–134.

Romanic, A.M., R.F. White, A.J. Arleth, E.H. Ohlstein and F.C. Barone. 1998. Matrix metalloproteinase expression increases after cerebral focal ischemia in rats—Inhibition of matrix metalloproteinase-9 reduces infarct size. Stroke 29: 1020–1030.

Rosell, A. and E.H. Lo. 2008. Multiphasic roles for matrix metalloproteinases after stroke. Curr. Opin. Pharmacol. 8: 82–89.

Rosenberg, G.A., E.Y. Estrada and J.E. Dencoff. 1998. Matrix metalloproteinases and TIMPs are associated with blood-brain barrier pening after reperfusion in rat brain. Stroke 29: 2189–2195.

Rudic, R.D., E.G. Shesely, N. Maeda, O. Smithies, S.S. Segal and W.C. Sessa. 1998. Direct evidence for the importance of endothelium-derived nitric oxide in vascular remodeling. J. Clin. Invest. 101: 731–736.

Sahni, A., A.A. Khorana, R.B. Baggs, H. Peng and C.W. Francis. 2006. FGF-2 binding to fibrin(ogen) is required for augmented angiogenesis. Blood 107: 126–131.

Sandoval, K.E. and K.A. Witt. 2008. Blood-brain barrier tight junction permeability and ischemic stroke. Neurobiol. Dis. 32: 200–219.

Saunders, W.B., B.L. Bohnsack, J.B. Faske, N.J. Anthis, K.J. Bayless, K.K. Hirschi and G.E. Davis. 2006. Coregulation of vascular tube stabilization by endothelial cell TIMP-2 and pericyte TIMP-3. J. Cell Biol. 175: 179–191.

Savitz, S.I., M. Chopp, R. Deans, S.T. Carmichael, D. Phinney and L. Wechsler. 2011. Stem cell therapy as an emerging paradigm for stroke (STEPS) II. Stroke 42: 825–829.

Schabitz, W.R., R. Laage, G. Vogt, W. Koch, R. Kollmar, S. Schwab, D. Schneider, G.F. Hamann, M. Rosenkranz, R. Veltkamp, J.B. Fiebach, W. Hacke, J.C. Grotta, M. Fisher and A. Schneider. 2010. AXIS: a trial of intravenous granulocyte colony-stimulating factor in acute ischemic stroke. Stroke 41: 2545–2551.

Schachtrup, C., J.K. Ryu, M.J. Helmrick, E. Vagena, D.K. Galanakis, J.L. Degen, R.U. Margolis and K. Akassoglou. 2010. Fibrinogen triggers astrocyte scar formation by promoting the availability of active TGF-beta after vascular damage. J. Neurosci. 30: 5843–5854.

Schachtrup, C., P. Lu, L.L. Jones, J.K. Lee, J. Lu, B.D. Sachs, B. Zheng and K. Akassoglou. 2007. Fibrinogen inhibits neurite outgrowth via beta3 integrin-mediated phosphorylation of the EGF receptor. Proc. Natl. Acad. Sci. USA 104: 11814–11819.

Schaller, B. and R. Graf. 2004. Cerebral ischemia and reperfusion: the pathophysiologic concept as a basis for clinical therapy. J. Cereb. Blood Flow Metab. 24: 351–371.

Schaper, W. 2009. Collateral circulation: past and present. Basic Res. Cardiol. 104: 5–21.

Schaper, W. and D. Scholz. 2003. Factors regulating arteriogenesis. Arterioscler. Thromb. Vasc. Biol. 23: 1143–1151.

Schlueter, C., H. Weber, B. Meyer, P. Rogalla, K. Roser, S. Hauke and J. Bullerdiek. 2005. Angiogenetic signaling through hypoxia: HMGB1: an angiogenetic switch molecule. Am. J. Pathol. 166: 1259–1263.

Schoch, H.J., S. Fischer and H.H. Marti. 2002. Hypoxia-induced vascular endothelial growth factor expression causes vascular leakage in the brain. Brain 125: 2549–2557.

Scholz, D., W.J. Cai and W. Schaper. 2001. Arteriogenesis, a new concept of vascular adaptation in occlusive disease. Angiogenesis 4: 247–257.

Seevinck, P.R., L.H. Deddens and R.M. Dijkhuizen. 2010. Magnetic resonance imaging of brain angiogenesis after stroke. Angiogenesis 13: 101–111.

Sehara, Y., T. Hayashi, K. Deguchi, H. Zhang, A. Tsuchiya, T. Yamashita, V. Lukic, M. Nagai, T. Kamiya and K. Abe. 2007. Potentiation of neurogenesis and angiogenesis by G-CSF after focal cerebral ischemia in rats. Brain Res. 1151: 142–149.

Semenza, G.L. 2000. Surviving ischemia: adaptive responses mediated by hypoxia-inducible factor 1. J. Clin. Invest. 106: 809–812.

Sharp, F.R. and M. Bernaudin. 2004. HIF1 and oxygen sensing in the brain. Nat. Rev. Neurosci. 5: 437–448.

Shen, L.H., Y. Li and M. Chopp. 2010. Astrocytic endogenous glial cell derived neurotrophic factor production is enhanced by bone marrow stromal cell transplantation in the ischemic boundary zone after stroke in adult rats. Glia 58: 1074–1081.

Shen, L.H., Y. Li, J. Chen, L.H. Shen, Y. Li, J. Chen, A. Zacharek, Q. Gao, A. Kapke, M. Lu, K. Raginski, P. Vanguri, A. Smith and M. Chopp. 2007a. Therapeutic benefit of bone marrow stromal cells administered 1 month after stroke. J. Cereb. Blood Flow Metab. 27: 6–13.

Shen, L.H., Y. Li, J. Chen, Y. Cui, C. Zhang, A. Kapke, M. Lu, S. Savant-Bhonsale and M. Chopp. 2007b. One-year follow-up after bone marrow stromal cell treatment in middle-aged female rats with stroke. Stroke 38: 2150–2156.

Shen, Q., S.K. Goderie, L. Jin, N. Karanth, Y. Sun, N. Abramova, P. Vincent, K. Pumiglia and S. Temple. 2004. Endothelial cells stimulate self-renewal and expand neurogenesis of neural stem cells. Science 304: 1338–1340.

Shi, H. 2009. Hypoxia inducible factor 1 as a therapeutic target in ischemic stroke. Curr. Med. Chem. 16: 4593–4600.

Shibata, M., S.R. Kumar, A. Amar, J.A. Fernandez, F. Hofman, J.H. Griffin and B.V. Zlokovic. 2001. Antiinflammatory, antithrombotic, and neuroprotective effects of activated protein C in a murine model of focal ischemic stroke. Circulation 103: 1799–1805.

Shintani, S., T. Murohara, H. Ikeda, T. Ueno, T. Honma, A. Katoh, K. Sasaki, T. Shimada, Y. Oike and T. Imaizumi. 2001. Mobilization of endothelial progenitor cells in patients with acute myocardial infarction. Circulation 103: 2776–2779.

Shuaib, A., K.R. Lees, P. Lyden, J. Grotta, A. Davalos, S.M. Davis, H.C. Diener, T. Ashwood, W.W. Wasiewski and U. Emeribe. 2007. SAINT II Trial Investigators: NXY-059 for the treatment of acute ischemic stroke. N. Engl. J. Med. 357: 562–571.

Shyu, W.C., S.Z. Lin, H.I. Yang, Y.S. Tzeng, C.Y. Pang, P.S. Yen and H. Li. 2004. Functional recovery of stroke rats induced by granulocyte colony-stimulating factor-stimulated stem cells. Circulation 110: 1847–1854.

Shyu, W.C., S.Z. Lin, M.F. Chiang, C.Y. Su and H. Li. 2006. Intracerebral peripheral blood stem cell (CD34+) implantation induces neuroplasticity by enhancing beta1 integrin-mediated angiogenesis in chronic stroke rats. J. Neurosci. 26: 3444–3453.

Silver, B., S. McCarthy, M. Lu, P. Mitsias, A.N. Russman, A. Katramados, D.C. Morris, C.A. Lewandowski and M. Chopp. 2009. Sildenafil treatment of subacute ischemic stroke: a safety study at 25-mg daily for 2 weeks. J. Stroke Cerebrovasc. Dis. 18: 381–383.

Silver, J. and J.H. Miller. 2004. Regeneration beyond the glial scar. Nat. Rev. Neurosci. 5: 146–156.

Slevin, M., J. Krupinski, A. Slowik, P. Kumar, A. Szczudlik and J. Gaffney. 2000. Serial measurement of vascular endothelial growth factor and transforming growth factor-beta1 in serum of patients with acute ischemic stroke. Stroke 31: 1863–1870.

Slevin, M., P. Kumar, J. Gaffney and J. Krupinski. 2006. Can angiogenesis be exploited to improve stroke outcome? Mechanisms and therapeutic potential. Clin. Sci. (Lond.) 111: 171–183.

Snapyan, M., M. Lemasson, M.S. Brill, M. Blais, M. Massouh, J. Ninkovic, C. Gravel, F. Berthod, M. Gotz, P.A. Barker, A. Parent and A. Saghatelyan. 2009. Vasculature guides migrating neuronal precursors in the adult mammalian forebrain via brain-derived neurotrophic factor signaling. J. Neurosci. 29: 4172–4188.

Sofroniew, M.V. and H.V. Vinters. 2010. Astrocytes: biology and pathology. Acta Neuropathol. 119: 7–35.

Stiver, S.I., X. Tan, L.F. Brown, E.T. Hedley-Whyte and H.F. Dvorak. 2004. VEGF-A angiogenesis induces a stable neovasculature in adult murine brain. J. Neuropathol. Exp. Neurol. 63: 841–55.

Strecker, J.K., S. Sevimli, M. Schilling, R. Klocke, S. Nikol, A. Schneider and W.R. Schabitz. 2010. Effects of G-CSF treatment on neutrophil mobilization and neurological outcome after transient focal ischemia. Exp. Neurol. 222: 108–113.

Stromhaug, P.E. and D.J. Klionsky. 2001. Approaching the molecular mechanism of autophagy. Traffic 2: 524–531.

Sun, Y., K. Jin, L. Xie, J. Childs, X.O. Mao, A. Logvinova and D.A. Greenberg. 2003. VEGF-induced neuroprotection, neurogenesis and angiogenesis after focal cerebral ischemia. J. Clin. Invest. 111: 1843–1851.

Taguchi, A., T. Matsuyama, H. Moriwaki, T. Hayashi, K. Hayashida, K. Nagatsuka, K. Todo, K. Mori, D.M. Stern, T. Soma and H. Naritomi. 2004. Circulating CD34-positive cells provide an index of cerebrovascular function. Circulation 109: 2972–2975.

Taguchi, A., T. Soma, H. Tanaka, T. Kanda, H. Nishimura, H. Yoshikawa, Y. Tsukamoto, H. Iso, Y. Fujimori, D.M. Stern, H. Naritomi and T. Matsuyama. 2004. Administration of CD34+ cells after stroke enhances neurogenesis via angiogenesis in a mouse model. J. Clin. Invest. 114: 330–338.

Tanaka, H., S. Takai, D. Jin, K. Furubayashi, N. Inoue, Y. Kajimoto, S. Miyatake, T. Kuroiwa and M. Miyazaki. 2007. Inhibition of matrix metalloproteinase-9 activity by trandolapril after middle cerebral artery occlusion in rats. Hypertens. Res. 30: 469–475.

Thiyagarajan, M., J. Fernandez, S. Lane, J. Griffin and B. Zlokovic. 2008. Activated protein C promotes neovascularization and neurogenesis in postischemic brain via protease-activated receptor 1. J. Neurosci. 28: 12788–12797.

Thored, P., A. Arvidsson, E. Cacci, H. Ahlenius, T. Kallur, V. Darsalia, C.T. Ekdahl, Z. Kokaia and O. Lindvall. 2006. Persistent production of neurons from adult brain stem cells during recovery after stroke. Stem Cells 24: 739–747.

Thored, P., J. Wood, A. Arvidsson, J. Cammenga, Z. Kokaia and O. Lindvall. 2007. Long-term neuroblast migration along blood vessels in an area with transient angiogenesis and increased vascularization after stroke. Stroke 38: 3032–3039.

Thwaites, J.W., V. Reebye, P. Mintz, N. Levicar and N. Habib. 2012. Cellular replacement and regenerative medicine therapies in ischemic stroke. Regen. Med. 7: 387–395.

Treutiger, C.J., G.E. Mullins, A.S. Johansson, A. Rouhiainen, H.M. Rauvala, H. Erlandsson-Harris, U. Andersson, H. Yang, K.J. Tracey, J. Andersson and J.E. Palmblad. 2003. High mobility group 1 B-box mediates activation of human endothelium. J. Intern. Med. 254: 375–385.

Troidl, C., K. Troidl, W. Schierling, W.J. Cai, H. Nef, H. Mollmann, S. Kostin, S. Schimanski, L. Hammer, A. Elsässer, T. Schmitz-Rixen and W. Schaper. 2009. Trpv4 induces collateral vessel growth during regeneration of the arterial circulation. J. Cell. Mol. Med. 13: 2613–2621.

Tsai, S.Y., C.M. Papadopoulos, M.E. Schwab and G.L. Kartje. 2011. Delayed anti-Nogo-A therapy improves function after chronic stroke in adult rats. Stroke 42: 186–190.

Turu, M.M., M. Slevin, S. Matou, D. West, C. Rodríguez, A. Luque, M. Grau-Olivares, L. Badimon, J. Martinez-Gonzalez and J. Krupinski. 2008. C-reactive protein exerts angiogenic effects on vascular endothelial cells and modulates associated signalling pathways and gene expression. BMC Cell Biol. 9: 47.

Ucuzian, A.A., A.A. Gassman, A.T. East and H.P. Greisler. 2010. Molecular mediators of angiogenesis. J. Burn Care Res. 31: 158–175.

Ukai, R., O. Honmou, K. Harada, K. Houkin, H. Hamada and J.D. Kocsis. 2007. Mesenchymal stem cells derived from peripheral blood protects against ischemia. J. Neurotrauma 24: 508–520.

van Hinsbergh, V.W. and P. Koolwijk. 2008. Endothelial sprouting and angiogenesis: matrix metallo-proteinases in the lead. Cardiovasc. Res. 78: 203–212.

van Royen, N., J.J. Piek, W. Schaper and W.F. Fulton. 2010. A critical review of clinical arteriogenesis research. J. Am. Coll. Cardiol. 55: 17–25.

Wagner, D.C., A. Deten, W. Härtig, J. Boltze and A. Kranz. 2012. Changes in T2 relaxation time after stroke reflect clearing processes. Neuroimage 61: 780–785.

Wang, G.L. and G.L. Semenza. 1993. Characterization of hypoxia-inducible factor 1 and regulation of DNA binding activity by hypoxia. J. Biol. Chem. 268: 21513–21518.

Wang, G., Q. Guo, M. Hossain, V. Fazio, E. Zeynalov, D. Janigro, M.R. Mayberg and S. Namura. 2009. Bone marrow-derived cells are the major source of MMP-9 contributing to blood-brain barrier dysfunction and infarct formation after ischemic stroke in mice. Brain Res. 1294: 183–192.

Wang, G.L., B.H. Jiang, E.A. Rue and G.L. Semenza. 1995. Hypoxia-inducible factor 1 is a basic-helix-loop-helix-PAS heterodimer regulated by cellular O_2 tension. Proc. Natl. Acad. Sci. USA 92: 5510–5514.

Wang, J. and R. Milner. 2006. Fibronectin promotes brain capillary endothelial cell survival and proliferation through alpha5beta1 and alphavbeta3 integrins via MAP kinase signaling. J. Neurochem. 96: 148–159.

Wang, L., Z.G. Zhang, R.L. Zhang, S.R. Gregg, A. Hozeska-Solgot, Y. LeTourneau, Y. Wang and M. Chopp. 2006. Matrix metalloproteinase 2 (MMP2) and MMP9 secreted by erythropoietin-activated endothelial cells promote neural progenitor cell migration. J. Neurosci. 26: 5996–6003.

Wang, Y., X.L. Deng, X.H. Xiao and B.X. Yuan. 2007. A nonsteroidal anti-inflammatory agent provides significant protection during focal ischemic stroke with decreased expression of matrix metalloproteinases. Curr. Neurovasc. Res. 4: 176–183.

Wang, Z., L.K. Tsai, J. Munasinghe, Y. Leng, E.B. Fessler, F. Chibane, P. Leeds and D.M. Chuang. 2012. Chronic valproate treatment enhances postischemic angiogenesis and promotes functional recovery in a rat model of ischemic stroke. Stroke 43: 2430–2436.

Wei, L., J.P. Erinjeri, C.M. Rovainen and T.A. Woolsey. 2001. Collateral growth and angiogenesis around cortical stroke. Stroke 32: 2179–2184.

Weintraub, M.I. 2006. Thrombolysis (tissue plasminogen activator) in stroke: A medicolegal quagmire. Stroke 37: 1917–1922.

Weis, S.M. and D.A. Cheresh. 2005. Pathophysiological consequences of VEGF-induced vascular permeability. Nature 437: 497–504.

Werner, N., J. Priller, U. Laufs, M. Endres, M. Böhm, U. Dirnagl and G. Nickenig. 2002. Bone marrow-derived progenitor cells modulate vascular reendothelialization and neointimal formation: effect of 3-hydroxy-3-methylglutaryl coenzyme a reductase inhibition. Arterioscler. Thromb.Vasc. Biol. 22: 1567–1572.

Wiessner, C., J. Gehrmann, D. Lindholm, R. Topper, G.W. Kreutzberg and K.A. Hossmann. 1993. Expression of transforming growth factor-beta 1 and interleukin-1 beta mRNA in rat brain following transient forebrain ischemia. Acta Neuropathol. 86: 439–446.

Xiong, Y., A. Mahmood and M. Chopp. 2010. Angiogenesis, neurogenesis and brain recovery of function following injury. Curr. Opin. Investig. Drugs 11: 298–308.

Yamashita, K., U. Gerken, P. Vogel, K. Hossmann and C. Wiessner. 1999. Biphasic expression of TGF-beta1 mRNA in the rat brain following permanent occlusion of the middle cerebral artery. Brain Res. 836: 139–145.

Yamashita, T., M. Ninomiya, P. Hernández Acosta, J.M. García-Verdugo, T. Sunabori, M. Sakaguchi, K. Adachi, T. Kojima, Y. Hirota, T. Kawase, N. Araki, K. Abe, H. Okano and K. Sawamoto. 2006. Subventricular zone-derived neuroblasts migrate and differentiate into mature neurons in the post-stroke adult striatum. J. Neurosci. 26: 6627–6636.

Yamashita, T. and K. Abe. 2012. Mechanisms of endogenous endothelial repair in stroke. Curr. Pharm. Design 18: 3649–3652.

Yang, Y. and A. Rosenberg. 2011. Blood-brain barrier breakdown in acute and chronic cerebrovascular disease. Stroke 42: 3323–3328.

Yanqing, Z., L. Yu-Min, Q. Jian, X. Bao-Guo and L. Chuan-Zhen. 2006. Fibronectin and neuroprotective effect of granulocyte colony-stimulating factor in focal cerebral ischemia. Brain Res. 1098: 161–169.

Yong, V.W. 2005. Metalloproteinases: mediators of pathology and regeneration in the CNS. Nat. Rev. Neurosci. 6: 931–944.

Zacharek, A., A. Shehadah, J. Chen, X. Cui, C. Roberts, M. Lu and M. Chopp. 2010. Comparison of bone marrow stromal cells derived from stroke and normal rats for stroke treatment. Stroke 41: 524–530.

Zhang, L., R.L. Zhang, Y. Wang, C. Zhang, Z.G. Zhang, H. Meng and M. Chopp. 2005a. Functional recovery in aged and young rats after embolic stroke: treatment with a phosphodiesterase type 5 inhibitor. Stroke 36: 847–852.

Zhang, R.L., Z.G. Zhang, and M. Chopp. 2005b. Neurogenesis in the adult ischemic brain: generation, migration, survival, and restorative therapy. Neuroscientist 11: 408–416.

Zhang, L., Z. Zhang, R.L. Zhang, Y. Cui, M.C. LaPointe, B. Silver and M. Chopp. 2006. Tadalafil, a long-acting type 5 phosphodiesterase isoenzyme inhibitor, improves neurological functional recovery in a rat model of embolic stroke. Brain Res. 1118: 192–198.

Zhang, R., L. Wang, L. Zhang, J. Chen, Z. Zhu, Z. Zhang and M. Chopp. 2003. Nitric oxide enhances angiogenesis via the synthesis of vascular endothelial growth factor and cGMP after stroke in the rat. Circ. Res. 92: 308–313.

Zhang, R., Y. Wang, L. Zhang, Z. Zhang, W. Tsang, M. Lu and M. Chopp. 2002a. Sildenafil (Viagra) induces neurogenesis and promotes functional recovery after stroke in rats. Stroke 33: 2675–2680.

Zhang, Z.G., L. Zhang, Q. Jiang and M. Chopp. 2002b. Bone marrow-derived endothelial progenitor cells participate in cerebral neovascularization after focal cerebral ischemia in the adult mouse. Circ. Res. 90: 284–288.

Zhang, R.L., M. Chopp, S.R. Gregg, Y. Toh, C. Roberts, Y. Letourneau, B. Buller, L. Jia, S.P Nejad Davarani and Z.G. Zhang. 2009. Patterns and dynamics of subventricular zone neuroblast migration in the ischemic striatum of the adult mouse. J. Cereb. Blood Flow Metab. 29: 1240–1250.

Zhang, R.L., Y. LeTourneau, S.R. Gregg, Y. Wang, Y. Toh, A.M. Robin, Z.G. Zhang and M. Chopp. 2007. Neuroblast division during migration toward the ischemic striatum: a study of dynamic migratory and proliferative characteristics of neuroblasts from the subventricular zone. J. Neurosci. 27: 3157–3162.

Zhang, Y. and B.A. Barres. 2010. Astrocyte heterogeneity: an underappreciated topic in neurobiology. Curr. Opin. Neurobiol. 20: 588–594.

Zhang, Z.G. and M. Chopp. 2009. Neurorestorative therapies for stroke: underlying mechanisms and translation to the clinic. Lancet Neurol.8: 491–500.

Zhang, Z.G., L. Zhang, Q. Jiang, R. Zhang, K. Davies, C. Powers, N. van Bruggen and M. Chopp. 2000. VEGF enhances angiogenesis and promotes blood-brain barrier leakage in the ischemic brain. J. Clin. Invest. 106: 829–838.

Zhao, B.Q., S. Wang, H.Y.Kim, H. Storrie, B.R. Rosen, D.J. Mooney, X.Y. Wang and E.H. Lo. 2006. Role of matrix metalloproteinases in delayed cortical responses after stroke. Nat. Med. 12: 441–445.

Ziegelhoeffer, T., B. Fernandez, S. Kostin, M. Heil, R. Voswinckel, A. Helisch and W. Schaper. 2004. Bone marrow-derived cells do not incorporate into the adult growing vasculature. Circ. Res. 94: 230–238.

Zlokovic, B.V. 2008. The blood-brain barrier in health and chronic neurodegenerative disorders. Neuron 57: 178–201.

Zlokovic, B.V., C.L. Zhang, D. Liu, J. Fernandez, J.H. Griffin and M. Chopp. 2005. Functional recovery after embolic stroke in rodents by activated protein C. Ann. Neurol. 58: 474–477.

Zorner, B. and M.E. Schwab. 2010. Anti-Nogo on the go: from animal models to a clinical trial. Ann. N. Y. Acad. Sci.1198: E22–E34.

CHAPTER 9

Impairment of Energy Metabolism and Vascular Function in Alzheimer's Disease

Victoria Campos Peña,[1] *Bryan V. Phillips Farfán*[2,a] and *Karla G. Carvajal Aguilera*[2,b,*]

9.1 Introduction

Chronic degenerative diseases have recently become one of the main health problems worldwide, particularly those affecting the nervous system. All these disorders are characterized by degeneration of specific cell populations and include multiple ailments in aging humans. Alzheimer's disease (AD) is without a doubt the most frequent cause of dementia, representing approximately 50–60 per cent of all cases. Its high prevalence puts this pathology among the principal causes of death in adults, after cancer and cardiovascular diseases.

There are many risk factors and pathological changes that contribute to AD and vascular dementia, such as insulin resistance (IR), dyslipidemias and hypertension, among others (Martins et al. 2006, Duron and Hanon

[1] Laboratorio Experimental de Enfermedades Neurodegenerativas, Instituto Nacional de Neurología y Neurocirugía, Insurgentes Sur 3877, Col. La Fama, México D.F., 14269, México. Email: neurovcp@ymail.com

[2] Laboratorio de Nutrición Experimental, Instituto Nacional de Pediatría, Insurgentes Sur 3700C, Col. Insurgentes Cuicuilco, México D.F., 04530, México.

[a] Email: bvphillips@yahoo.com

[b] Email: karla_ca@yahoo.com

* Corresponding author

2008, Milionis et al. 2008, Bhat 2010, Kerola et al. 2011). These factors are also associated to a clinical entity known as the metabolic syndrome (MS; see below). Many authors have reviewed the relationship between AD and the factors that are associated to the MS (Pasinetti and Eberstein 2008, Frisardi et al. 2010, Farooqui et al. 2012). However, a risk factor that is often overlooked is endothelial dysfunction generated by oxidative stress, which both causes and is the result of an impairment of energy metabolism. These factors importantly participate in the pathogenesis of AD and the MS. Thus, this chapter emphasizes the impairment of energy metabolism that leads to endothelial dysfunction and their relationship to both diseases.

9.2 Alzheimer's Disease

At the clinical level, AD is characterized by a progressive loss of memory and other cognitive functions. Although several clinical tools exist for its diagnosis, such as the mini-mental state examination (Folstein et al. 1975) and the clinical dementia rating, the definitive diagnosis of AD can only be established after postmortem examination of the affected brain areas. Histochemical studies have revealed the presence of two lesion types that define AD at the pathological level: neuritic plaques (NP) and neurofibrillary tangles (NFT, Fig. 1) (Pearson et al. 1985, Terry and Katzman 1991, Roberts et al. 1993).

Although AD has been classified as a "non-vascular" dementia, in the last years much evidence has been found which indicates that the presence of vascular alterations is tightly associated with AD development. It is known that AD patients exhibit evident vascular pathology, as well as microglial and astrocyte activation that lead to a local neuro-inflammatory process. In the same manner, the neurovascular functional unit is strongly compromised affecting the function and permeability of the blood-brain barrier (BBB). Under normal circumstances, central nervous system (CNS) functionality is mostly regulated by the vascular flow that transports the necessary nutrients for neuronal activity (Girouard and Iadecola 2006). AD patients show important changes in the microcirculation of the brain that include a reduction of capillary density, an increase of endothelial pinocytosis, a decrease in the number of mitochondria, collagen accumulation in the basement membrane and loss of tight junctions resulting in BBB failure (Iadecola 2004, Zlokovic et al. 2005). These types of alterations are associated to oxidative stress and abnormal accumulation of proteins within blood vessels or the brain parenchyma, which ultimately generate toxic damage altering neuronal homeostasis and function contributing to the degenerative process. Amyloid beta (Aβ) accumulation within intra-cerebral arteries leads to amyloid cerebral angiopathy, observed in more than 80 per cent of AD

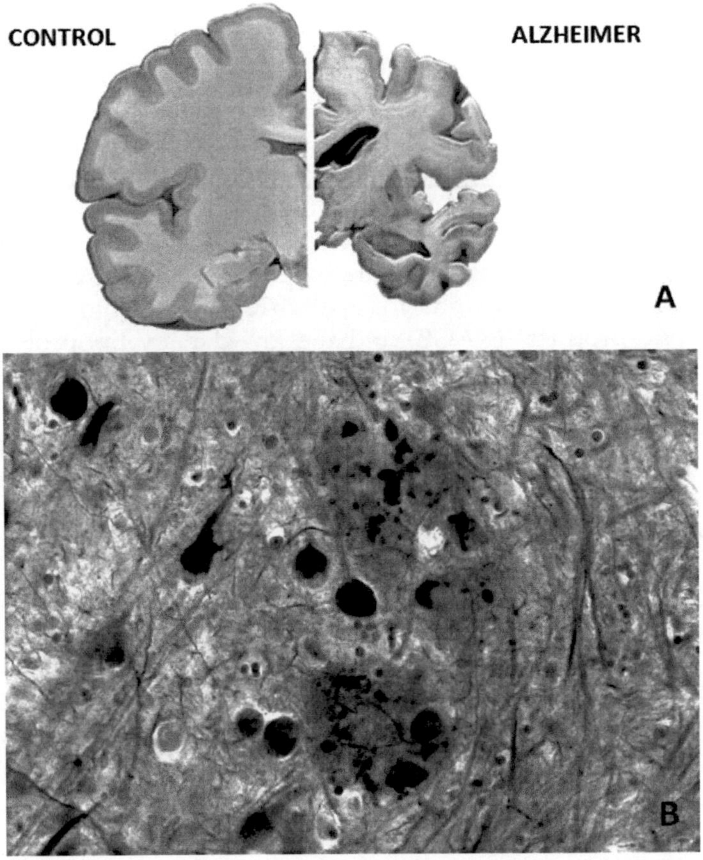

Figure 1. Appearance of a brain affected by Alzheimer's disease (AD). (A) Macroscopic appearance compared to a normal aged brain. Note the shrinkage of the brain as a whole, as well as the reduction in the thickness of the cerebral cortex and the enlarged ventricles. (B) Microscopic appearance of a doubly stained section from an Alzheimer's brain. Darkly stained neurons are filled by neurofibrillary tangles (Bielschowsky), while the extracellular reddish stain corresponds to neuritic plaques identified with an antibody (6F/3D).

Color image of this figure appears in the color plate section at the end of the book.

patients (Jellinger and Attems 2008, Grinberg et al. 2012). The presence of Aβ in the small arteries and arterioles has as a consequence vascular cell atrophy favoring the rupture of blood vessel walls, which originates intra-cerebral hemorrhages in approximately 30 per cent of AD patients (Viswanathan and Greenberg 2011, Hernandez-Guillamon et al. 2012). Similarly, the presence of mutations in the amyloid precursor protein (APP) gene or duplication of its chromosomic locus results in early onset AD, amyloid cerebral angiopathy and severe intra-cerebral hemorrhages (Rovelet-Lecrux et al. 2006).

9.3 Aβ

At the current moment, the most accepted theory of AD pathogenesis is the Aβ hypothesis, which suggests that the accumulation of the Aβ peptide and the formation of NP in the patient's brains is the fundamental event that leads to the eventual development of the pathology (Mullane and Williams 2013); while the other pathological findings (NFT and neuronal death) are supposed to be secondary events that occur after amyloid aggregation. In the brain, the levels of amyloid are regulated by an active process that transports it across the BBB (Sharma et al. 2012). The receptor for advanced glycation end-products (RAGE) and the low-density lipoprotein (LDL) receptor-related protein 1 (LRP1) participate importantly in amyloid peptide entry and exit through the BBB (Sharma et al. 2012); thus, any alteration in the function of these proteins results in the pathological accumulation of amyloid. Both molecules are fundamental for barrier function, since they regulate brain protein homeostasis. LRP1, of the LDL receptor family, has scavenger and signaling activity besides handling amyloid transport out of the brain for its elimination.

The Aβ peptide is derived from the proteolytic cleavage of the APP (Fig. 2) (Kang et al. 1987), a trans-membrane type I glycoprotein (Kang et al. 1987), whose gene is localized in chromosome 21 (Yoshikai et al. 1991, Lamb et al. 1993). Aβ peptide formation is carried out by β- and γ-secretase, localized in the luminal or extracellular space. The peptides that are formed vary from 39 to 43 amino-acid residues (Aβ39, 40, 42, 43); although Aβ40 is the most abundant, Aβ42 is the most hydrophobic and is found in a greater proportion of the observed NP in AD patients (Jarrett et al. 1993).

The damage caused by Aβ is determined mostly by the level of peptide aggregation, in other words, by the conversion of the non-toxic monomeric form to highly toxic oligomeric forms (Klein et al. 2001, Walsh and Selkoe 2004, Crews and Masliah 2010). Progressive accumulation of these oligomers causes the formation of insoluble fibers that finally result in development of NP. *In vitro* studies have demonstrated that the direct administration of Aβ to cell cultures has a neurotoxic effect, since it increases oxidative stress and induces apoptosis (Forloni 1993, Loo et al. 1993). Oxidative stress has been observed in the early stages of AD and has been directly associated to Aβ accumulation. The reactive oxygen species (ROS) that are generated affect membrane proteins, mitochondrial DNA, lipids, cytoplasmic proteins (Mecocci et al. 1994, Christen 2000) and also contribute to the vascular damage observed in AD patients.

Most of the activity of β- and γ-secretase, as well as APP is localized in the endosomal compartment and the trans-Golgi network; thus most Aβ is formed in these subcellular domains and is later secreted by exocytosis. This excreted amyloid, that forms the extracellular aggregates, can enter neurons

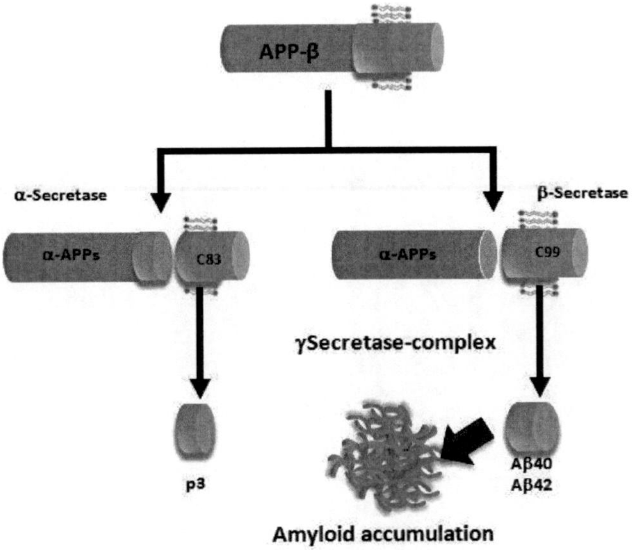

Figure 2. Proteolytic cleavage of the amyloid precursor protein (APP-β). The enzyme α-secretase cleaves APP within the amyloid beta (Aβ) domain and releases APPsα, precluding amyloid formation. The subsequent action of γ-secretase results in the release of a short peptide containing a C-terminal region of Aβ (p3). Under normal conditions it is the principal form of APP processing. The enzymes β- and γ-secretase produce the Aβ peptide that accumulates to generate neuritic plaques. β-secretase activity initiates amyloid production, cleaving at the amino terminus of the amyloid sequence to generate the secreted APPsβ ectodomain and the membrane-bound carboxyl-terminal fragment C99. The final step in amyloid production occurs when the C99 fragment undergoes cleavage by γ-secretase.

Color image of this figure appears in the color plate section at the end of the book.

or endothelial cells through several transporters and membrane receptors, such as the acetylcholine receptor, N-methyl-D-aspartate receptor, LRP1 and RAGE (Fig. 3) (LaFerla et al. 2007, Sakono and Zako 2010). Recently, it has been proposed that intracellular buildup of amyloid precedes the formation of NP; suggesting that amyloid accumulation inside cells is an early event in the progression of the disease. In fact, it has been demonstrated that the intra-neuronal levels of Aβ decrease as extracellular plaques accumulate (Gouras et al. 2000, Mori et al. 2002).

Destabilization of intracellular membranes is another mechanism that could contribute to the deposition of Aβ inside cells. Aβ formed in the different compartments can buildup within lysosomes and rupture their membranes, liberating the peptide within the cytosol where it could aggregate (Yang et al. 1998). The mechanism through which these monomers come together to form oligomers is still unknown. Similarly, the mechanisms of toxicity produced by the oligomeric forms are not entirely clear; however,

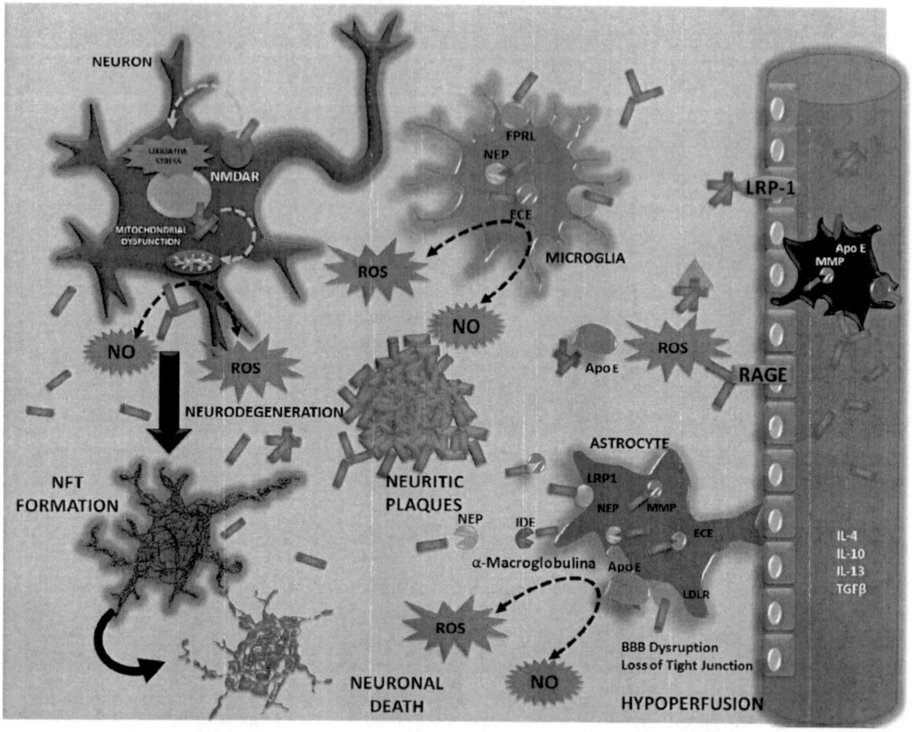

Figure 3. Amyloid beta (Aβ) toxicity in the central nervous system (CNS). Endothelin converting enzyme (ECE), insulin degrading enzyme (IDE), matrix metalloproteinase (MMP) and neutral endopeptidase (NEP) all actively participate in Aβ degradation. However, their activity is reduced by pathological conditions; inducing Aβ peptide accumulation. The presence of Aβ aggregates leads to the extensive production of reactive oxygen species (ROS) and severe endothelial alterations, which cause chronic hypo-perfusion favoring neuronal damage. ROS and nitric oxide (NO) produce mitochondrial dysfunction as well as activation of astrocytes and microglia. The latter generate a chronic pro-inflammatory environment that exacerbates the formation of neuritic plaques (NP) and neurofibrillary tangles (NFT) that finally result in neuronal death. ApoE: apolipoprotein E, BBB: blood-brain barrier, FPRL: formyl peptide receptor like, IL-4: interleukin 4, IL-10: interleukin 10, IL-13: interleukin 13, LDLR: low density lipoprotein receptor, LPR1: low-density lipoprotein receptor-related protein 1, NMDAR: N-methyl D-aspartate receptor, RAGE: receptor for advanced glycation end products, TGFβ: transforming growth factor-beta.

Color image of this figure appears in the color plate section at the end of the book.

it has been proposed that they may inhibit the proteasomal degradation system, which would favor neuronal death.

Numerous authors (Yatin et al. 1999, Butterfield et al. 2001, Boyd-Kimball et al. 2004, Butterfield and Boyd-Kimball 2005, Butterfield and Sultana 2011) have suggested that an increase in Aβ accumulation could importantly contribute to endothelial vascular alterations, reducing blood flow and affecting vascular auto-regulation. These alterations have been observed in animal models that over-express APP and are tightly associated with an increase in oxidative stress (Butterfield et al. 2001).

9.4 Oxidative Stress and Vascular Alterations

The presence of oxidative stress in AD has been documented in recent years. This has been evident in the brains of AD patients, which show augmented levels of oxidized proteins, elevated levels of protein nitrosylation and carbonylation, lipid peroxidation, RNA and DNA oxidation, as well as sugar modifications and the presence of ROS. These types of damage can be caused by the presence and accumulation of Aβ; *in vivo* models have demonstrated that expression of the wild-type or mutated human protein induces oxidative stress and protein modification (Drake et al. 2003, LaFontaine et al. 2002). The mechanisms through which Aβ can cause all these changes have not been elucidated. However, it has been established that methionine 35 in the Aβ42 peptide sequence is very important for its oxidative and neurotoxic properties (Butterfield and Bush 2004).

Recent studies have demonstrated that Aβ can cause important endothelial vascular alterations. This damage has been observed in transgenic mice that over-express the human APP protein, which show severe vascular alterations even before the formation of amyloid deposits (Iadecola et al. 1999, Zhang et al. 1997), indicating that the sole over-production of Aβ can cause endothelial damage. These alterations include changes in blood pressure and a reduction in cerebral blood-flow (Niwa et al. 2002); probably due to a reduction of endothelial-derived relaxing factors, such as nitric oxide and some bradykinins, suggesting that endothelial vascular alterations might be an early event in AD pathogenesis and amyloid peptides may contribute to this damage. Blood flow transports the necessary nutrients for cellular function and also importantly contributes to Aβ elimination. Thus, alterations in blood flow may exacerbate Aβ accumulation and favor buildup of NP.

Finally, Aβ also contributes to oxidative stress which generates vascular injury. Several models have demonstrated that amyloid induces ROS overproduction, further increasing damage. This increase in ROS is also due to a reduction in the activity of anti-oxidant enzymes, which increases protein oxidation (Habib et al. 2010). This also results from a decreased function of the proteosomal system, which alters catabolism of oxidized proteins, increasing damage and favoring the neurodegenerative processes observed in AD patients.

9.5 Metabolic Syndrome and Alzheimer's Disease

MS is a heterogeneous pathology; its clinical manifestations include IR, dyslipidemias, abdominal obesity and arterial hypertension. MS is an important risk factor for the development of cardiac and vascular diseases. MS itself, as well as obesity and IR are risk factors for dementia in elderly people, especially AD, since endothelial vascular damage is produced during MS development, leading to a reduction in brain irrigation. The pathogenesis of MS is complex, but it is clear that a prevailing common denominator is a disequilibrium in energy production and utilization, which in the long run alters the function of vital systems such as the CNS (Carvajal-Aguilera and Carrillo 2009).

MS increases the risk of suffering clinical dementia and has become a risk factor that augments the likelihood of developing neurodegenerative diseases such as AD. It is possible that each of the factors that constitute MS are involved in the development of neurodegenerative processes that result in the appearance of dementia—the principal characteristic of AD—and other neuronal alterations, such as the loss of motor control and other basic functions (Milionis et al. 2008).

In the last few years there have been a growing number of studies about the relationship between dementia and metabolic diseases such as diabetes, obesity, hypertension and dyslipidemias. The etiologic heterogeneity of these diseases and their comorbidities pose challenges in determining the relationship between them; however, IR is a common factor among them. IR consists of the incapacity of cells to respond to the hormonal stimulus, especially in skeletal and cardiac muscle, adipose tissue and certainly the CNS. In the brain, it is believed that IR importantly participates in APP and Tau protein metabolism, since it increases Aβ accumulation and buildup of NP (Milionis et al. 2008). Many studies suggest that type II diabetes is a consequence of IR, which is characterized by glucose intolerance, obesity and hyperinsulinemia. Type II diabetes increases the risk of developing neurodegenerative diseases, principally AD (Ho et al. 2004). A high body mass index, one of the principal factors that characterize IR, has been strongly associated with loss of cognitive capacity and AD development (Gustafson et al. 2003). It has also been demonstrated that obesity leads to hypercholesterolemia, which is linked to atrophy of the hippocampus as well as problems with learning and memory, suggesting that obesity can contribute to cognitive dysfunction. The mechanisms by which these risk factors contribute to neurodegeneration could be related to the endothelial injury caused by the metabolic disturbances linked to these pathologies.

On the other hand, IR causes a decrease in glucose utilization by different tissues. The brain uses glucose, provided by the cerebral blood flow, as the principal source of energy production. It is well known that the energy metabolism of neurons is deteriorated in brains affected by AD and that this energy deficit can be attributed to changes in insulin-dependent glucose uptake and damage to the different proteins that participate in its metabolism (Bosco et al. 2011). Positron emission tomography performed in AD patients shows a progressive reduction of glucose metabolism as well as diminished blood flow in the parietal and temporal lobes, which correlates with the severity of dementia in these patients (Friedland et al. 1983). The first evidence of a possible relationship between AD and alterations in glucose transport was provided by studies using cultured neurons, these studies found that Aβ alters glucose uptake, which in turn causes a decrease in adenosine triphosphate levels (Copani et al. 1991, Mark et al. 1997). Recent studies in obese patients with associated dyslipidemias suggested that the reduction in glucose transport may also be due to the reaction between 4-hydroxinonenal, a neurotoxic aldehyde product of polyunsaturated fatty acid oxidation, and the glucose transporter GLUT3 (Manzano-Leon et al. 2006).

The participation of obesity as a factor that contributes to AD development is also related to the excessive production of inflammatory cytokines by adipose tissue; among them leptin, tumor necrosis factor (TNFα) and adiponectin, as well as interleukin β and 6. The metabolic alterations produced by obesity and MS are influenced by the level of cytokine production (Recasens et al. 2004). The participation of leptin in the development of neurodegeneration is controversial. It has been demonstrated that leptin is neuro-protective, since it inhibits formation of NP. Thus, it is hard to explain why neurodegeneration is increased in patients suffering from MS, since it usually includes hyperleptinemia. This may be partially explained by resistance to this hormone in peripheral tissues, causing augmented secretion of this peptide, as well as diminished leptin transport across the BBB (Pasinetti and Eberstein 2008). Similarly, it has been demonstrated that TNFα is over-expressed in the adipose tissue of obese insulin-resistant rodents and humans. This cytokine is also elevated in the brains of AD patients and adults with mild cognitive impairment, consistent with the idea that increased TNFα—associated with obesity, IR and hyperinsulinemia—can cause an elevation in the cerebral accumulation of Aβ or increased neurodegenerative processes (Craft 2009). In fact, it has been shown that a higher secretion of these cytokines by adipose tissue leads to chronic inflammation of the vascular wall, causing endothelial dysfunction within the brain and augmenting neuronal damage.

Other MS components that could participate in AD development are the dyslipidemias, such as hypertriglyceridemia and hypercholesterolemia. The brain is the organ that contains the greatest amount of cholesterol in the body (25 per cent). Cholesterol is the principal component of myelin and the membranes of neurons and glia (Dietschy and Turley 2001). It also has a very important participation in the development and maintenance of neuronal plasticity, making it very important for CNS function (Pfrieger 2003). Most of the cholesterol present in the brain is synthetized *in situ*, since it is incapable of passing through the BBB. The brain does not eliminate cholesterol directly; excess cholesterol is converted to 24S-hydroxicholesterol, which can easily pass through the BBB and arrive at the liver for its elimination (Bjorkhem et al. 1997, Bjorkhem et al. 1998). Most cholesterol is found within special regions of the plasma membrane known as "lipid rafts", which are highly dynamic and structured micro-domains that contain high levels of cholesterol and sphingolipids. The lipid rafts also contain a select group of proteins that perform different neurological functions including signaling, cellular adhesion and axonal growth (Paratcha and Ibanez 2002, Golub et al. 2004, Kamiguchi 2006).

In the last few years, several models have been developed that try to explain how the levels of cholesterol could affect the metabolism of Aβ. It is noteworthy that a fraction of total APP and the secretases responsible for its processing are anchored to the plasmatic membrane, thus their localization, half-life and metabolism could be modified by alterations in membrane fluidity generated by changes in cholesterol levels (Puglielli et al. 2003). A cholesterol-rich diet distinctly elevates the production of amyloid in rabbit hippocampal neurons (Sparks et al. 1994). The induction of hypercholesterolemia in a transgenic model augmented the levels of Aβ in the CNS; Aβ buildup directly correlated to the increase in the total levels of cholesterol (Refolo et al. 2000). These results have been abundantly documented by different researchers throughout the world (Levin-Allerhand et al. 2002, Shie et al. 2002, Oksman et al. 2006, Thirumangalakudi et al. 2008). A very important aspect in the relationship between the production of amyloid and hypercholesterolemia resides in the activity of γ-secretase. It is known that the activity of this enzyme complex is dependent on cholesterol levels. Depletion of this lipid completely inhibits the activity of γ-secretase (Wahrle et al. 2002). Undoubtedly, cholesterol redistribution prominently contributes to APP processing and by extension to the formation and accumulation of Aβ (Vetrivel and Thinakaran 2010).

Hypertension is another MS factor implicated in AD. It has been proposed that cardiovascular diseases elevate the risk of developing subcortical white matter lesions commonly found in AD. Hypertension has a positive correlation with NP and NFT disease burden and increased vascular permeability in the brain accompanied by protein extravasation through endothelial cell junctions (Milionis et al. 2008). Endothelial vascular function in peripheral and coronary circulation is altered by IR and type II diabetes; however, the effects of these illnesses on cerebral arteries are practically unexplored. In MS, it has been shown that the increase in circulating cytokines produced by the adipose tissue leads to a chronic inflammatory state in the vascular wall, producing endothelial dysfunction of cerebral blood-vessels (Guzik et al. 2007).

IR generates cerebrovascular dysfunction in genetically obese rats, mediated by oxidative stress (Erdos et al. 2004) and activation of protein kinase C (PKC). PKC belongs to a family of enzymes that participate in several cellular functions including growth, differentiation, ionic channel regulation, neuronal plasticity and memory. Several reports have identified alterations of PKC in the brain and in fibroblasts of AD patients; these changes in PKC may be an early event in AD pathology (Etcheberrigaray and Bhagavan 1999).

9.6 Alterations in Mitochondrial Metabolism and AD

Currently there is evidence supporting the important participation of mitochondrial dysfunction and Aβ in AD pathogenesis. High concentrations of the Aβ peptide increase mitochondrial ROS production. Interestingly, inhibition of energy metabolism alters APP processing, favoring the production of Aβ peptide fragments that can easily buildup, accelerating the formation of the lesions that characterize AD (Pasinetti and Eberstein 2008).

An important mitochondrial enzyme associated to AD is creatine kinase (CK), which has four isoforms expressed only in highly energy demanding tissues or cells, such as the muscle and brain. In the brain, cytosolic CK exists as a homodimer known as brain type creatine kinase (CK-BB). Mitochondria express both dimeric and octameric isoenzymes—the latter is known as ubiquitous mitochondrial creatine kinase (uMtCK)-found in the inter-membrane space. The function of this enzyme is fundamental for the efficiency of energy metabolism in brain cells; in addition it also initiates and regulates apoptosis or necrosis of these cells. It has been shown that mice lacking the expression of CK-BB show anomalous hippocampal mossy fiber connections, delayed seizure development, reduced open-field habituation and slower spatial learning (Jost et al. 2002). Mice lacking expression of CK-BB and uMtCK show diminished body weight, extremely poor spatial learning and reduced sound-induced startle reflex responses (Streijger et al. 2005).

CK is crucial for cellular energy metabolism in the brain. Any perturbation in the function of this enzyme could alter normal processing of APP and lead to accumulation of Aβ. CK function may be altered in AD, leading to deficits in the maintenance of optimal energy levels and altered energy supply within glia, neurons and synapses (Burklen et al. 2006). It has been demonstrated that the molecular structure of the different CK isoforms makes them highly susceptible to oxidative stress (Speer et al. 2005, Burklen et al. 2006), which is enhanced in both MS and AD. Although the redox state of CK and its contribution to the development of neurodegeneration have not been studied in detail, its activity is most likely diminished due to oxidation of some of its aminoacid residues. In this sense, it has been reported that the experimental activation of the CK energy circuit, by supplementation of creatine in the diet, notably improves cellular integrity of cultured neurons subjected to toxic manipulations (Andres et al. 2005).

9.7 Therapeutic Options for AD and MS

Therapies for AD and MS are quite limited, as most of them are still actively being studied and developed. Most target a single component of these diseases, which is unlikely to be very helpful, since they are quite complex multi-factorial pathologies. The successful treatment of AD and MS will probably require multiple drugs affecting different etiological aspects of these diseases.

A preventive strategy is to reduce the impairment of energy metabolism by diminishing the quantity of energy generated (caloric restriction) and/or consuming the energy that is available (exercise). Exercise diminished amyloid pathology as well as improved learning and memory in mouse models of AD (Wolf et al. 2006, Parachikova et al. 2008). Caloric restriction also reduces amyloid burden and enhances cognitive performance in an AD mouse model (Halagappa et al. 2007). Dietary nutrients including, but not limited to, polyphenols (resveratrol, curcumin) and omega-3 fatty acids (docosahexaenoic acid, eicosapentaenoic acid) show promise as possible therapeutic agents for AD and MS. The polyphenols have anti-inflammatory and antioxidant (free radical scavenger) effects, while docosahexaenoic acid reduces oxidative damage (Pasinetti and Eberstein 2008). Another non-pharmacological option for the treatment of MS is laparoscopic surgery, which shows some benefit in reducing weight, dyslipidemia and IR (Pontiroli et al. 2002).

Since IR is a common important factor of these pathologies, a logical treatment option is insulin therapy. Insulin administration has been shown to enhance cognition, memory and neurotransmitter function (Messier and Teutenberg 2005). However, it probably requires augmented glucose accessibility and may not assist patients when APP-Aβ42 levels are highly increased (Galasko 2003). Intranasal insulin is a similar safe therapeutic option that efficiently elevates insulin levels in the brain, enhances memory and improves the ratio of APP-Aβ40 to Aβ42 in the cerebrospinal fluid. However, only apolipoprotein E-ε4 negative patients were helped by this treatment (Reger et al. 2008).

An analogous approach consists of increasing insulin secretion by using incretins. Glucagon-like peptide-1 (GLP-1) is an incretin secreted by intestinal cells after a meal. In addition to its gastrointestinal effects, GLP-1 encourages neurite growth, is neuroprotective and may decrease Aβ build-up (Li 2007, Liu et al. 2007, Biswas et al. 2008, Holscher 2010). However, GLP-1 is rapidly degraded by the enzyme dipeptidyl peptidase-4 (DPP-4); making it impractical for clinical use. This has prompted the use of similar molecules with a longer half-life or the reduction of DPP-4 action. DPP-4 inhibition diminishes oxidative or nitrosative stress, memory deficits, inflammation and Aβ accumulation in a mouse model of AD (D'Amico et al. 2010). Liraglutide (a GLP-1 receptor agonist) augments hippocampal neurogenesis, as well as diminishing inflammation and accumulation of Aβ in an AD model; thus, it averts memory deficits, synaptic loss and declining synaptic plasticity (McClean et al. 2011).

The modulation of peroxisome proliferator-activated receptor (PPAR) function is yet another therapeutic option. Although PPARδ agonists reduce the expression of APP and buildup of Aβ (de la Monte et al. 2006), PPARγ agonists have been studied more. The main effect of PPARγ agonists, such as thiazolidinediones, is to sensitize the cells to insulin and thus recover its function (Kaundal and Sharma 2010). They have been shown to reduce deficits in learning and memory in animal models (Pedersen et al. 2006) and enhance cognition in humans (Landreth 2007). However, rosiglitazone mostly only helped apolipoprotein E-ε4 negative subjects in keeping their cognitive function intact (Risner et al. 2006).

Metformin is a biguanide which dramatically reduces hepatic glucose production and has a host of other peripheral effects that make it useful for the treatment of MS, such as increasing binding of insulin to its receptor and greater insulin receptor activity (Matthaei et al. 1991). Metformin also augments glucose uptake and utilization in different cell types (Matthaei et al. 1991, Hundal et al. 1992, Hamann et al. 1993). In combination with insulin it decreases Aβ plaque burden and reduces down-regulation of the insulin receptor by Aβ oligomers (Chen et al. 2009).

9.8 Final Remarks and Proposal

IR produces hypertension and vascular damage via intervening steps (Fig. 4). These vascular risk factors participate importantly in the development of chronic-degenerative diseases, such as AD and MS. These illnesses are characterized by alterations of endothelial function caused by extensive formation of ROS which damage DNA, RNA, proteins and cell membranes, producing cell death. The vascular pathology observed in AD includes pervasive accumulation of Aβ peptides in blood-vessel walls, vascular fibrosis and alterations of endothelial cells which result in chronic hypo-perfusion. This, in turn, causes irreversible neuronal damage, since the brain does not have oxygen reserves. Aβ peptides also lead to production

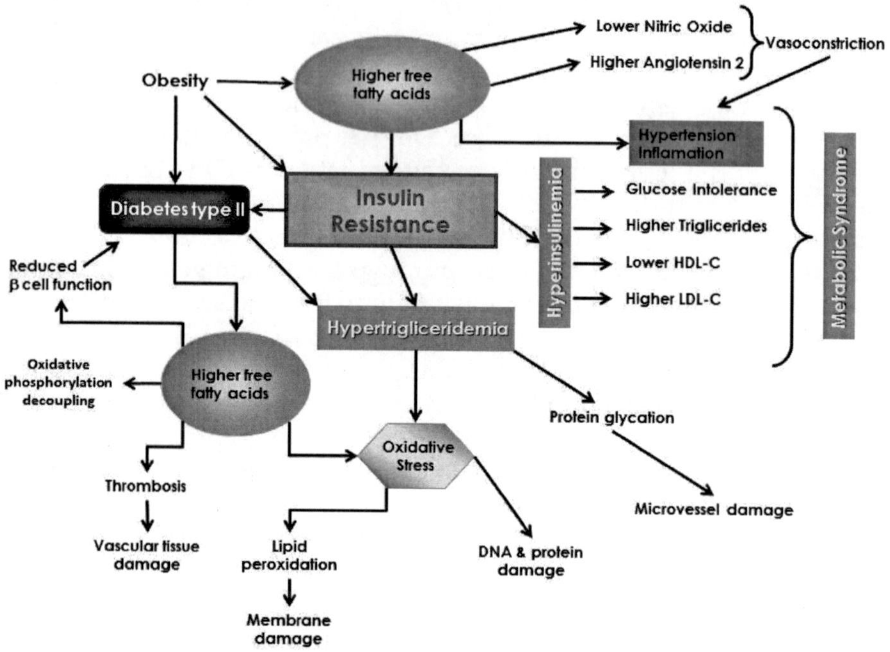

Figure 4. Schematic representation of the mechanisms through which insulin resistance produces hypertension and vascular damage. Lipid disorders, such as hypertriglyceridemia, high LDL (low density lipoprotein) and low HDL (high density lipoprotein) may trigger oxidative stress; which in turn results in protein and DNA oxidation, leading to further cellular damage. In addition, insulin resistance directly affects cellular function in the pancreas, liver, muscle, the endothelium and the brain.

Color image of this figure appears in the color plate section at the end of the book.

of ROS which contribute to the reduced blood flow and, consequently, to deregulation of energy metabolism. In addition, Aβ peptides generate atrophy and rupture of small vessels, thus aggravating the signs of dementia. These factors lead to cerebral hypo-perfusion and endothelial dysfunction; thus causing breakdown of BBB homeostasis and allowing damage to the CNS by free radicals and cytokines. The diminished blood flow favors the accumulation of Aβ due to altered expression of transporters, such as RAGE and LRP which regulate the entry and exit of the Aβ peptide through the BBB. RAGE controls re-entry of the circulating amyloid into the brain and mediates the inflammatory response; while LRP is the principal receptor that transports Aβ through the BBB for elimination in the periphery. Therefore, changes in the homeostasis or function of these proteins result in Aβ accumulation, exacerbating neurodegenerative damage.

Given all that was discussed above, our proposal is as follows: an early event in the pathogenesis of AD and/or MS is the impairment of energy metabolism and consequent generation of free radicals, free radicals in turn attack and impair the function and integrity of many proteins and the structure of the cells. The impairment of energy metabolism, together with IR and inflammation are three of the most important triggers of MS and AD.

References

Andres, R.H., A.D. Ducray, A. Perez-Bouza, U. Schlattner, A.W. Huber, S.H. Krebs, R.W. Seiler, T. Wallimann and H.R. Widmer. 2005. Creatine supplementation improves dopaminergic cell survival and protects against MPP+ toxicity in an organotypic tissue culture system. Cell Transplant. 14: 537–550.

Bhat, N.R. 2010. Linking cardiometabolic disorders to sporadic Alzheimer's disease: a perspective on potential mechanisms and mediators. J. Neurochem. 115: 551–562.

Biswas, S.C., J. Buteau and L.A. Greene. 2008. Glucagon-like peptide-1 (GLP-1) diminishes neuronal degeneration and death caused by NGF deprivation by suppressing Bim induction. Neurochem. Res. 33: 1845–1851.

Bjorkhem, I., D. Lutjohann, O. Breuer, A. Sakinis and A. Wennmalm. 1997. Importance of a novel oxidative mechanism for elimination of brain cholesterol. Turnover of cholesterol and 24(S)-hydroxycholesterol in rat brain as measured with 18O2 techniques *in vivo* and in vitro. J. Biol. Chem. 272: 30178–30184.

Bjorkhem, I., D. Lutjohann, U. Diczfalusy, L. Stahle, G. Ahlborg and J. Wahren. 1998. Cholesterol homeostasis in human brain: turnover of 24S-hydroxycholesterol and evidence for a cerebral origin of most of this oxysterol in the circulation. J. Lipid Res. 39: 1594–1600.

Bosco, D., A. Fava, M. Plastino, T. Montalcini and A. Pujia. 2011. Possible implications of insulin resistance and glucose metabolism in Alzheimer's disease pathogenesis. J. Cell. Mol. Med. 15: 1807–1821.

Boyd-Kimball, D., H.M. Abdul, T. Reed, R. Sultana and D.A. Butterfield. 2004. Role of phenylalanine 20 in Alzheimer's amyloid beta-peptide (1-42)-induced oxidative stress and neurotoxicity. Chem. Res. Toxicol. 17: 1743–1749.

Burklen, T.S., U. Schlattner, R. Homayouni, K. Gough, M. Rak, A. Szeghalmi and T. Wallimann. 2006. The creatine kinase/creatine connection to Alzheimer's disease: CK-inactivation, APP-CK complexes and focal creatine deposits. J. Biomed. Biotechnol. 35936.

Butterfield, D.A. and A.I. Bush. 2004. Alzheimer's amyloid beta-peptide (1-42): involvement of methionine residue 35 in the oxidative stress and neurotoxic properties of this peptide. Neurobiol. Aging 25: 563–568.

Butterfield, D.A. and D. Boyd-Kimball. 2005. The critical role of methionine 35 in Alzheimer's amyloid beta-peptide (1-42)-induced oxidative stress and neurotoxicity. Biochim. Biophys. Acta 1703: 149–156.

Butterfield, D.A. and R. Sultana. 2011. Methionine-35 of aβ(1-42): importance for oxidative stress in Alzheimer's disease. J. Amino Acids 198430.

Butterfield, D.A., J. Drake, C. Pocernich and A. Castegna. 2001. Evidence of oxidative damage in Alzheimer's disease brain: central role for amyloid beta-peptide. Trends Mol. Med. 7: 548–554.

Carvajal-Aguilera, K.G. and S. Carrillo-Sanchez. 2009. Señales moleculares que modulan el metabolismo energético: implicaciones en el desarrollo de obesidad, diabetes y cardiopatías. Mensaje Bioquimico 33: 71–82.

Chen, Y., K. Zhou, R. Wang, Y. Liu, Y.D. Kwak, T. Ma, R.C. Thompson, Y. Zhao, L. Smith, L. Gasparini, Z. Luo, H. Xu and F.F. Liao. 2009. Antidiabetic drug metformin (GlucophageR) increases biogenesis of Alzheimer's amyloid peptides via up-regulating BACE1 transcription. Proc. Natl. Acad. Sci. USA 106: 3907–3912.

Christen, Y. 2000. Oxidative stress and Alzheimer disease. Am. J. Clin. Nutr. 71: 621S–629S.

Copani, A., J.Y. Koh and C.W. Cotman. 1991. Beta-amyloid increases neuronal susceptibility to injury by glucose deprivation. Neuroreport 2: 763–765.
Craft, S. 2009. The role of metabolic disorders in Alzheimer's disease and vascular dementia: Two roads converged? 66: 300–305.
Crews, L. and E. Masliah. 2010. Molecular mechanisms of neurodegeneration in Alzheimer's disease. Hum. Mol. Genet. 19: R12–R20.
D'Amico, M., C. Di Filippo, R. Marfella, A.M. Abbatecola, F. Ferraraccio, F. Rossi and G. Paolisso. 2010. Long-term inhibition of dipeptidyl peptidase-4 in Alzheimer's prone mice. Exp. Gerontol. 45: 202–207.
de la Monte, S.M., M. Tong, N. Lester-Coll, M. Plater Jr. and J.R. Wands. 2006. Therapeutic rescue of neurodegeneration in experimental type 3 diabetes: relevance to Alzheimer's disease. J. Alzheimers Dis. 10: 89–109.
Dietschy, J.M. and S.D. Turley. 2001. Cholesterol metabolism in the brain. Curr. Opin. Lipidol. 12: 105–112.
Drake, J., C.D. Link and D.A. Butterfield. 2003. Oxidative stress precedes fibrillary deposition of Alzheimer's disease amyloid h-peptide (1-42) in a transgenic caenorhabditis elegans model. Neurobiol. Aging 24: 415–420.
Duron, E. and O. Hanon. 2008. Vascular risk factors, cognitive decline, and dementia. Vasc. Health Risk Manag. 4: 363–381.
Erdos, B., J.A. Snipes, A.W. Miller and D.W. Busija. 2004. Cerebrovascular dysfunction in Zucker obese rats is mediated by oxidative stress and protein kinase C. Diabetes 53: 1352–1359.
Etcheberrigaray, R. and S. Bhagavan. 1999. Ionic and signal transduction alterations in Alzheimer's disease: relevance of studies on peripheral cells. Mol. Neurobiol. 20: 93–109.
Farooqui, A.A., T. Farooqui, F. Panza and V. Frisardi. 2012. Metabolic syndrome as a risk factor for neurological disorders. Cell. Mol. Life Sci. 69: 741–762.
Folstein, M.F., S.E. Folstein and P.R. McHugh. 1975. "Mini-mental state". A practical method for grading the cognitive state of patients for the clinician. J. Psychiatr. Res. 12: 189–198.
Forloni, G. 1993. beta-Amyloid neurotoxicity. Funct. Neurol. 8: 211–225.
Friedland, R.P., T.F. Budinger, E. Ganz, Y. Yano, C.A. Mathis, B. Koss, B.A. Ober, R.H. Huesman and S.E. Derenzo. 1983. Regional cerebral metabolic alterations in dementia of the Alzheimer type: positron emission tomography with [18F]Fluorodeoxyglucose. J. Comput. Assist. Tomogr. 7: 590–598.
Frisardi, V., V. Solfrizzi, D. Seripab, C. Capurso, A. Santamato, D. Sancarlo, G. Vendemiale, A. Pilotto and F. Panza. 2010. Metabolic-cognitive syndrome: A cross-talk between metabolic syndrome and Alzheimer's disease. Ageing Res. Rev. 9: 399–417.
Galasko, D. 2003. Insulin and Alzheimer's disease: an amyloid connection. Neurology 60: 1886–1887.
Girouard, H. and C. Iadecola. 2006. Neurovascular coupling in the normal brain and in hypertension, stroke, and Alzheimer disease. J. Appl. Physiol. 100: 328–335.
Golub, T., S. Wacha and P. Caroni. 2004. Spatial and temporal control of signaling through lipid rafts. Curr. Opin. Neurobiol. 14: 542–550.
Gouras, G.K., J. Tsai, J. Naslund, B. Vincent, M. Edgar, F. Checler, J.P. Greenfield, V. Haroutunian, J.D. Buxbaum, H. Xu, P. Greengard and N.R. Relkin. 2000. Intraneuronal Abeta42 accumulation in human brain. Am. J. Pathol. 156: 15–20.
Grinberg, L.T., A.D. Korczyn and H. Heinsen. 2012. Cerebral amyloid angiopathy impact on endothelium. Exp. Gerontol. 47: 838–842.
Gustafson, D., E. Rothenberg, K. Blennow, B. Steen and I. Skoog. 2003. An 18-year follow-up of overweight and risk of Alzheimer disease. Arch. Intern. Med. 163: 1524–1528.
Guzik, T.J., P.J. Marvar, M. Czesnikiewicz-Guzik and R. Korbut. 2007. Perivascular adipose tissue as a messenger of the brain-vessel axis: role in vascular inflammation and dysfunction. J. Physiol. Pharmacol. 58: 591–610.
Habib, L.K., M.T. Lee and J. Yang. 2010. Inhibitors of catalase-amyloid interactions protect cells from beta-amyloid-induced oxidative stress and toxicity. J. Biol. Chem. 285: 38933–38943.

Halagappa, V.K., Z. Guo, M. Pearson, Y. Matsuoka, R.G. Cutler, F.M. LaFerla and M.P. Mattson. 2007. Intermittent fasting and caloric restriction ameliorate age-related behavioral deficits in the triple-transgenic mouse model of Alzheimer's disease. Neurobiol. Dis. 26: 212–220.

Hamann, A., H. Benecke, H. Greten and S. Matthaei. 1993. Metformin increases glucose transporter protein and gene expression in human fibroblasts. Biochem. Biophys. Res. Commun. 196: 382–387.

Hernandez-Guillamon, M., P. Delgado, A. Penalba, D. Rodriguez-Luna, C.A. Molina, A. Rovira, J. Alvarez-Sabin, M. Boada and J. Montaner. 2012. Plasma β-amyloid levels in cerebral amyloid angiopathy-associated hemorrhagic stroke. Neurodegener. Dis. 10: 320–323.

Ho, L., W. Qin, P.N. Pompl, Z. Xiang, J. Wang, Z. Zhao, Y. Peng, G. Cambareri, A. Rocher, C.V. Mobbs, P.R. Hof and G.M. Pasinetti. 2004. Diet-induced insulin resistance promotes amyloidosis in a transgenic mouse model of Alzheimer's disease. FASEB J. 18: 902–904.

Holscher, C. 2010. Incretin analogues that have been developed to treat type 2 diabetes hold promise as a novel treatment strategy for Alzheimer's disease. Recent Pat. CNS Drug Discov. 5: 109–117.

Hundal, H.S., T. Ramlal, R. Reyes, L.A. Leiter and A. Klip. 1992. Cellular mechanism of metformin action involves glucose transporter translocation from an intracellular pool to the plasma membrane in L6 muscle cells. Endocrinology 131: 1165–1173.

Iadecola, C. 2004. Neurovascular regulation in the normal brain and in Alzheimer's disease. Nat. Rev. Neurosci. 5: 347–360.

Iadecola, C., F. Zhang, K. Niwa, C. Eckman, S.K. Turner, E. Fischer, S. Younkin, D.R. Borchelt, K.K. Hsiao and G.A. Carlson. 1999. SOD1 rescues cerebral endothelial dysfunction in mice overexpressing amyloid precursor protein. Nat. Neurosci. 2: 157–161.

Jarrett, J.T., E.P. Berger and P.T. Lansbury Jr. 1993. The carboxy terminus of the beta amyloid protein is critical for the seeding of amyloid formation: implications for the pathogenesis of Alzheimer's disease. Biochemistry 32: 4693–4697.

Jellinger, K.A. and J. Attems. 2008. Prevalence and impact of vascular and Alzheimer pathologies in Lewy body disease. Acta Neuropathol. 115: 427–436.

Jost, C.R., C.E. van der Zee, H.J. in 't Zandt, F. Oerlemans, M. Verheij, F. Streijger, J. Fransen, A. Heerschap, A.R. Cools and B. Wieringa. 2002. Creatine kinase B-driven energy transfer in the brain is important for habituation and spatial learning behaviour, mossy fibre field size and determination of seizure susceptibility. Eur. J. Neurosci. 15: 1692–1706.

Kamiguchi, H. 2006. The region-specific activities of lipid rafts during axon growth and guidance. J. Neurochem. 98: 330–335.

Kang, J., H.G. Lemaire, A. Unterbeck, J.M. Salbaum, C.L. Masters, K.H. Grzeschik, G. Multhaup, K. Beyreuther and B. Muller-Hill. 1987. The precursor of Alzheimer's disease amyloid A4 protein resembles a cell-surface receptor. Nature 325: 733–736.

Kaundal, R.K. and S.S. Sharma. 2010. Peroxisome proliferator activated receptor gamma agonists as neuroprotective agents. Drug News Perspect. 23: 241–256.

Kerola, T., R. Kettunen and T. Nieminen. 2011. The complex interplay of cardiovascular system and cognition: how to predict dementia in the elderly? Int. J. Cardiol. 150: 123–129.

Klein, W.L., G.A. Krafft and C.E. Finch. 2001. Targeting small Abeta oligomers: the solution to an Alzheimer's disease conundrum? Trends Neurosci. 24: 219–224.

LaFerla, F.M., K.N. Green and S. Oddo. 2007. Intracellular amyloid-beta in Alzheimer's disease. Nat. Rev. Neurosci. 8: 499–509.

LaFontaine, M.P. Mattson and D.A. Butterfield. 2002. Oxidative stress in synaptosomal proteins from mutant presenilin-1 knock-in mice: implications for familial Alzheimer's disease. Neurochem. Res. 27: 417–421.

Lamb, B.T., S.S. Sisodia, A.M. Lawler, H.H. Slunt, C.A. Kitt, W.G. Kearns, P.L. Pearson, D.L. Price and J.D. Gearhart. 1993. Introduction and expression of the 400 kilobase amyloid precursor protein gene in transgenic mice. Nat. Genet. 5: 22–30.

Landreth, G. 2007. Therapeutic use of agonists of the nuclear receptor PPARgamma in Alzheimer's disease. Curr. Alzheimer Res. 4: 159–164.

Levin-Allerhand, J.A., C.E. Lominska and J.D. Smith. 2002. Increased amyloid levels in APPSWE transgenic mice treated chronically with a physiological high-fat high-cholesterol diet. J. Nutr. Health Aging 6: 315–319.

Li, L. 2007. Is Glucagon-like peptide-1, an agent treating diabetes, a new hope for Alzheimer's disease? Neurosci. Bull. 23: 58–65.

Liu, J., F. Yin, X. Zheng, J. Jing and Y. Hu. 2007. Geniposide, a novel agonist for GLP-1 receptor, prevents PC12 cells from oxidative damage via MAP kinase pathway. Neurochem. Int. 51: 361–369.

Loo, D.T., A. Copani, C.J. Pike, E.R. Whittemore, A.J. Walencewicz and C.W. Cotman. 1993. Apoptosis is induced by beta-amyloid in cultured central nervous system neurons. Proc. Natl. Acad. Sci. USA 90: 7951–7955.

Manzano-Leon, N., B. Delgado-Coello, M. Guaderrama-Diaz and J. Mas-Oliva. 2006. Beta-adaptin: key molecule for microglial scavenger receptor function under oxidative stress. Biochem. Biophys. Res. Commun. 351: 588–594.

Mark, R.J., Z. Pang, J.W. Geddes, K. Uchida and M.P. Mattson. 1997. Amyloid beta-peptide impairs glucose transport in hippocampal and cortical neurons: involvement of membrane lipid peroxidation. J. Neurosci. 17: 1046–1054.

Martins, I.J., E. Hone, J.K. Foster, S.I. Sunram-Lea, A. Gnjec, S.J. Fuller, D. Nolan, S.E. Gandy and R.N. Martins. 2006. Apolipoprotein E, cholesterol metabolism, diabetes, and the convergence of risk factors for Alzheimer's disease and cardiovascular disease. Mol. Psychiatry 11: 721–736.

Matthaei, S., A. Hamann, H.H. Klein, H. Benecke, G. Kreymann, J.S. Flier and H. Greten. 1991. Association of metformin's effect to increase insulin-stimulated glucose transport with potentiation of insulin-induced translocation of glucose transporters from intracellular pool to plasma membrane in rat adipocytes. Diabetes 40: 850–857.

McClean, P.L., V. Parthsarathy, E. Faivre and C. Holscher. 2011. The diabetes drug liraglutide prevents degenerative processes in a mouse model of Alzheimer's disease. J. Neurosci. 31: 6587–6594.

Mecocci, P., U. MacGarvey and M.F. Beal. 1994. Oxidative damage to mitochondrial DNA is increased in Alzheimer's disease. Ann. Neurol. 36: 747–751.

Messier, C. and K. Teutenberg. 2005. The role of insulin, insulin growth factor, and insulin-degrading enzyme in brain aging and Alzheimer's disease. Neural Plast. 12: 311–328.

Milionis, H.J., M. Florentin and S. Giannopoulos. 2008. Metabolic syndrome and Alzheimer's disease: a link to a vascular hypothesis? CNS Spectr. 13: 606–613.

Mori, E., K. Lee, M. Yasuda, M. Hashimoto, H. Kazui, N. Hirono and M. Matsui. 2002. Accelerated hippocampal atrophy in Alzheimer's disease with apolipoprotein E epsilon4 allele. Ann. Neurol. 51: 209–214.

Mullane, K. and M. Williams. 2013. Alzheimer's therapeutics: continued clinical failures question the validity of the amyloid hypothesis-but what lies beyond? Biochem. Pharmacol. 85: 289–305.

Niwa, K., V.A. Porter, K. Kazama, D. Cornfield, G.A. Carlson and C. Iadecola. 2001. A beta-peptides enhance vasoconstriction in cerebral circulation. Am. J. Physiol. Heart Circ. Physiol. 281: H2417–H2424.

Oksman, M., H. Iivonen, E. Hogyes, Z. Amtul, B. Penke, I. Leenders, L. Broersen, D. Lutjohann, T. Hartmann and H. Tanila. 2006. Impact of different saturated fatty acid, polyunsaturated fatty acid and cholesterol containing diets on beta-amyloid accumulation in APP/PS1 transgenic mice. Neurobiol. Dis. 23: 563–572.

Parachikova, A., K.E. Nichol and C.W. Cotman. 2008. Short-term exercise in aged Tg2576 mice alters neuroinflammation and improves cognition. Neurobiol. Dis. 30: 121–129.

Paratcha, G. and C.F. Ibanez. 2002. Lipid rafts and the control of neurotrophic factor signaling in the nervous system: variations on a theme. Curr. Opin. Neurobiol. 12: 542–549.

Pasinetti, G.M. and J.A. Eberstein. 2008. Metabolic syndrome and the role of dietary lifestyles in Alzheimer's disease. J. Neurochem. 106: 1503–1514.

Pearson, R.C., M.M. Esiri, R.W. Hiorns, G.K. Wilcock and T.P. Powell. 1985. Anatomical correlates of distribution of the pathological changes in the neocortex in Alzheimer`s disease. Proc. Natl. Acad. Sci. USA 82: 4531–4534.

Pedersen, W.A., P.J. McMillan, J.J. Kulstad, J.B. Leverenz, S. Craft and G.R. Haynatzki. 2006. Rosiglitazone attenuates learning and memory deficits in Tg2576 Alzheimer mice. Exp. Neurol. 199: 265–273.

Pfrieger, F.W. 2003. Cholesterol homeostasis and function in neurons of the central nervous system. Cell. Mol. Life Sci. 60: 1158–1171.

Pontiroli, A.E., P. Pizzocri, M.C. Librenti, P. Vedani, M. Marchi, E. Cucchi, C. Orena, M. Paganelli, M. Giacomelli, G. Ferla and F. Folli. 2002. Laparoscopic adjustable gastric banding for the treatment of morbid (grade 3) obesity and its metabolic complications: a three-year study. J. Clin. Endocrinol. Metab. 87: 3555–3561.

Puglielli, L., R.E. Tanzi, and D.M. Kovacs. 2003. Alzheimer's disease: the cholesterol connection. Nat. Neurosci. 6: 345–351.

Recasens, M., W. Ricart and J.M. Fernandez-Real. 2004. Obesity and inflammation. Rev. Med. Univ. Navarra 48: 49–54.

Refolo, L.M., B. Malester, J. LaFrancois, T. Bryant-Thomas, R. Wang, G.S. Tint, K. Sambamurti, K. Duff and M.A. Pappolla. 2000. Hypercholesterolemia accelerates the Alzheimer's amyloid pathology in a transgenic mouse model. Neurobiol. Dis. 7: 321–331.

Reger, M.A., G.S. Watson, P.S. Green, L.D. Baker, B. Cholerton, M.A. Fishel, S.R. Plymate, M.M. Cherrier, G.D. Schellenberg, W.H. Frey 2nd and S. Craft. 2008. Intranasal insulin administration dose-dependently modulates verbal memory and plasma amyloid-beta in memory-impaired older adults. J. Alzheimers Dis. 13: 323–331.

Risner, M.E., A.M. Saunders, J.F. Altman, G.C. Ormandy, S. Craft, I.M. Foley, M.E. Zvartau-Hind, D.A. Hosford, A.D. Roses and Rosiglitazone in Alzheimer's Disease Study Group. 2006. Efficacy of rosiglitazone in a genetically defined population with mild-to moderate Alzheimer's disease. Pharmacogenomics J. 6: 246–254.

Roberts, G.W., M. Nash, P.G. Ince, M.C. Royston and S.M. Gentleman. 1993. On the origin of Alzheimer disease: a hypothesis. Neuroreport 4: 7–9.

Rovelet-Lecrux, A., D. Hannequin, G. Raux, N. Le Meur, A. Laquerriere, A. Vital, C. Dumanchin, S. Feuillette, A. Brice, M. Vercelletto, F. Dubas, T. Frebourg and D. Campion. 2006. APP locus duplication causes autosomal dominant early-onset Alzheimer disease with cerebral amyloid angiopathy. Nat. Genet. 38: 24–26.

Sakono, M. and T. Zako. 2010. Amyloid oligomers: formation and toxicity of Abeta oligomers. FEBS J. 277: 1348–1358.

Sharma, H.S., R.J. Castellani, M.A. Smith and A. Sharma. 2012. The blood-brain barrier in Alzheimer's disease: novel therapeutic targets and nanodrug delivery. Int. Rev. Neurobiol. 102: 47–90.

Shie, F.S., L.W. Jin, D.G. Cook, J.B. Leverenz and R.C. LeBoeuf. 2002. Diet-induced hypercholesterolemia enhances brain A beta accumulation in transgenic mice. Neuroreport 13: 455–459.

Sparks, D.L., S.W. Scheff, J.C. Hunsaker 3rd, H. Liu, T. Landers and D.R. Gross. 1994. Induction of Alzheimer-like beta-amyloid immunoreactivity in the brains of rabbits with dietary cholesterol. Exp. Neurol. 126: 88–94.

Speer, O., N. Back, T. Buerklen, D. Brdiczka, A. Koretsky, T. Wallimann and O. Eriksson. 2005. Octameric mitochondrial creatine kinase induces and stabilizes contact sites between the inner and outer membrane. Biochem. J. 385: 445–450.

Streijger, F., F. Oerlemans, B.A. Ellenbroek, C.R. Jost, B. Wieringa and C.E. van der Zee. 2005. Structural and behavioural consequences of double deficiency for creatine kinases BCK and UbCKmit. Behav. Brain Res. 157: 219–234.

Terry, R.D. and R. Katzman. 1983. Senile dementia of the Alzheimer type. Ann. Neurol. 14: 497–506.

Thirumangalakudi, L., A. Prakasam, R. Zhang, H. Bimonte-Nelson, K. Sambamurti, M.S. Kindy and N.R. Bhat. 2008. High cholesterol-induced neuroinflammation and amyloid precursor

protein processing correlate with loss of working memory in mice. J. Neurochem. 106: 475–485.
Vetrivel, K.S. and G. Thinakaran. 2010. Membrane rafts in Alzheimer's disease beta-amyloid production. Biochim. Biophys. Acta 1801: 860–867.
Viswanathan, A. and S.M. Greenberg. 2011. Cerebral amyloid angiopathy in the elderly. Ann. Neurol. 70: 871–880.
Wahrle, S., P. Das, A.C. Nyborg, C. McLendon, M. Shoji, T. Kawarabayashi, L.H. Younkin, S.G. Younkin and T.E. Golde. 2002. Cholesterol-dependent gamma-secretase activity in buoyant cholesterol-rich membrane microdomains. Neurobiol. Dis. 9: 11–23.
Walsh, D.M. and D.J. Selkoe. 2004. Oligomers on the brain: the emerging role of soluble protein aggregates in neurodegeneration. Protein Pept. Lett. 11: 213–228.
Wolf, S.A., G. Kronenberg, K. Lehmann, A. Blankenship, R. Overall, M. Staufenbiel and G. Kempermann. 2006. Cognitive and physical activity differently modulate disease progression in the amyloid precursor protein (APP)-23 model of Alzheimer's disease. Biol. Psychiatry 60: 1314–1323.
Yang, A.J., D. Chandswangbhuvana, L. Margol and C.G. Glabe. 1998. Loss of endosomal/lysosomal membrane impermeability is an early event in amyloid Abeta1-42 pathogenesis. J. Neurosci. Res. 52: 691–698.
Yatin, S.M., S. Varadarajan, C.D. Link and D.A. Butterfield. 1999. *In vitro* and *in vivo* oxidative stress associated with Alzheimer's amyloid beta-peptide (1-42). Neurobiol. Aging 20: 325-330.
Yoshikai, S., H. Sasaki, K. Doh-ura, H. Furuya and Y. Sakaki. 1991. Genomic organization of the human-amyloid beta-protein precursor gene. Gene 102: 291–292.
Zhang, F., C. Eckman, S. Younkin, K.K. Hsiao and C. Iadecola. 1997. Increased susceptibility to ischemic brain damage in transgenic mice overexpressing the amyloid precursor protein. J. Neurosci. 17: 7655–7661.
Zlokovic, B.V., R. Deane, J. Sallstrom, N. Chow and J.M. Miano. 2005. Neurovascular pathways and Alzheimer amyloid beta-peptide. Brain Pathol. 15: 78–83.

Color Plate Section

Chapter 1

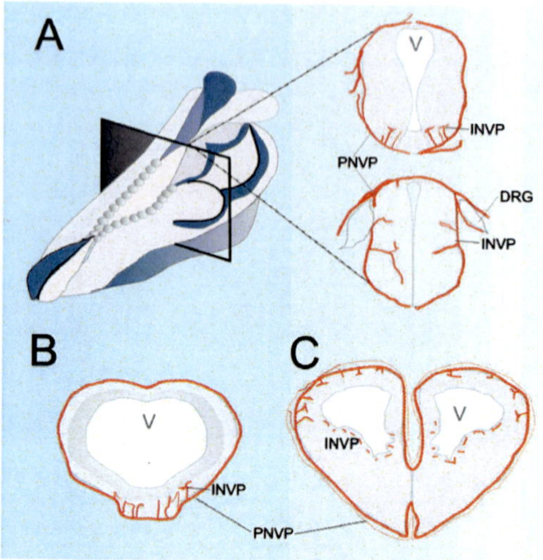

Figure 1. Early brain vascularization follows a stereotypical pattern. Neural tube closure instructs perineural vascular plexus (PNVP) formation, which starts at the rhombencephalic-cervical boundary (A, *left*). Intraneural vascular plexus (INVP) development is modulated by neuromere-specific local cues as demonstrated by its entry points into the central nervous system. PNVP-derived vessels sprout into the rhombencephalon (A, *top right*) and mesencephalon (B) ventrally while they enter the spinal cord laterally (A, *bottom right*). In the prosencephalon, pial and ventricular vessels develop independently (C). DRG= dorsal root ganglion, V= ventricle.

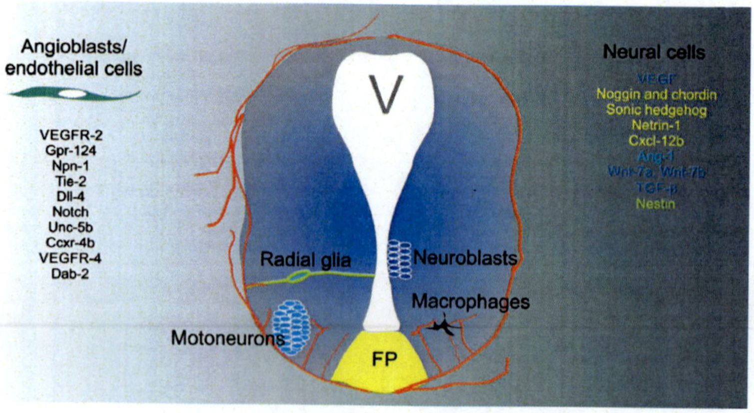

Figure 2. Neural tube-derived cues regulate perineural vascular plexus (PNVP) formation and vessel entry into the central nervous system (i.e., intraneural vascular plexus development). Some of the molecules expressed in angioblasts or endothelial cells are shown on the left, while neural tissue-derived signals are shown on the right (color code indicates the specific cell type that produces the protein). FP= floor plate, V= ventricle. For abbreviations of the molecules see text.

Chapter 2

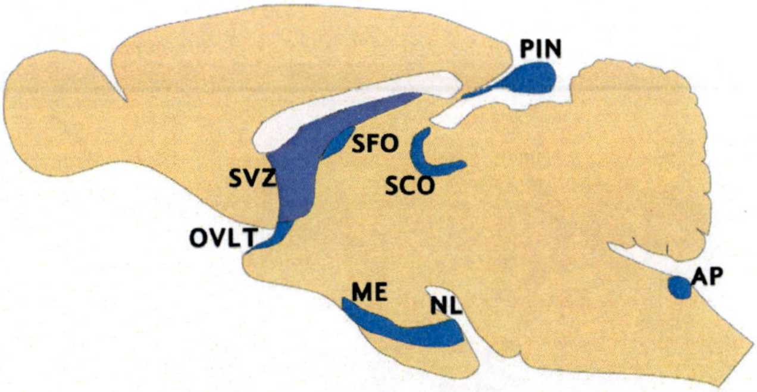

Figure 1. Schematic representation of a sagittal section through the rat brain. The areas highlighted with blue or purple lack tight junctions between endothelial cells (Blue: circumventricular organs, Purple: subventricular zone). AP = area postrema; ME = median eminence; NL = neural lobe of the pituitary; OVLT = vascular organ of the lamina terminalis; PIN = pineal organ; SCO = subcomissural organ; SFO = subfornical organ; SVZ = subventricular zone.

Chapter 3

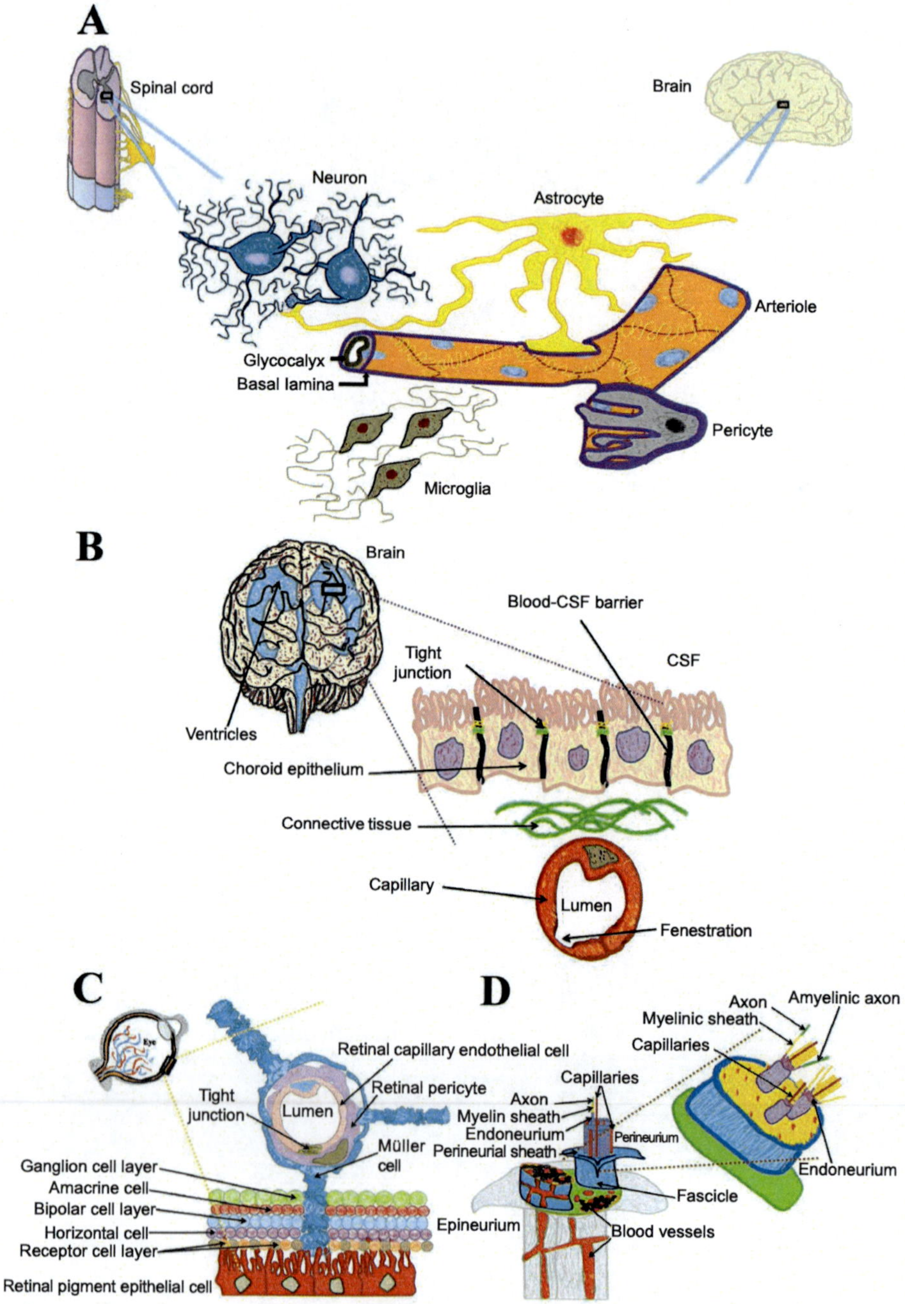

Figure 1. Blood-neural barriers. A) Blood-brain and blood-spinal cord barriers, B) Blood-cerebrospinal fluid barrier, C) Blood-retinal barrier and D) Blood-nerve barrier. CSF = cerebrospinal fluid.

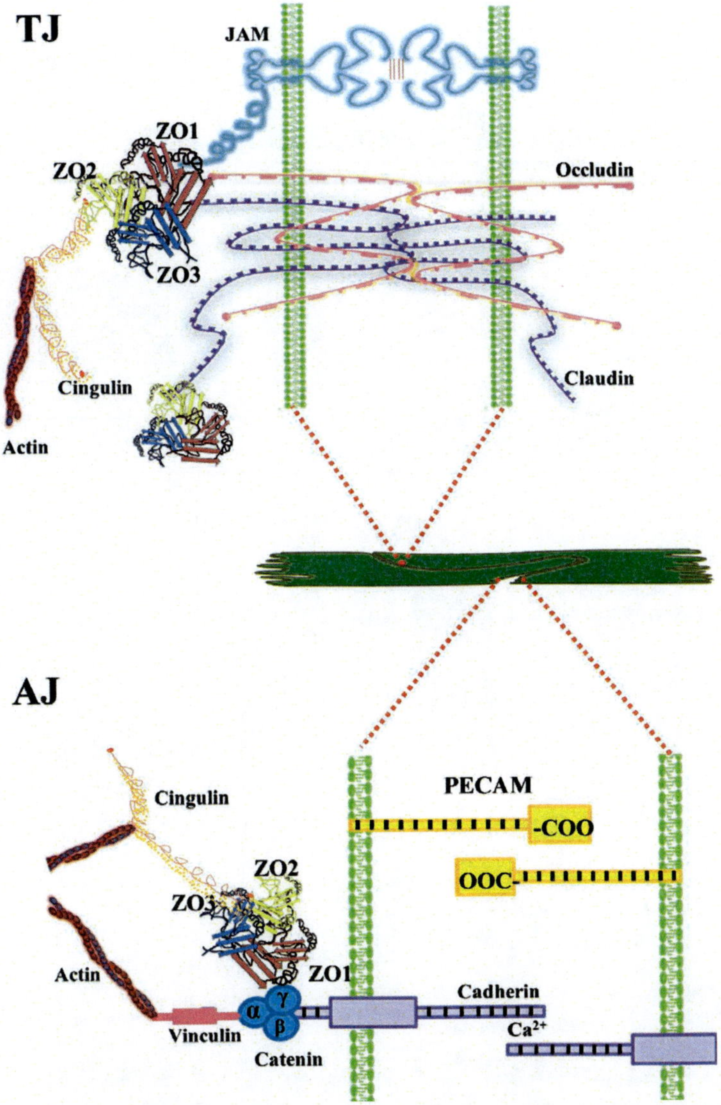

Figure 2. The interendothelial junctional complex. The image shows tight junctions (TJ) in the apical region and adherens junctions (AJ) in the basolateral region of brain endothelial cells. Abbreviations: ZO-1 to 3 = Zonula occludens 1, 2 and 3; JAM = junctional adhesion molecules; PECAM = platelet endothelial cell adhesion molecule.

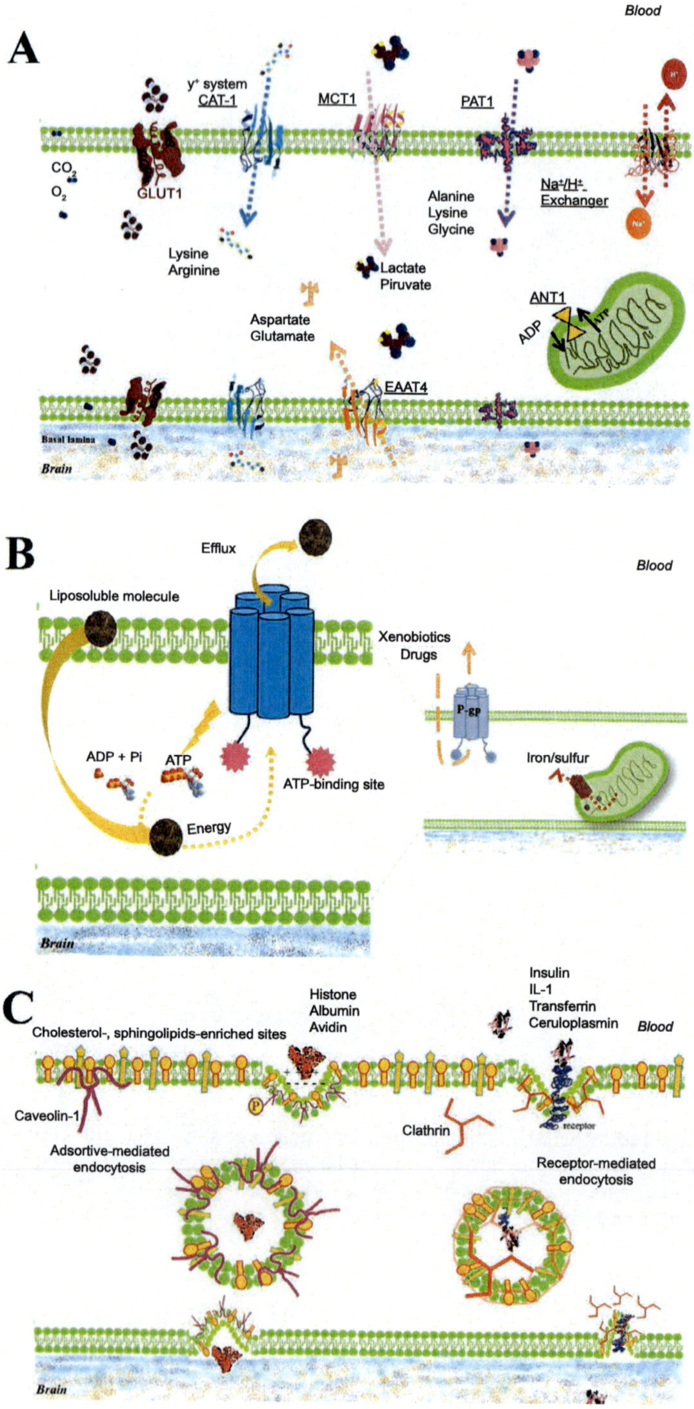

Figure 3. contd....

Chapter 4

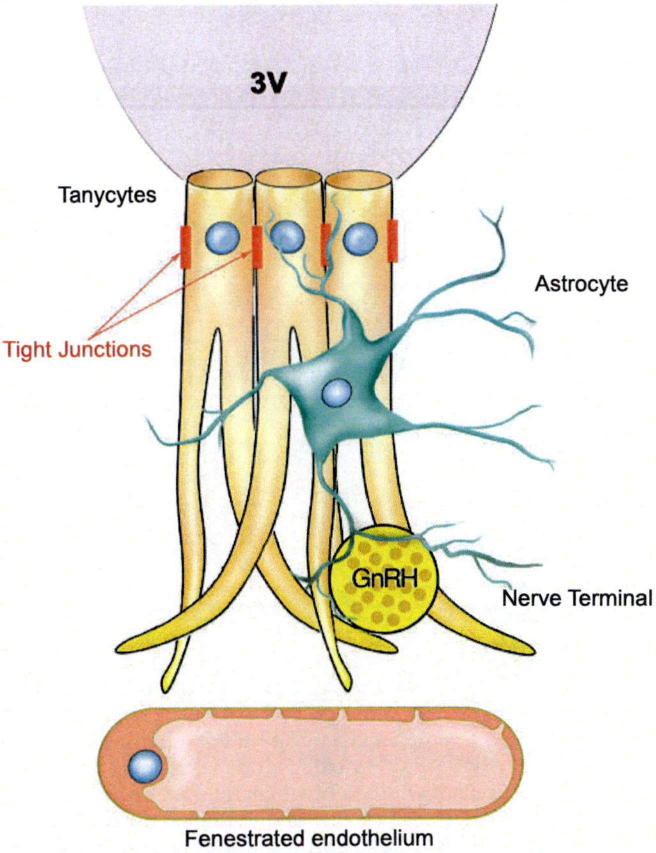

Figure 1. Schematic representation of the cell types (tanycytes, astrocytes and endothelial cells) and neuronal elements (neuroendocrine terminals) that reside within the hypothalamic median eminence. The median eminence of the hypothalamus is the brain structure that establishes the floor of the third ventricle (3V). It is one of the circumventricular organs of the brain, capable of conveying information from the brain to the periphery via the release of neurohormones into the circulation and conversely sensing information reaching the brain via the bloodstream. GnRH= gonadotropin-releasing hormone. From (Prevot et al. 2010a) with permission.

Figure 3. contd.

Figure 3. Specialized carrier systems in brain endothelial cells. A) Solute carrier systems for glucose, amino acids and ion exchange; B) ATP-binding cassette transporters for the efflux of xenobiotics and drugs; C) adsorptive and receptor-mediated endocytosis. Abbreviations: GLUT-1 = Glucose transporter 1; CAT-1 = cationic amino acid transporter (y+ system); MCT1 = monocarboxylic acid transporter; LAT-1 = large amino acid transporter 1; EAAT-4 = excitatory amino acid transporter; ANT-1 = mitochondrial adenine nucleotide translocator; Na^+/H^+ = sodium proton pump; P-gp = P glycoprotein; IL-1 = interleukin 1; ATP = adenosine triphosphate; ADP = adenosine biphosphate; Pi = inorganic phosphate.

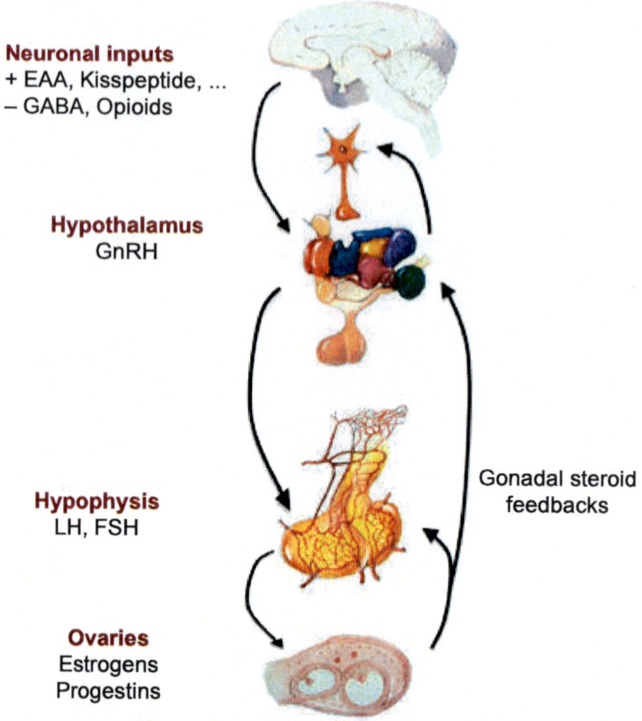

Figure 2. Neuronal control of the pituitary gonadal axis. Hypothalamic gonadotropin-releasing hormone (GnRH) neurons are the final common pathway for central control of gonatropin secretion, they are subjected to a complex array of excitatory and inhibitory trans-synaptic inputs that modulate their activity. GnRH neuroendocrine neurons project to the median eminence, where they make contact with the basal lamina and open into the pericapillary space of the primary hypophyseal portal plexus. Upon reaching the pituitary portal system, GnRH goes to the pituitary to stimulate the synthesis and secretion of pituitary gonadotropins: luteinizing hormone (LH) and follicle stimulating homone (FSH). Blood-borne LH and FSH act on target cells in the gonads (here the ovaries) to direct production of gametes, as well as the secretion of steroid hormones. Within the brain, gonadal steroids influence GnRH secretion via neuroendocrine feedback loops. EAA: excitatory amino acids, GABA: gamma-aminobutyric acid. From (Prevot 2010, Prevot et al. 2010b) with permission.

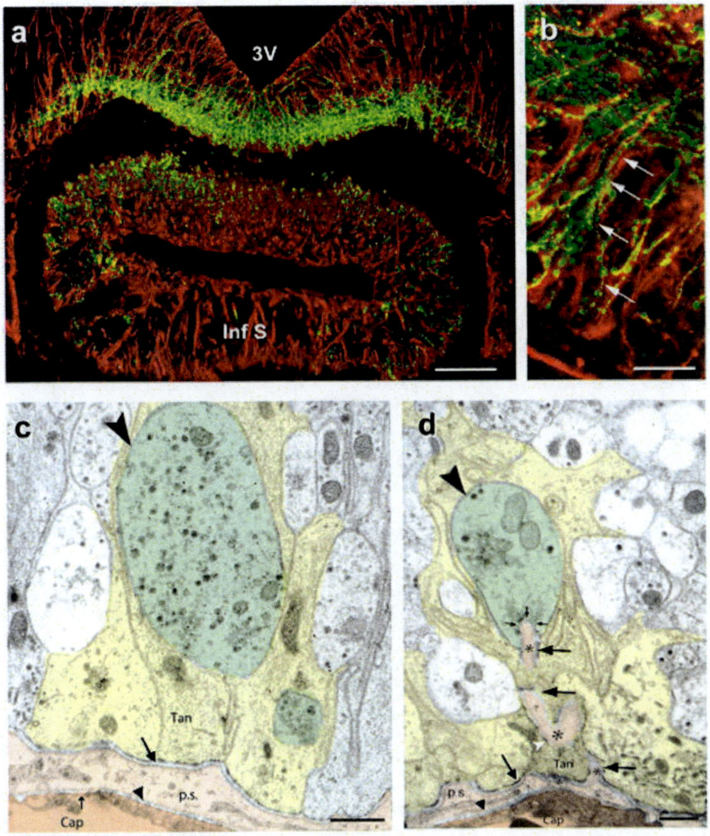

Figure 3. Close association between gonadotropin-releasing hormone (GnRH) nerve terminals and tanycytic processes in the median eminence of the hypothalamus. a) Fluorescent micrograph of a frontal section showing dense GnRH-immunoreactivity (green) and tanycytic processes visualized by their immunoreactivity for vimentin (red) in the external zone of the median eminence. 3V: third ventricle, Inf S: infundibular stem. b) High magnification image showing GnRH axon fibers (green) alongside tanycytic processes (red) projecting to the capillary zone of the median eminence (arrows). c & d) Electron micrographs which illustrate the dynamic changes occurring in the external zone of the median eminence that control the direct access of GnRH nerve terminals to the pericapillary space during the estrous cycle in the rat. Cap: capillaries, Tan: tanacyte endfeet, PS: pericapillary space. c) Electron micrograph of GnRH-immunoreactive terminals (large arrowhead, green) in the external zone of the median eminence in close proximity to fenestrated capillaries (red) of the portal vasculature. During most stages of the reproductive cycle, GnRH nerve terminals (labeled with 15-nm gold particles) are entirely embedded in tanycytic endfeet (yellow), which prevent them from contacting the pericapillary space (pink) delineated by the parenchymatous basal lamina (arrow). Arrowhead: basal lamina of the endothelium, short arrows: endothelial fenestration. Scale bar: 0.5 μm. d) On the day of proestrus, the time when the pre-ovulatory surge of GnRH and LH occurs, a significant fraction of GnRH nerve endings (large arrowhead, green) directly contact the pericapillary space (pink). Scale bar: 0.5 μm. From (Prevot et al. 1998, Prevot et al. 1999a) with permission.

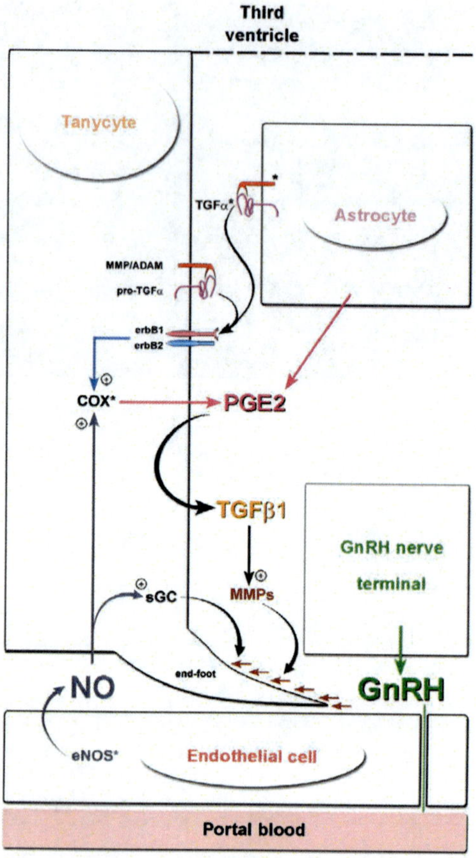

Figure 4. Morphological plasticity of glial coverage of gonadotropin-releasing hormone (GnRH) neurons in the median eminence. Under conditions of low gonadotropin output such as in diestrus, tanycytes envelop neuro-secretory GnRH nerve terminals and, through their endfeet, prevent them from directly contacting the perivascular space and releasing GnRH into the vascular compartment. Transforming growth factor alpha (TGFα), which is produced by tanycytes and astrocytes, activates epidermal growth factor receptor (erbB) 1 and 2 heterodimers to stimulate prostaglandin E2 (PGE_2) production and the subsequent PGE_2-dependent release of TGFβ1 from tanycytes. Astrocytes also constitute a source of PGE_2 in response to epidermal growth factor (EGF) peptides. Tanycyte-derived TGFβ1 acts in an autocrine manner to induce retraction of tanycytic processes through metalloproteinase (MMP)-mediated degradation of the extracellular matrix. Endothelial cells also regulate the morphological plasticity of tanycytes via the release of nitric oxide (NO), which diffuses towards tanycytes to stimulate the enzymatic activity of cyclooxygenase (COX) and the NO-sensitive soluble guanylyl cyclase (sGC). Notably, the mechanism by which the activation of sGC promotes actin cytoskeleton remodeling in tanycytes remains unknown. The retraction of tanycytic endfeet enables GnRH nerve endings to directly contact the pericapillary space (i.e., the space delineated by the parenchymatous basal lamina on one side and by the endothelial basal lamina on the other side; not shown). The released GnRH then enters the portal blood via the fenestrations of endothelial cells. Asterisks indicate the molecular components shown to be positively regulated by gonadal steroids in primary cultures of astrocytes, tanycytes and endothelial cells (see main text for details). From (Sharif et al. 2013) with permission.

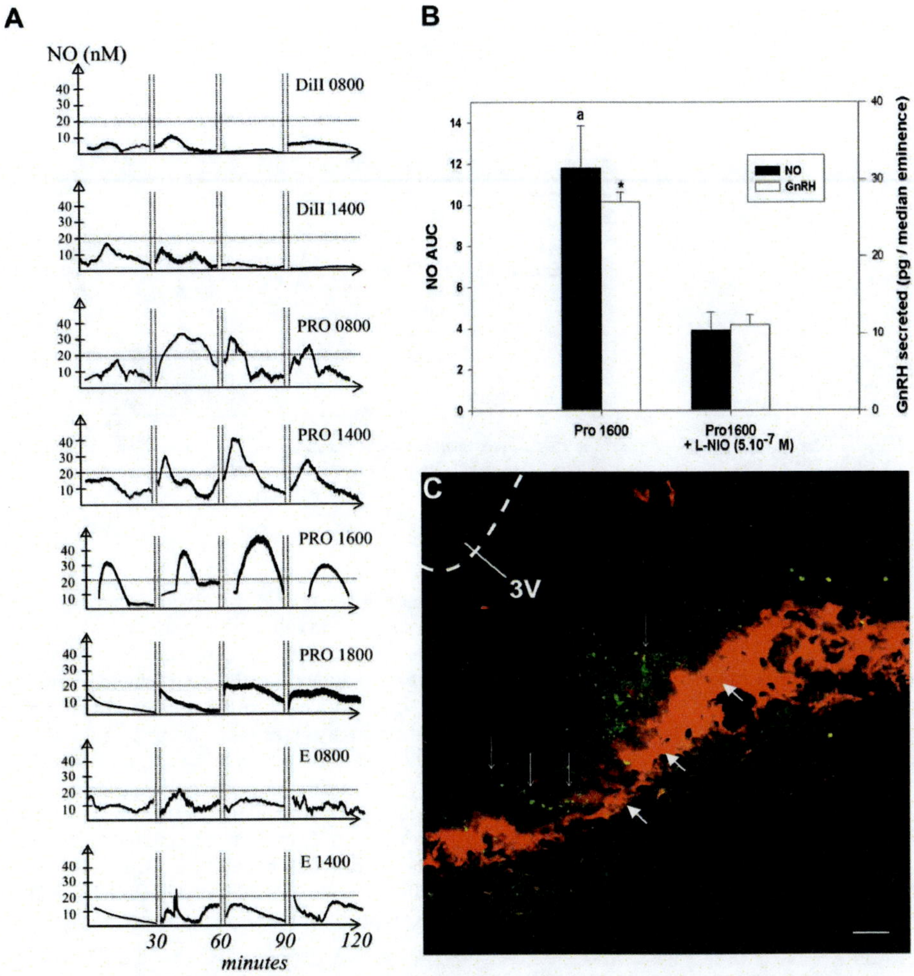

Figure 5. In the median eminence of the hypothalamus, endothelial nitric oxide (NO) secretion may represent one of the synchronizing cues that coordinates gonadotropin-releasing hormone (GnRH) release from GnRH neuroendocrine terminals, that are distributed over two mm within the median eminence, allowing the occurrence of functionally meaningful episodes of GnRH secretion. A) Real time amperometric measurement of spontaneous NO release from median eminence explants at different stages of the rat estrous cycle. DiII: diestrus II, PRO: proestrus, E: estrus. B) On the afternoon of proestrus, the pre-ovulatory GnRH-NO release is blocked with L-NIO, an NO synthase (NOS) inhibitor selective for endothelial NOS (eNOS) at 0.5 µM. * and a: significantly different from treated samples, p<0:05. AUC: area under the curve during a 30 min period. C) Photomicrograph showing GnRH axonal fibers in the external zone of the median eminence (green fluorescence, arrows) in close apposition to eNOS-immunoreactive portal vasculature (red fluorescence, arrowheads). 3V: third ventricle. The dotted lines outline the third ventricle. Scale bar: 75 µm. From (Prevot et al. 2000a, Knauf et al. 2001) with permission.

Chapter 6

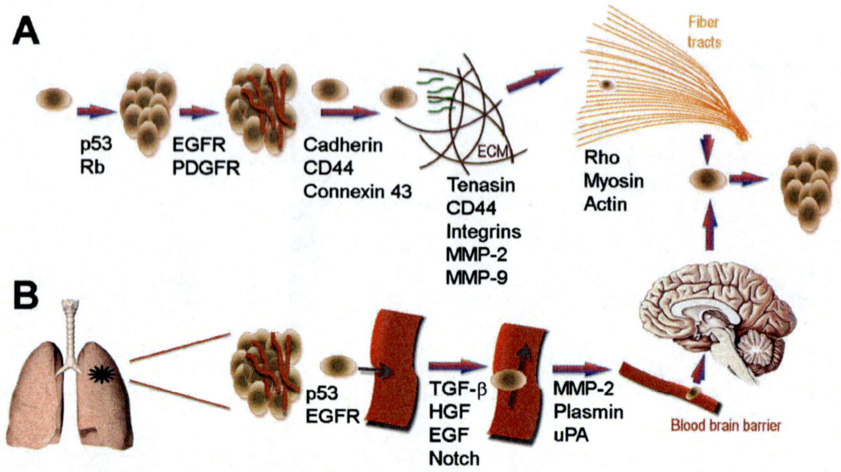

Figure 2. Cell invasion in primary and secondary brain tumors. A) Cell intrinsic genetic alterations of the p53 and retinoblastoma (Rb) pathways lead to cell proliferation in low grade gliomas (LGG). Transformation from LGG to glioblastoma multiforme can occur by mutation or amplifications of epidermal growth factor receptor (EGFR) and platelet derived growth factor receptor (PDGFR) pathways; these changes increase cell proliferation motility and invasiveness of tumor cells by activation of different pathways (see text). Brain parenchyma invasion by glioma cells involves detachment of cells from the primary tumor, adhesion to and degradation of the extracellular matrix (ECM) by tumor cells and cell motility along fiber tracts to settle in another site and continue proliferating. B) Metastatic brain tumors need cell invasion mechanisms to colonize the brain parenchyma. Tumor cells detached from the primary tumor interact and degrade the ECM in order to reach blood vessels. Blood flow transports tumoral cells to the brain parenchyma, where they settle and proliferate. The main molecules that have been implicated in each step of cell proliferation and invasion are shown (for abbreviations of molecules see text).

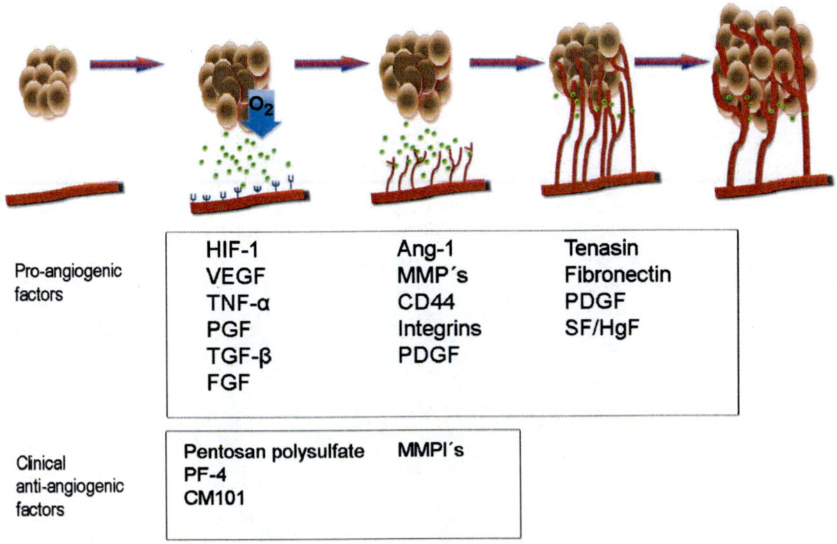

Figure 3. Angiogenesis in primary and secondary brain tumors. Angiogenesis is necessary for tumor growth when its size exceeds the capacity of the original blood vessels to nourish and oxygenate. A decrease in tumor oxygen levels triggers the imbalance between proangiogenic and antiangiogenic factors. In response to hypoxia, tumor cells secrete pro-angiogenic factors that reach the receptors on the endothelial cells of preexisting blood vessels. Formation of new vessels involves remodeling and indeed dissolution of preexisting ones. New vessels surround the tumor providing nutrients and oxygen to cells and maintaining tumor growth. For abbreviations see text.

Chapter 7

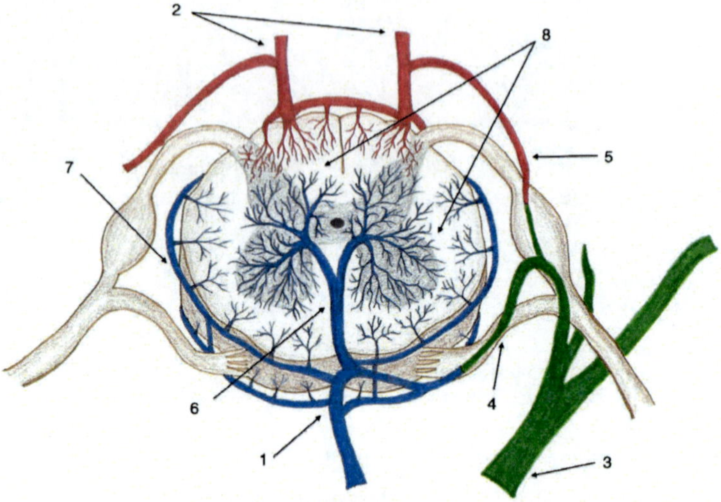

Figure 1. Arterial blood supply and microvasculature of the human spinal cord. The figure separates the anterior arterial system (blue), the posterior arterial system (red) and the segmental arterial system (green). (1) Anterior spinal artery (ASA); (2) Posterior spinal arteries (PSA); (3) Spinal branch from the segmental arteries of the aorta; (4) Anterior radiculomedullary artery; (5) Posterior radiculomedullary artery; (6) Central or sulcal artery; (7) Vasocorona or pial arterial plexus; (8) Watershed zone between central and peripheral arterial systems.

Chapter 9

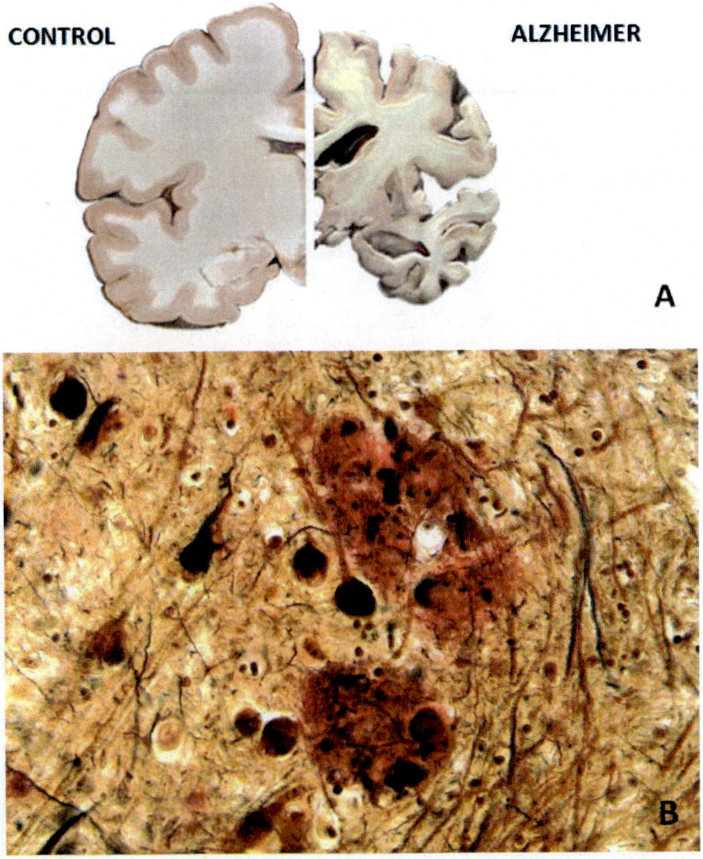

Figure 1. Appearance of a brain affected by Alzheimer's disease (AD). (A) Macroscopic appearance compared to a normal aged brain. Note the shrinkage of the brain as a whole, as well as the reduction in the thickness of the cerebral cortex and the enlarged ventricles. (B) Microscopic appearance of a doubly stained section from an Alzheimer's brain. Darkly stained neurons are filled by neurofibrillary tangles (Bielschowsky), while the extracellular reddish stain corresponds to neuritic plaques identified with an antibody (6F/3D).

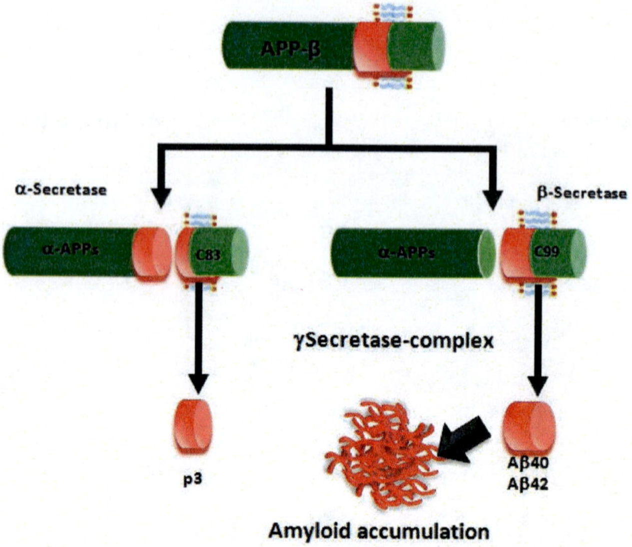

Figure 2. Proteolytic cleavage of the amyloid precursor protein (APP-β). The enzyme α-secretase cleaves APP within the amyloid beta (Aβ) domain and releases APPsα, precluding amyloid formation. The subsequent action of γ-secretase results in the release of a short peptide containing a C-terminal region of Aβ (p3). Under normal conditions it is the principal form of APP processing. The enzymes β- and γ-secretase produce the peptide that accumulates to generate neuritic plaques. β-secretase activity initiates amyloid production, cleaving at the amino terminus of the amyloid sequence to generate the secreted APPsβ ectodomain and the membrane-bound carboxyl-terminal fragment C99. The final step in amyloid production occurs when the C99 fragment undergoes cleavage by γ-secretase.

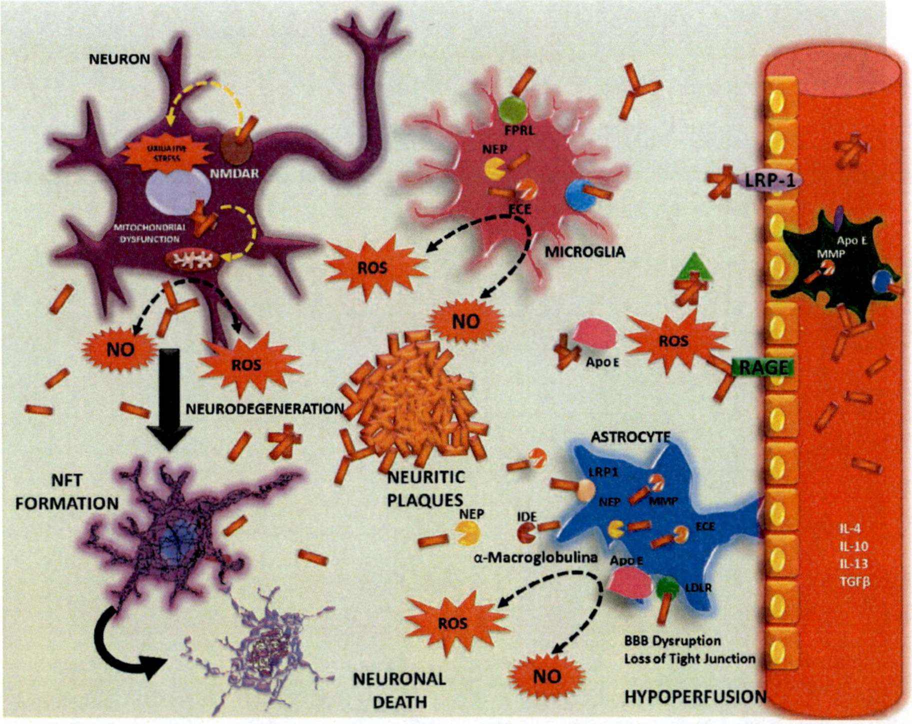

Figure 3. Amyloid beta (Aβ) toxicity in the central nervous system (CNS). Endothelin converting enzyme (ECE), insulin degrading enzyme (IDE), matrix metalloproteinase (MMP) and neutral endopeptidase (NEP) all actively participate in Aβ degradation. However, their activity is reduced by pathological conditions; inducing Aβ peptide accumulation. The presence of Aβ aggregates leads to the extensive production of reactive oxygen species (ROS) and severe endothelial alterations, which cause chronic hypo-perfusion favoring neuronal damage. ROS and nitric oxide (NO) produce mitochondrial dysfunction as well as activation of astrocytes and microglia. The latter generate a chronic pro-inflammatory environment that exacerbates the formation of neuritic plaques (NP) and neurofibrillary tangles (NFT) that finally result in neuronal death. ApoE: apolipoprotein E, BBB: blood-brain barrier, FPRL: formyl peptide receptor like, IL-4: interleukin 4, IL-10: interleukin 10, IL-13: interleukin 13, LDLR: low density lipoprotein receptor, LPR1: low-density lipoprotein receptor-related protein 1, NMDAR: N-methyl D-aspartate receptor, RAGE: receptor for advanced glycation end products, TGFβ: transforming growth factor-beta.

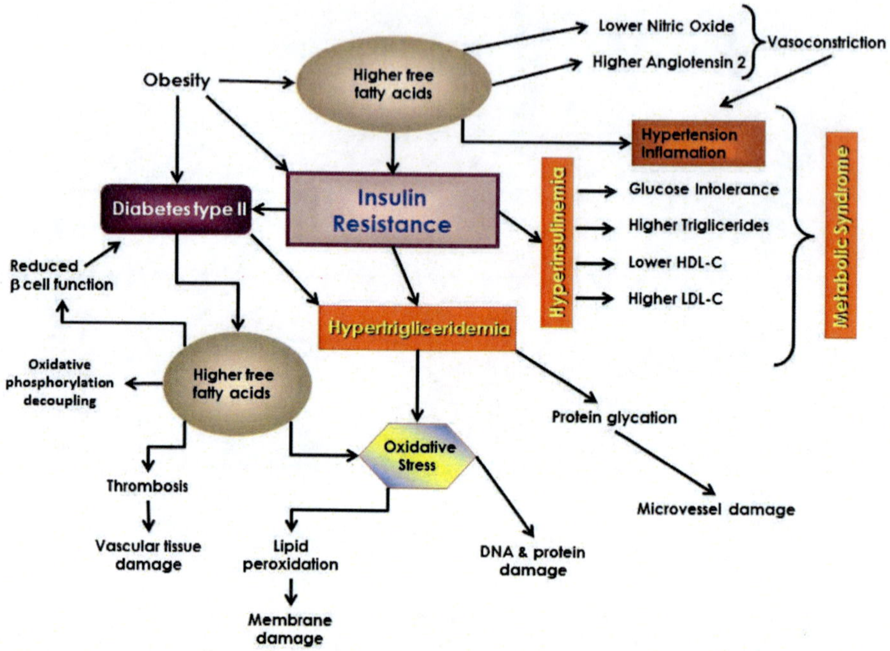

Figure 4. Schematic representation of the mechanisms through which insulin resistance produces hypertension and vascular damage. Lipid disorders, such as hypertriglyceridemia, high LDL (low density lipoprotein) and low HDL (high density lipoprotein) may trigger oxidative stress; which in turn results in protein and DNA oxidation, leading to further cellular damage. In addition, insulin resistance directly affects cellular function in the pancreas, liver, muscle, the endothelium and the brain.